AF553484

COMPUTATIONAL BIOMECHANICS

By

Dr. M. Prakash

Dept. of Zoology
M.M.H. Post Graduate College
Ghaziabad (U.P.)
(India)

DISCOVERY PUBLISHING HOUSE PVT. LTD.
NEW DELHI-110 002

First Published-2009

ISBN: 978-81-8356-475-5

Published by:

DISCOVERY PUBLISHING HOUSE PVT. LTD.
4831/24, Ansari Road, Prahlad Street
Darya Ganj, New Delhi-110002 (India)
Phone: 23279245 • Fax: 91-11-23253475
E-mail: dphbooks@rediffmail.com • dphtemp@indiatimes.com
web: www.discoverypublishinggroup.com

Printed at:

Sachin Printers
Delhi

Preface

The present title "*Computational Biomechanics*" is an exciting new field bringing together life scientists, mathematicians, computer scientists and engineers to explore a new and deeper understanding of biological systems. The present text has been compilated for computer science and mathematics graduate and level undergraduate students. It will also be of interest to molecular biologists interested in bioinformatics.

The rationable of the book is to present algorithmic ideas in computational biology and to show how they are connected to molecular biology and to biotechnology. To achieve this goal, the book has a substantial "computating biology without formulas" component that presents biological motivation and computational ideas in a simple way. This simplified presentation of biology and computing aims to make the book accessible to computer scientists entering this new areas and to biologists who do not have sufficient background for more involved computational techniques. Every chapter has an introductory section that describes both computational and biological ideas without any formulas. Thus book concentrate on computational ideas rather than details of the algorithms and makes special efforts to prevent there ideas in a simple way.

To make the work more comprehensive and informative, the author has consulted many authoritative books, research journals, abstracts, monographs etc. He is grateful to all those great scholars whose work are cited or substantially reproduced.

There can be no claim to originality except in the manner of treatment and much of the information has been obtained from the books and scientific journals available in the different libraries.

The author expresses his thanks to his friends and colleagues whose continue inspirations have initiated him to bring out this book.

The author expresses his gratitude to Mr. Wasan and staff of M/s Discovery Publishing House Pvt. Ltd. for their whole hearted cooperation in the publication of this book.

In the mean time, the author will remain sincerely responsible for any shortcomings of the book and be grateful to the readers for their suggestions and constructive criticism for the continuous betterment of the book. He takes this opportunity to appeal to the readers to send their suggestions straightaway to his Publisher.

Author

Contents

INTRODUCTION

1

The first hints of the chemical basis of life were noted approximately 150 years ago. Leading up to this initial awareness were a series of insights that living organisms comprise a hierarchy of structures: organs, which are composed of individual cells, which are themselves formed of organelles of different chemical compositions, and so on. From this realization and the observation that nonviable extracts from organisms such as yeast could by themselves catalyze chemical reactions, it became clear that life itself was the result of a complex combination of individual chemicals and chemical reactions. These advances stimulated investigations into the nature of the molecules responsible for biochemical reactions, culminating in the discovery of the genetic code and the molecular structure of deoxyribonucleic acid (DNA) in the early 1950s by Watson and Crick. One of the most fascinating aspects of their discovery was that an understanding of the mechanism by which the genetic code functioned could not be achieved until knowledge of the three-dimensional (3D) structure of DNA was attained.

The discovery of the structure of DNA and its relationship to DNA function had a tremendous impact on all subsequent biochemical investigations, basically defining the paradigm of modern biochemistry and molecular biology. This established the primary importance of molecular structure for an understanding of the function of biological molecules and the need to investigate the relationship between structure and function in order to advance our understanding of the fundamental processes of life. As the molecular structure of DNA was being elucidated, scientists made significant contributions to revealing the structures of proteins and enzymes. Sanger resolved the primary sequence of insulin in 1953, followed by that of an enzyme, ribonuclease A, 10 years later. The late 1950s saw the first high resolution 3D structures of proteins, myoglobin and hemoglobin, as determined by Kendrew *et al.* and Perutz *et al.* respectively, followed by the first 3D structure of an enzyme, lysozyme, by Phillips and coworkers in 1965. Since then, the structures of a very large number of proteins and other biological molecules have been determined.

There are currently over 10,000 3D structures of proteins available along with several hundred DNA and RNA structures and a number of protein-nucleic acid complexes. Prior to the elucidation of the 3D structure of proteins via experimental methods, theoretical approaches made significant inroads toward understanding protein structure. One of the most significant contributions was made by Pauling and Corey in 1951, when they predicted the existence of the main elements of secondary structure in proteins, the α-helix and β-sheet. Their prediction was soon confirmed by Perutz who made the first glimpse of the secondary structure at low resolution.

This landmark work by Pauling and Corey marked the dawn of theoretical studies of biomolecules. It was followed by prediction of the allowed conformations of amino acids, the basic building block of proteins, in 1963 by Ramachandran *et al*. This work, which was based on simple hard-sphere models, indicated the potential of computational approaches as tools for understanding the atomic details of biomolecules. Energy minimization algorithms with an explicit potential energy function followed readily to assist in the refinement of model structures of peptides by Scheraga and of crystal structures of proteins by Levitt and Lifson. The availability of the first protein structures determined by X-ray crystallography led to the initial view that these molecules were very rigid, an idea consistent with the lock-and-key model of enzyme catalysis.

Detailed analysis of protein structures, however, indicated that proteins had to be flexible in order to perform their biological functions. For example, in the case of myoglobin and hemoglobin, there is no path for the escape of O_2 from the heme-binding pocket in the crystal structure; the protein must change structure in order for the O_2 to be released. This and other realizations lead to a rethinking of the properties of proteins, which resulted in a more dynamic picture of protein structure. Experimental methods have been developed to investigate the dynamic properties of proteins; however, the information content from these studies is generally isotropic in nature, affording little insight into the atomic details of these fluctuations.

Atomic resolution information on the dynamics of proteins as well as other biomolecules and the relationship of dynamics to function is an area where computational studies can extend our knowledge beyond what is accessible to experimentalists. The first detailed microscopic view of atomic motions in a protein was provided in 1977 via a molecular dynamics (MD) simulation of bovine pancreatic trypsin inhibitor by McCammon *et al.* This work, marking the beginning of modern computational biochemistry and biophysics, has been followed by a large number of theoretical investigations of many complex biomolecular systems. It is this large body of work, including the numerous methodological advances in computational studies of biomolecules over the last decade, that largely motivated the production of the present book.

BIOMOLECULES

Although the dynamic nature of biological molecules has been well accepted for over 20 years, the extent of that flexibility, as manifested in the large structural changes that biomolecules can undergo, has recently become clearer due to the availability of experimentally determined structures of the same biological molecules in different environments. For example, the enzyme triosephosphate isomerase contains an 11 amino acid residue loop that moves by more than 7 Å following the binding of substrate, leading to a catalytically competent structure. In the enzyme cytosine-5-methyltransferase, a loop containing one of the catalytically essential residues undergoes a large conformational change upon formation of the DNA-coenzyme-protein complex, leading to some residues changing position by over 20 Å. DNA, typically envisioned in the canonical B form, has been shown to undergo significant distortions upon binding to proteins. Bending of 90° has been seen in the CAP-DNA complex, and binding of the TATA box binding protein to the TATAAAA consensus sequence leads to the DNA assuming a unique conformation referred to as the TA form.

Even though experimental studies can reveal the end points associated with these conformational transitions, these methods typically cannot access structural details of the pathway between the end points. Such information is directly accessible via computational approaches. Computational approaches can be used to investigate the energetics associated with changes in both conformation and chemical structure. An example is afforded by the conformational transitions discussed in the preceding paragraph. Conformational free energy differences and barriers can be calculated and then directly compared with experimental results. Overviews of these methods are included.

Recent advances in techniques that combine quantum mechanical (QM) approaches with molecular mechanics (MM) now allow for a detailed understanding of processes involving bond breaking and bond making and how enzymes can accelerate those reactions. Gives a detailed overview of the implementation and current status of QM/MM methods. The ability of computational biochemistry to reveal the microscopic events controlling reaction rates and equilibrium at the atomic level is one of its greatest strengths. Biological membranes provide the essential barrier between cells and the organelles of which cells are composed. Cellular membranes are complicated extensive biomolecular sheet-like structures, mostly formed by lipid molecules held together by cooperative noncovalent interactions.

A membrane is not a static structure, but rather a complex dynamical two-dimensional liquid crystalline fluid mosaic of oriented proteins and lipids. A number of experimental approaches can be used to investigate and characterize biological membranes. However, the complexity of membranes is such that experimental data remain very difficult to interpret at the microscopic level. In recent years, computational studies of membranes based on detailed atomic models, as summarized, have greatly increased the ability to interpret experimental data, yielding a much-improved picture of the structure and dynamics of lipid bilayers and the relationship of those properties to membrane function. Computational approaches are now being used to facilitate the experimental determination of macromolecular structures by aiding in structural refinement based on either nuclear magnetic resonance (NMR) or X-ray data.

The current status of the application of computational methods to the determination of biomolecular structure and dynamics is presented. Computational approaches can also be applied in situations where experimentally determined structures are not available. With the rapid advances in gene technology, including the human genome project, the ability of computational approaches to accurately predict 3D structures based on primary sequence represents an area that is expected to have a significant impact. Prediction of the 3D structures of proteins can be performed via homology modeling or threading methods. Related to this is the area of protein folding. As has been known since the seminal experimental refolding studies of ribonuclease A in the 1950s, the primary structure of many proteins dictates their 3D structure. Accordingly, it should be possible "in principle" to compute the 3D structure of many proteins based on knowledge of just their primary sequences. Although this has yet to be achieved on a wide scale, considerable efforts are being made to attain this goal, as overviewed. Drug design and development is another area of research where computational biochemistry and biophysics are having an ever-increasing impact.

Computational approaches can be used to aid in the refinement of drug candidates, systematically changing a drug's structure to improve its pharmacological properties, as well as in the identification of novel lead compounds. The latter can be performed via the identification of compounds with a high potential for activity from available databases of chemical compounds or via de novo drug design approaches, which build totally novel ligands into the binding sites of target molecules. Techniques used for these types of studies are presented. In addition to aiding in the design of compounds that target specific molecules, computational approaches offer the possibility of being able to improve the ability of drugs to access their targets in the body. These gains will be made through an understanding of the energetics associated with the crossing of lipid membranes and using the information to rationally enhance drug absorption rates.

As evidenced by the recent contribution of computational approaches in the development of inhibitors of the HIV protease, many of which are currently on the market, it can be expected that these methods will continue to have an increasing role in drug design and development. Clearly, computational and theoretical studies of biological molecules have advanced significantly in recent years and will progress rapidly in the future. These advances have been partially fueled by the ever-increasing number of available structures of proteins, nucleic acids, and carbohydrates, but at the same time significant methodological improvements have been made in the area of physics relevant to biological molecules.

These advances have allowed for computational studies of biochemical processes to be performed with greater accuracy and under conditions that allow for direct comparison with experimental studies. Examples include improved force fields, treatment of long-range atom-atom interactions, and a variety of algorithmic advances, as covered. The combination of these advances with the exponential increases in computational resources has greatly extended and will continue to expand the applicability of computational approaches to biomolecules.

SCOPE OF THE BOOK

The overall scope of this book is the implementation and application of available theoretical and computational methods toward understanding the structure, dynamics, and function of biological molecules, namely proteins, nucleic acids, carbohydrates, and membranes. The large number of computational tools already available in computational chemistry preclude covering all topics, as Schleyer et al. are doing in *The Encyclopedia of Computational Chemistry*. Instead, we have attempted to create a book that covers currently available theoretical methods applicable to biomolecular research along with the appropriate computational applications. We have designed it to focus on the area of biomolecular computations with emphasis on the special requirements associated with the treatment of macromolecules. In combination, the book should serve as a useful reference for both theoreticians and experimentalists in all areas of biophysical and biochemical research. Its content represents progress made over the last decade in the area of computational biochemistry and biophysics. Books by Brooks *et al.* and McCammon and Harvey are recommended for an overview of earlier developments in the field. Although efforts have been made to include the most recent advances in the field along with the underlying fundamental concepts, it is to be expected that further advances will be made even as this book is being published.

TOWARD A NEW ERA

The 1998 Nobel Prize in Chemistry was given to John A. Pople and Walter Kohn for their work in the area of quantum chemistry, signifying the widespread acceptance of computation as a valid tool for investigating chemical phenomena. With its extension to bimolecular systems, the range of possible applications of computational chemistry was greatly expanded. Though still a relatively young field, computational biochemistry and biophysics is now pervasive in all aspects of the biological sciences. These methods have aided in the interpretation of experimental data, and will continue to do so, allowing for the more rational design of new experiments, thereby facilitating investigations in the biological sciences. Computational methods will also allow access to information beyond that obtainable via experimental techniques. Indeed, computer-based approaches for the study of virtually any chemical or biological phenomena may represent the most powerful tool now available to scientists, allowing for studies at an unprecedented level of detail. It is our hope that the present book will help expand the accessibility of computational approaches to the vast community of scientists investigating biological systems.

DYNAMICS OF BIOLOGICAL MOLECULES

2

One of the major uses of molecular simulation is to provide useful theoretical interpretation of experimental data. Before the advent of simulation this had to be done by directly comparing experiment with analytical (mathematical) models. The analytical approach has the advantage of simplicity, in that the models are derived from first principles with only a few, if any, adjustable parameters. However, the chemical complexity of biological systems often precludes the direct application of meaningful analytical models or leads to the situation where more than one model can be invoked to explain the same experimental data. Computer simulation gets around this problem by allowing more complicated, detailed, and realistic models of biological systems to be investigated.

However, the price to pay for this is a degree of mathematical defeat, as the extra complexity prevents analytical solution of the relevant equations. We must therefore rely on numerical methods and hope that when allied with sufficient computer power they will lead to converged solutions of the equations we wish to solve. The simulation methods used involve empirical potential energy functions of the molecular mechanics type, and the equations of motion are solved using either stepwise integration of the full anharmonic function by molecular dynamics (MD) simulation or, for vibrating systems, by normal mode analysis involving a harmonic approximation to the potential function. In the case of molecular dynamics simulations of the dynamic properties of proteins, convergence is particularly difficult to achieve, because motions occur in proteins on time scales that are the same as or longer than the presently accessible time scale (nanoseconds). In the future things should improve, due to more efficient algorithms and the continuing rapid increase in computer power.

Moreover, many dynamical phenomena of physical and biological interest occur on the subnanosecond time scale and thus can already be well sampled. The examples given in this chapter are mainly on this time scale. Computer simulation can be used to provide a stepping stone between experiment and the simplified analytical descriptions of the physical behavior of biological systems. But before gaining the right to do this, we must first validate a simulation by direct comparison with experiment. To do this we must compare physical quantities that are measurable or derivable from measurements with the same quantities derived from simulation. If the quantities agree, we then have some justification for using the detailed information present in the simulation to interpret the experiments. The spectroscopic techniques that have been most frequently used to investigate biomolecular dynamics are those that are commonly available in laboratories, such as nuclear magnetic resonance (NMR), fluorescence, and Mossbauer spectroscopy.

Here we examine scattering of X-ray and neutron radiation. Neutrons and X-rays share the property of being found in expensive sources not commonly available in the laboratory. Neutrons are produced by a nuclear reactor or "spallation" source. X-ray experiments are routinely performed using intense synchrotron radiation, although in favorable cases laboratory sources may also be used.

The X-ray and neutron scattering processes provide relatively direct spatial information on atomic motions via determination of the wave vector transferred between the photon/neutron and the sample; this is a Fourier transform relationship between wave vectors in reciprocal space and position vectors in real space.

Neutrons, by virtue of the possibility of resolving their energy transfers, can also give information on the time dependence of the motions involved. The comparison with experiment can be made at several levels. The first, and most common, is in the comparison of "derived" quantities that are not directly measurable, for example, a set of average crystal coordinates or a diffusion constant. A comparison at this level is convenient in that the quantities involved describe directly the structure and dynamics of the system. However, the obtainment of these quantities, from experiment and/or simulation, may require approximation and model-dependent data analysis. For example, to obtain experimentally a set of average crystallographic coordinates, a physical model to interpret an electron density map must be imposed.

To avoid these problems the comparison can be made at the level of the measured quantities themselves, such as diffraction intensities or dynamic structure factors. A comparison at this level still involves some approximation. For example, background corrections have to made in the experimental data reduction. However, fewer approximations are necessary for the structure and dynamics of the sample itself, and comparison with experiment is normally more direct. This approach requires a little more work on the part of the computer simulation team, because methods for calculating experimental intensities from simulation configurations must be developed. The comparisons made here are of experimentally measurable quantities. Having made the comparison with experiment one may then make an assessment as to whether the simulation agrees sufficiently well to be useful in interpreting the experiment in detail.

In cases where the agreement is not good, the determination of the cause of the discrepancy is often instructive. The errors may arise from the simulation model or from the assumptions used in the experimental data reduction or both. In cases where the quantities examined agree, the simulation can be decomposed so as to isolate the principal components responsible for the observed intensities. Sometimes, then, the dynamics involved can be described by a simplified concept derived from the simulation.

BASIC EQUATIONS RELATING ATOMIC POSITIONS TO X-RAY AND NEUTRON SCATTERING

Scattering experiments involve processes in which incident particles (X-rays or neutrons) with wave vector $\vec{k}_i$ (energy E_i) interact with the sample and emerge with wave vector $\vec{k}_f$ (energy E_f), obeying the conservation laws for momentum and energy,

$$\mathbf{k}_i - \mathbf{k}_f = \mathbf{Q} \quad ...(1)$$

$$E_i - E_f = \hbar\omega \quad ...(2)$$

The vectors $\mathbf{k}_i$, $\mathbf{k}_f$, and $\mathbf{Q}$ define the scattering geometry.

We first examine the relationship between particle dynamics and the scattering of radiation in the case where both the energy and momentum transferred between the sample and the incident radiation are measured. Linear response theory allows dynamic structure factors to be written in terms of equilibrium fluctuations of the sample. For neutron scattering from a system of identical particles, this is

$$S_{\text{coh}}(\vec{Q},\omega) = \frac{1}{2\pi}\iint dt\, d^3r e^{i(\vec{Q}\cdot\vec{r}-\omega t)} G(\vec{r},t) \quad ...(3)$$

$$S_{\text{inc}}(\vec{Q},\omega) = \frac{1}{2\pi}\iint dt\, d^3r e^{i(\vec{Q}\cdot\vec{r}-\omega t)} G_s(\vec{r},t) \quad ...(4)$$

where $\bar{Q}$ is the scattering wave vector, ω the energy transfer, and the subscripts coh and inc refer to coherent and incoherent scattering, discussed later. $G_s(\bar{r}, t)$ and $G(\bar{r}, t)$ are van Hove correlation functions, which, for a system of N particles undergoing classical dynamics, are defined as follows:

$$G(\vec{r},t) = \frac{1}{N}\sum_i \left\langle \delta(\vec{r} - \vec{R}_i(t) + \vec{R}_j(0)) \right\rangle \quad ...(5)$$

$$G_s(\vec{r},t) = \frac{1}{N}\sum_i \left\langle \delta(\vec{r} - \vec{R}_i(t) + \vec{R}_j(0)) \right\rangle \quad ...(6)$$

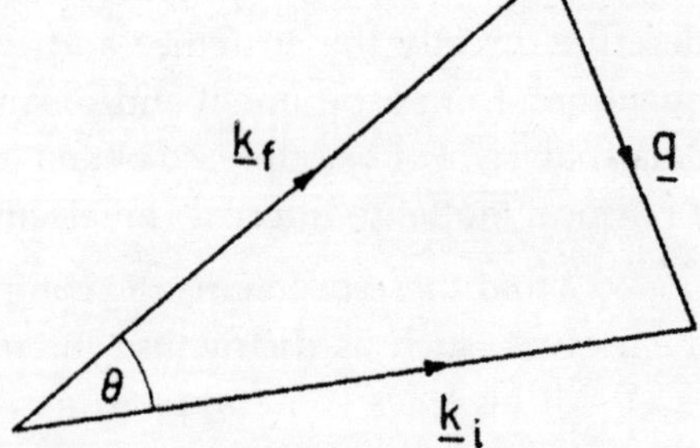

Fig. 2.1. Scattering vector triangle.

where $\vec{R}_i(t)$ is the position vector of the fth scattering nucleus and $\langle\ldots\rangle$ indicates an ensemble average.

$G(\bar{r}, t)$ is the probability that, given a particle at the origin at time $t - 0$, any particle (including the original particle) is at $\bar{r}$ at time t. $G_s(\bar{r}, t)$ is the probability that, given a particle at the origin at time $t = 0$, the same particle is at $\bar{r}$ at time t.

Equation (3) has an analogous form in X-ray scattering, where the scattered intensity is given as

$$|F(\vec{Q},\omega)|^2 = \frac{1}{2\pi}\iint d^3r\, dtP(\vec{r},t)\, e^{i(\vec{Q}.\vec{r}-\omega t)} \quad ...(7)$$

where $P(\bar{r}, t)$ is the spatiotemporal Patterson function given by

$$P(\bar{r},t) = \iint d\vec{R}\, dt\, \rho(\bar{R},t)\, \rho(\vec{r}+\vec{R}, t+\tau) \quad ...(8)$$

and $\rho(\bar{r}, t)$ is the time-dependent electron density. Unfortunately, X-ray photons with wavelengths corresponding to atomic distances have energies much higher than those associated with thermal fluctuations. For example, an X-ray photon of 1.8 Å wavelength has an energy of 6.9 keV corresponding to a temperature of 8×10^7 K. Until very recently X-ray detectors were not sensitive enough to accurately measure the minute fractional energy changes associated with molecular fluctuations, so the practical exploitation of Eq. (7) was difficult. However, the use of new third-generation synchrotron sources has enabled useful inelastic X-ray scattering experiments to be performed on, for example, glass-forming liquids and liquid water.

Although inelastic X-ray scattering holds promise for the investigation of biological molecular dynamics, it is still in its infancy, so in the subsequent discussion X-ray scattering is examined only in the case in which inelastic and elastic scattering are indistinguishable experimentally.

SCATTERING BY CRYSTALS

In an X-ray crystallography experiment the instantaneous scattered intensity is given by

$$I_{hkl} = |F_{hkl}|^2 = \sum_{i=1}^{N}\sum_{j=1}^{N} f_i f_j^* \exp[i\vec{Q}\cdot(\vec{R}_i - \vec{R}_j)] \qquad ...(9)$$

where F_{hkl} is the structure factor, $\bar{R}_i$ is the position vector of atom i in the crystal, and f_i is the X-ray atomic form factor. For neutron diffraction f_i is replaced by the coherent scattering length.

In an experimental analysis it is not feasible to insert into Eq. (9) the atomic positions for all the atoms in the crystal for every instant in the time of the experiment. Rather, the intensity must be evaluated in terms of statistical relationships between the positions. A convenient approach is to consider a real crystal as a superposition of an ideal periodic structure with slight perturbations. When exposed to X-rays the real crystal gives rise to two scattering components: the set of Bragg reflections arising from the periodic structure, and scattering outside the Bragg spots (diffuse scattering) that arises from the structural perturbations:

$$I_{hkl} = I_{hkl}^{B} + I_{hkl}^{D} \qquad ...(10)$$

where I_{hkl}^{B} is the Bragg intensity found at integer values of h, k, and l, and I_{hkl}^{D} is the diffuse scattering, not confined to integer values of h, k, and l.

In terms of structure factors the various intensities are given by

$$I_{hkl} = |F_{hkl}|^2 \qquad ...(11)$$

$$I_{hkl}^{B} = |\langle F_{hkl}\rangle|^2 \qquad ...(12)$$

and

$$I_{hkl}^{D} = |\Delta F_{hkl}|^2 \qquad ...(13)$$

A. Bragg Diffraction

The Bragg peak intensity reduction due to atomic displacements is described by the well-known "temperature" factors. Assuming that the position $\vec{r}_i$ can be decomposed into an average position, $\langle \vec{r}_i \rangle$ and an infinitesimal displacement, $\vec{u}_i = \delta\vec{R}_i = \vec{R}_i - \langle \vec{R}_i \rangle$ then the X-ray structure factors can be expressed as follows:

$$F_{hkl} = \sum_{i=1}^{N} f_i(\vec{Q}) \exp[i\vec{Q}\cdot\langle \vec{R}_i \rangle \exp[W_i(Q)] \qquad ...(14)$$

where $W_i(Q) = -(1/3)\langle u_{i,Q}^2 \rangle Q^2$ and $\langle u_{i,Q}^2 \rangle$ is the mean-square fluctuation in the direction of Q. $W_i(Q)$ is the Debye-Waller factor.

Example: *Low Temperature Vibrations in Acetanilide*

Temperature factors are of interest to structural biologists mainly as a means of deriving qualitative information on the fluctuations of segments of a macromolecule. However, X-ray temperature factor analysis has drawbacks. One of the most serious is the possible presence of a static disorder contribution to the atomic fluctuations. This cannot be distinguished from the dynamic disorder due to the aforementioned absence of accurate energy analysis of the scattered X-ray photons. For quantitative work this problem can be avoided by choosing a system in which there is negligible static disorder and in which the harmonic approximation is valid. An example of such a system is acetanilide, ($C_6H_5-CONH-CH_3$), at 15 K. The structure of acetanilide is shown in Figure 2.2. In recent work a molecular mechanics force field was parametrized for this crystal, and normal mode analysis was performed in the full configurational space of the crystal *i.e.*, including all intramolecular and intermolecular degrees of freedom. As a quantitative test of the accuracy of the force field, anisotropic quantum mechanical mean-square displacements of the hydrogen atoms were calculated in each Cartesian direction as a sum over the phonon normal modes of the crystal and compared with experimental neutron diffraction temperature factors. The experimental and theoretical temperature factors are presented in Table 2.1. The values of the mean-square displacements are in excellent agreement.

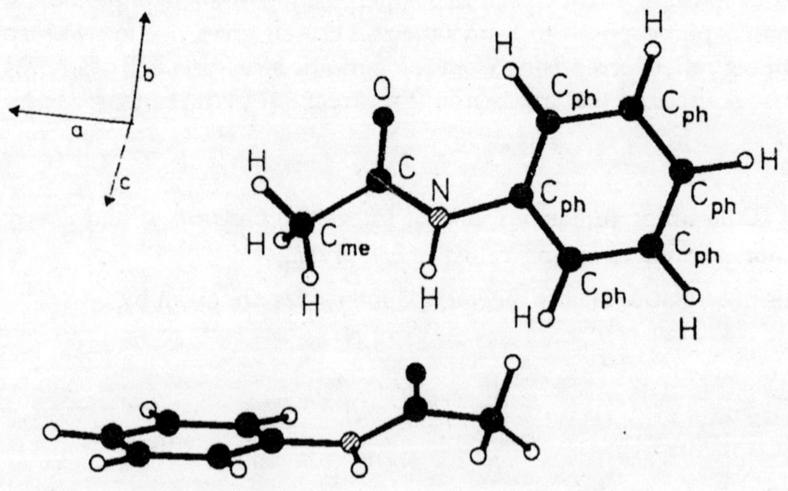

Fig. 2.2. Crystalline acetanilide.

B. X-Ray Diffuse Scattering

Any perturbation from ideal space-group symmetry in a crystal will give rise to diffuse scattering. The X-ray diffuse scattering intensity I^D_{hkl} at some point *(hkl)* in reciprocal space can be written as

$$I^D_{hkl} = N\sum_m \langle (F_n - \langle F \rangle)(F_{n+m} - \langle F \rangle^*) \rangle \exp(-\vec{Q}\cdot\vec{R}_m) \qquad ...(15)$$

where F_n is the structure factor of the nth unit cell and the sum Σ_m runs over the relative position vectors $\vec{R}_m$ between the unit cells. The correlation function $\langle (F_n - \langle F \rangle)(F_{n+m} - \langle F \rangle)^* \rangle$ is determined by correlations between atomic displacements.

If the diffuse scattering of dynamic origin contributes significantly to the measured scattering, it may provide information on the nature of correlated motions in biological macromolecules that may themselves be of functional significance. To examine this possibility it is necessary to construct dynamic models of the crystal, calculate their diffuse scattering, and compare with experimental results. The advent of high intensity synchrotron sources and image plate detectors has allowed good quality X-ray diffuse scattering images to be obtained from macromolecular crystals.

SERENA (scattering of X-rays elucidated by numerical analysis) is a program for calculating X-ray diffuse scattering intensities from configurations of atoms in molecular crystals. The configurations are conveniently derived from molecular dynamics simulations, although in principle any collection of configurations can be used. SERENA calculates structure factors from the individual configurations and performs the averaging required in Eq. (11).

Displacements correlated within unit cells but not between them lead to very diffuse scattering that is not associated with the Bragg peaks. This can be conveniently explored using present-day simulations of biological macromolecules. However, motions correlated over distances larger than the size of the simulation model will clearly not be included.

Table 2.1 Mean-Square Displacements of Hydrogen Atoms in Crystalline Acetanilide at 15 K[a]

I		*II*	*III*	*I*		*II*	*III*
Methyl H	*a*	0.0359	0.0366	Phenyl H_{para}	*a*	0.126	0.0118
	b	0.0366	0.0438		*b*	0.0273	0.0278
	c	0.0253	0.0258		*c*	0.0279	0.0258
isotropic		0.0326	0.0349	isotropic		0.0226	0.0218
Methyl H	*a*	0.0150	0.0160	Phenyl H_{meta}	*a*	0.0215	0.0220
	b	0.0482	0.0510		*b*	0.0189	0.0194
	c	0.0328	0.0336		*c*	0.0265	0.0246
isotropic		0.0320	0.0335	isotropic		0.0223	0.0220
Methyl H	*a*	0.0301	0.0286	Phenyl H_{meta}	*a*	0.0154	0.0162
	b	0.0162	0.0172		*b*	0.0218	0.0210
	c	0.0483	0.0606		*c*	0.0296	0.0256
isotropic		0.0315	0.0354	isotropic		0.0222	0.0209
Amide H	*a*	0.0178	0.0192	Phenyl H_{ontho}	*a*	0.0159	0.0160
	b	0.0110	0.0120		*b*	0.0189	0.0194
	c	0.0258	0.0300		*c*	0.0286	0.0250
isotropic		0.0182	0.0204	isotropic		0.0211	0.0201
				Phenyl H_{ontho}	*a*	0.0203	0.0200
					b	0.0170	0.0164
					c	0.0264	0.0254
				isotropic		0.0212	0.0206

* Column I: Hydrogen atoms of acetanilide. Phenyl hydrogens are named according to their position relative to the N substitution site, *a, b, c* refer to the crystallographic directions. Column II: Anisotropic (*a, b, c* crystallographic directions) and isotropic mean-square displacements (Å^2) from neutron diffraction data [9]. Column III: Anisotropic (*a, b, c* crystallographic directions) and isotropic mean-square displacements (Å^2) from harmonic analysis.

Example: *Correlated Motions in Lysozyme*

Ligand binding and cooperativity often require conformational change involving correlated displacements of atoms. A simple model for long distance transmission of information across a protein involves the activation and amplification of correlated motions that are present in the unperturbed protein. Although long-range correlated fluctuations are required for functional, dynamic information transfer, it is not clear to what extent they contribute to equilibrium thermal fluctuations in proteins. It is therefore important to know whether equilibrium motions in proteins can indeed be correlated over long distances or whether anharmonic and damping effects destroy such correlations. To examine the dynamic origins of X-ray diffuse scattering by proteins, experimental scattering was measured from orthorhombic lysozyme crystals and compared to patterns calculated using molecular simulation.

The diffuse scattering was found to be approximately reproduced by both normal modes and molecular dynamics. More recently, a molecular dynamics analysis was performed of the dynamics of a unit cell of orthorhombic lysozyme, including four protein molecules. Diffuse scattering calculated from the MD trajectory is compared in Fig. 2.3 with that calculated from trajectories of rigid bodies obtained by fitting to the full simulation. The full simulation scattering is reproduced by the approximate representation to an agreement factor (R-factor) of 6%.

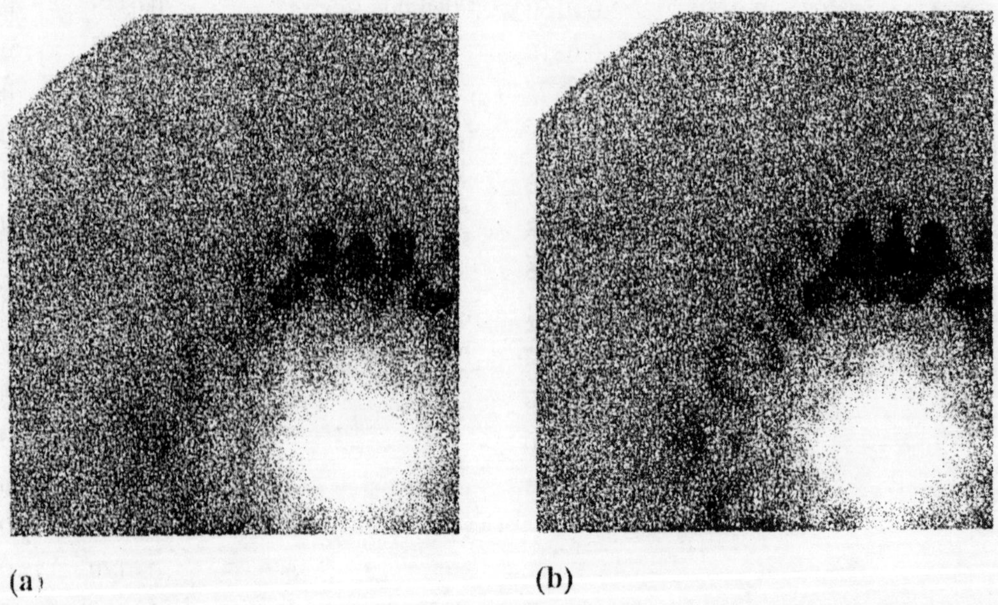

Fig. 2.3. Calculated X-ray diffuse scattering patterns from (*a*) a full molecular dynamics trajectory of orthorhombic hen egg white lysozyme and (*b*) a trajectory obtained by fitting to the full trajectory rigid-body side chains and segments of the backbone.

NEUTRON SCATTERING

In contrast to X-rays, the mass of the neutron is such that the energy exchanged in exciting or deexciting picosecond time scale thermal motions is a large fraction of the incident energy and can be measured relatively precisely. A thermal neutron of 1.8 Å wavelength has an energy of 25 meV corresponding to k_bT at 300 K. To further examine the neutron scattering case, we perform space Fourier transformation of the van Hove correlation functions [Eqs. (3) and (4)]:

$$S_{\text{coh}}(\vec{Q}, \omega) = \frac{1}{2}\int_{\infty}^{+\infty} dte^{i\omega t} I_{coh}(\vec{Q}, t) \quad ...(16)$$

$$I_{\text{coh}}(\vec{Q}. t) = \frac{1}{N}\sum_{i,j} b^{*}_{i,\text{coh}} b_{j,\text{coh}} e^{-\vec{Q}\cdot\vec{R}_i(0)} \left\langle e^{i\vec{Q}\cdot\vec{R}_j(t)} \right\rangle \quad ...(17)$$

$$I_{\text{inc}}(\vec{Q}. \omega) = \frac{1}{2\pi}\int_{\infty}^{+\infty} dte^{i\omega t} I_{\text{inc}}(\vec{Q}, t) \quad ...(18)$$

$$I_{\text{inc}}(\vec{Q}. t) = \frac{1}{N}\sum_{i} b^{2}_{i,\text{coh}} \left\langle e^{-i\vec{Q}\cdot\vec{R}_i(0)} e^{i\vec{Q}\cdot\vec{R}_i(t)} \right\rangle \quad ...(19)$$

Neutrons are scattered by the nuclei of the sample. Because of the random distribution of nuclear spins in the sample, the scattered intensity will contain a *coherent* part arising from the average neutron-nucleus potential and an *incoherent* part arising from fluctuations from the average. The coherent scattering arises from self- and cross-correlations of atomic motions, and the incoherent scattering, from single-atom motions. Each isotope has a coherent scattering length $b_{,\text{coh}}$ and an incoherent scattering length $b_{,\text{iinc}}$ that define the strength of the interaction between the nucleus of the atom and the neutron. We see from Eqs. (16) and that the coherent and incoherent dynamic structure factors are time Fourier transforms of the coherent and incoherent intermediate scattering functions, $I_{\text{coh}}(\vec{Q}, t)$; and $I_{\text{inc}}(\vec{Q}, t)$; these are time-correlation functions. $S_{\text{inc}}(\vec{Q}, \omega)$ and $S_{\text{coh}}(\vec{Q}, \omega)$ may contain elastic ($\omega = 0$) and inelastic ($\omega \neq 0$) parts. Elastic scattering probes correlations of atomic positions at long times, whereas the inelastic scattering process probes position correlations as a function of time.

A. Coherent Inelastic Neutron Scattering

The use of coherent neutron scattering with simultaneous energy and momentum resolution provides a probe of time-dependent pair correlations in atomic motions. Coherent inelastic neutron scattering is therefore particularly useful for examining lattice dynamics in molecular crystals and holds promise for the characterization of correlated motions in biological macro-molecules. A property of lattice modes is that for particular wave vectors there are well-defined frequencies; the relations between these two quantities are the phonon dispersion relations. Neutron scattering is presently the most effective technique for determining phonon dispersion curves. The following momentum conservation law is obeyed:

$$\mathbf{k}_i - \mathbf{k}_f = \mathbf{Q} = \boldsymbol{\tau} + \mathbf{q} \quad ...(20)$$

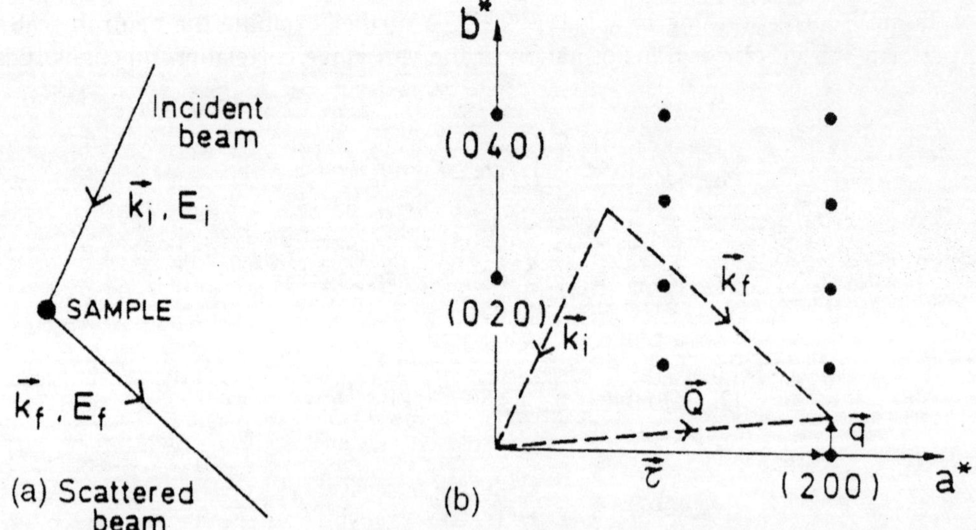

Fig. 2.4. Schematic vector diagrams illustrating the use of coherent inelastic neutron scattering to determine phonon dispersion relationships, (*a*) Scattering in real space; (*b*) a scattering triangle illustrating the momentum transfer, Q, of the neutrons in relation to the reciprocal lattice vector of the sample τ and the phonon wave vector, q. Heavy dots represent Bragg reflections.

The vibrational excitations have a wave vector q that is measured from a Brillouin zone center (Bragg peak) located at τ, a reciprocal lattice vector.

If the displacements of the atoms are given in terms of the harmonic normal modes of vibration for the crystal, the coherent one-phonon inelastic neutron scattering cross section can be analytically expressed in terms of the eigenvectors and eigenvalues of the harmonic analysis.

Example: *Lattice Vibrations in L-Alanine*

Zwitterionic L-alanine ($^{+}H_3N-C(CH_3)-CO_2-$) is a dipolar molecule that forms large well-ordered crystals in which the molecules form hydrogen-bonded columns. The strong interactions lead to the presence of well-defined intra- and intermolecular vibrations that can usefully be described using harmonic theory.

Coherent inelastic neutron scattering experiments have been combined with normal mode analyses with a molecular mechanics potential function to examine the collective vibrations in deuterated L-alanine. In Fig. 2.5 are shown experimental phonon frequencies ν_i (**q**) ($\nu = \omega/2\pi$) for several modes propagating along the crystallographic direction **b***. The solid lines represent the most probable paths for the dispersion curves ν_i(**q**). The theoretical dispersion curves are also given. The comparison between theory and experiment can be used to assess the accuracy with which the theory reproduces long-range interactions in the crystal.

B. Incoherent Neutron Scattering

Neutron scattering from nondeuterated organic molecules is dominated by incoherent scattering from the hydrogen atoms. This is largely because the incoherent scattering cross section ($4\pi b^2_{inc}$) of hydrogen is approximately 15 times greater than the total scattering cross section of carbon, nitrogen, or oxygen. The measured incoherent scattering thus essentially gives information on self-correlations of hydrogen atom motions. A program for calculating neutron scattering properties from molecular dynamics simulations has been published.

In practice, the measured incoherent scattering energy spectrum is divided into elastic, quasielastic, and inelastic scattering. Inelastic scattering arises from vibrations. Quasi-elastic scattering is typically Lorentzian or a sum of Lorentzians centered on $\omega = 0$ and arises from diffusive motions in the sample. Elastic scattering gives information on the self-probability distributions of the hydrogen atoms in the sample. It is assumed that the atomic position vectors can be decomposed into two contributions, one due to diffusive motion, $\vec{r}_{i,d}(t)$, and the other from vibrations, $\vec{u}_{i,v}(t)$, *i.e.*,

$$\vec{R}_i(t) = \vec{r}_{i,d}(t) + \vec{u}_{i,v}(t) \qquad ...(21)$$

Combining Eq. (21) with Eq. (19) and assuming that $\vec{r}_{i,d}(t)$ and $\vec{u}_{i,v}(t)$ are uncorrelated, one obtains

$$I_{\text{inc}}(\vec{Q}, t) = I_d(\vec{Q}, t) I_v(\vec{Q}, t) \qquad ...(22)$$

where $I_d(\vec{Q}, t)$ and $I_v(\vec{Q}, t)$ are obtained by substituting $\vec{R}_i(t)$ in Eq. (19) with $\vec{r}_{i,d}(t)$ and $\vec{u}_{i,v}(t)$, respectively.

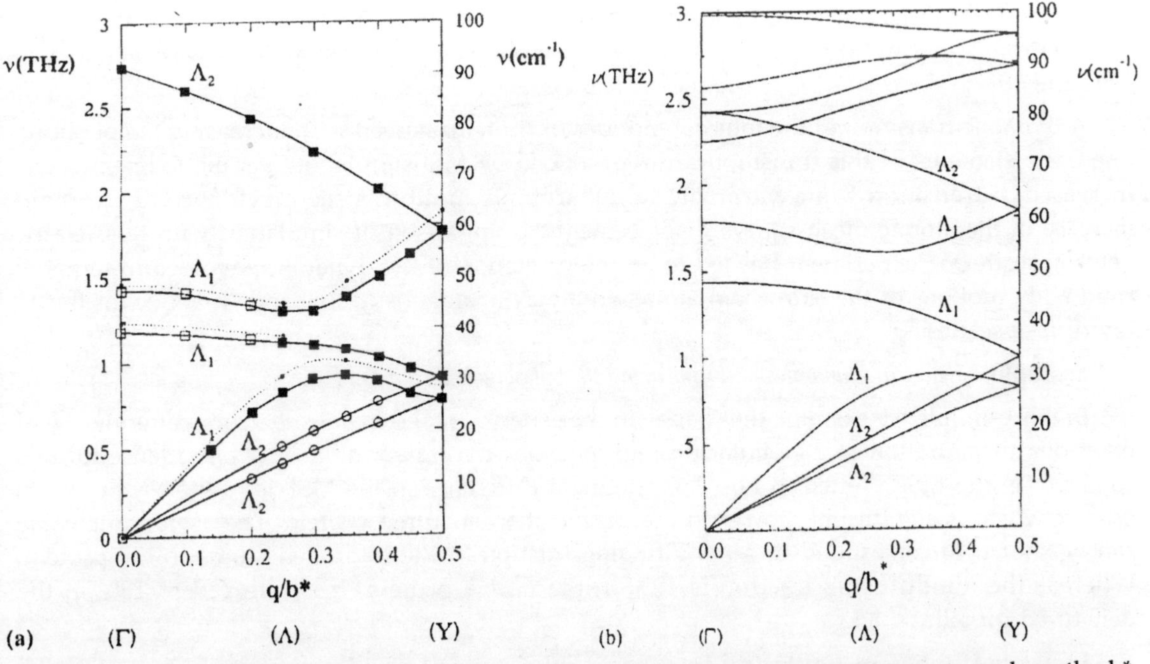

Fig. 2.5. **(*a*) Dispersion curves for crystalline zwitterionic L-alanine at room temperature along the b* crystallographic direction determined by coherent inelastic neutron scattering. The ○ and ■ symbols are associated with phonon modes observed in predominantly transverse and purely longitudinal configurations, respectively, *i.e.*, for vectors Q and q perpendicular and parallel to one another, respectively. They correspond to measurements performed around the strong Bragg reflections. The □ symbols are neutron data points obtained around the reciprocal lattice points in a mixed configuration. Solid lines indicate the most probable connectivity of the dispersion curves, and dashed lines correspond to the measurements performed at low temperature T = 100 K. (*b*) Theoretical dispersion curves for L-alanine determined from normal mode analysis.**

The Fourier transform of Eq. (22) gives

$$S(\vec{Q}, \omega) = S_d(\bar{Q}, \omega) \otimes S_v(\bar{Q}, \omega) \quad ...(23)$$

where $S_d(\bar{Q}, \omega)$ and $S_v(\bar{Q}, \omega)$ are obtained by Fourier transformation of $I_d(\bar{Q}, t)$ and $I_v(\bar{Q}, t)$ and the symbol $\otimes$ denotes the convolution product. Appropriate descriptions of $\vec{r}_{i,d}(t)$ and $\vec{u}_{i,v}(t)$ can be obtained from analytical theory or computer simulation.

$I_d(\vec{Q}, t)$ can be separated into time-dependent and time-independent parts as follows:

$$I_d(\vec{Q}, t) = A_0(\vec{Q}) + I_d'(\vec{Q}, t) \quad ...(24)$$

The elastic incoherent structure factor (EISF), $A_0(\bar{Q})$, is defined as

$$A_0(\bar{Q}) = \lim_{t \to \infty} I_d(\vec{Q}, t) = \int d^3 r e^{i\vec{q}\cdot\vec{r}} \lim_{t \to \infty} G_d(\vec{r}, t) \quad ...(25)$$

where $G_d(\vec{r}, t)$ is the contribution to the van Hove self-correlation function due to diffusive motion. $A_0(Q)$ is thus determined by the diffusive contribution to the space probability distribution of the hydrogen nuclei.

Direct experiment-simulation quasielastic neutron scattering comparisons have been performed for a variety of small molecule and polymeric systems. The combination of simulation and neutron scattering in the analysis of internal motions in globular proteins was reviewed in 1991 and 1997.

A dynamic transition in the internal motions of proteins is seen with increasing temperature. The basic elements of this transition are reproduced by MD simulation. As the temperature is increased, a transition from, harmonic to anharmonic motion is seen, evidenced by a rapid increase in the atomic mean-square displacements. Comparison of simulation with quasielastic neutron scattering experiment has led to an interpretation of the dynamics involved in terms of rigid-body motions of the side chain atoms, in a way analogous to that shown above for the X-ray diffuse scattering.

Example: *Change in Dynamics on Denaturing Phosphoglycerate Kinase*

In this example we examine the change in the experimental dynamic neutron scattering signal on strong denaturation of a globular protein, phosphoglycerate kinase (PGK). Evidence for this comes from the EISF plotted in Fig. 2.6. The main difference in the EISF is in the asymptote as $Q \to \infty$ which is significantly lower in the case of the denatured protein. The asymptotic value can be shown to correspond to a nondiffusing fraction of the hydrogen atoms in the protein. Whereas the nondiffusing fraction is 40% in the native protein, it is reduced to 18% in the denatured protein.

Inelastic Incoherent Scattering Intensity : For a system executing harmonic dynamics, the transform in Eq. (4) can be performed analytically and the result expanded in a power series over the normal modes in the sample. The following expression is obtained :

$$S_{inc}(\vec{Q}, \omega) = \sum_i b_{inc}^2 \exp[-2W_i(\vec{Q})]$$

$$\times \prod_\lambda \left[\sum_{n\lambda} \exp\left(\frac{n_\lambda \hbar \omega_\lambda \beta}{2} \right) I_{n\lambda} \left(\frac{\hbar (\vec{Q} \cdot \vec{e}_{\lambda,i})^2}{2M\omega_\lambda \sinh(\hbar\omega_\lambda \beta/2} \right) \delta\left(\omega - \sum_\lambda n_\lambda \omega_\lambda \right) \right] \quad ...(26)$$

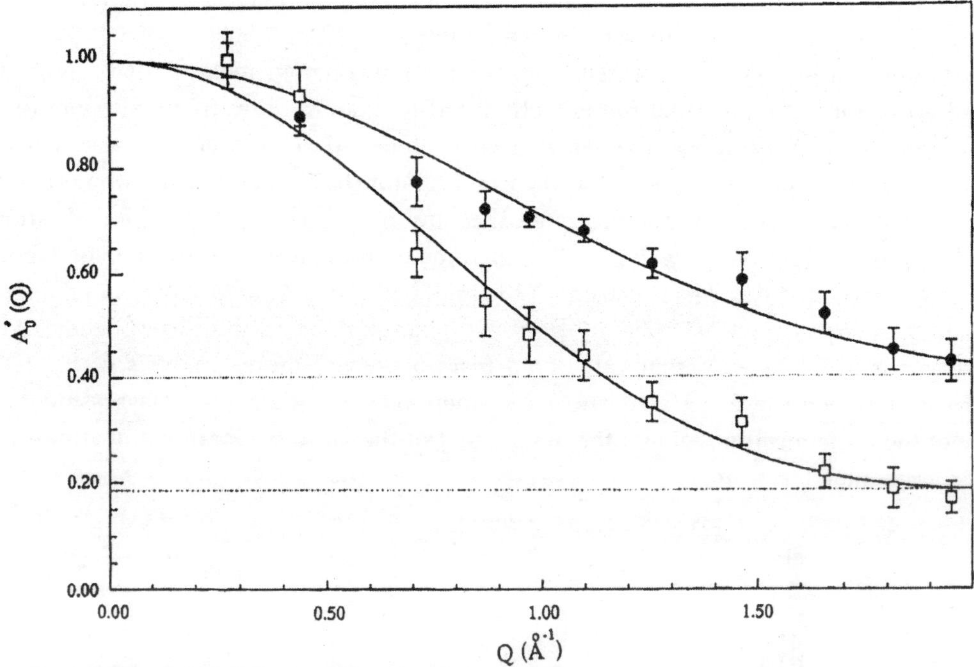

Fig. 2.6. Apparent elastic incoherent structure factor $A'_0(Q)$ for (□) denatured and (●) native phosphoglycerate kinase. The solid line represents the fit of a theoretical model in which a fraction of the hydrogens of the protein execute only vibrational motion (this fraction is given by the dotted line) and the rest undergo diffusion in a sphere.

In Eq. (26), M is the hydrogen mass, λ labels the mode, $\vec{e}_{\lambda,i}$ is the atomic eigenvector for hydrogen i in mode λ, and ω_λ is the mode angular frequency. n_λ is the number of quanta of energy $\hbar\omega_\lambda$ exchanged between the neutron and mode λ. $I_{n\lambda}$ is a modified Bessel function.

$W_i(\vec{Q})$ is the exponent of the Debye-Waller factor, $\exp[-2W_i(\vec{Q})]$, for hydrogen atom i and is given as follows:

$$2W_i(\vec{Q}) = \frac{1}{2NM}\sum_{\lambda}\frac{\hbar(\vec{Q}\cdot\vec{e}_{\lambda,i})^2}{\omega_\lambda[2n(\omega_\lambda)+1]} = Q^2\left\langle u_{Q,i}^2\right\rangle \qquad ...(27)$$

In Eq. (27), N is the number of modes, $n(\omega_\lambda)$ is the Bose occupancy, and $\left\langle u_{Q,i}^2\right\rangle$ is the mean-square displacement for atom i in the direction of $\vec{Q}$.

Equation (26) is an exact quantum mechanical expression for the scattered intensity. A detailed interpretation of this equation is given in Ref. 27. Inserting the calculated eigenvectors and eigenvalues into the equation allows the calculation of the incoherent scattering in the harmonic approximation for processes involving any desired number of quanta exchanged between the neutrons and the sample, *e.g.*, one-phonon scattering involving the exchange of one quantum of energy $\hbar\omega_\lambda$, two-phonon scattering, and so on.

The label λ in Eq. (26) runs over all the modes of the sample. In the case of an isolated molecule, λ runs over the $3N - 6$ normal modes of the molecule, where N is the number of atoms.

Example: *Vibrations in Staphylococcal Nuclease.*

Vibrations in proteins can be conveniently examined using normal mode analysis of isolated molecules. The results of such analyses indicate the presence of a variety of vibrations, with frequencies upward of a few inverse centimeters (cm^{-1}). Incoherent inelastic neutron scattering combined with normal mode analysis is well suited to examine low frequency vibrations in proteins. This is primarily due to the fact that large-amplitude displacements scatter neutrons strongly. Experiments on bovine pancreatic trypsin inhibitor (BPTI), combined with normal mode analysis of the isolated protein, demonstrated that low frequency underdamped vibrations do exist in the protein. More recently, the TFXA spectrometer at the Rutherford Appleton laboratory in Oxford was used to measure a spectrum of the high frequency local vibrations in the globular protein staphylococcal nuclease. Fig. 2.7 presents a comparison of the experimental dynamic structure factor at 25 K, with that calculated from a normal mode analysis of the protein. Comparison between the calculated and experimental profiles allows an assessment of the accuracy of the dynamical model and the assignment of the various vibrational features making up the experimental spectrum.

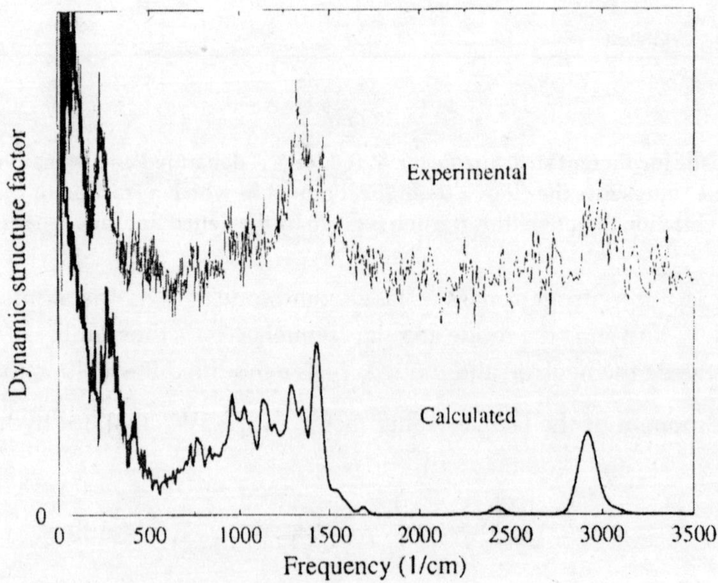

Fig. 2.7. Experimental and theoretical inelastic neutron scattering spectrum from staphylococcal nuclease at 25 K. The experimental spectrum was obtained on the TFXA spectrometer at Oxford. The calculated spectrum was obtained from a normal mode analysis of the isolated molecule.

Applications of Molecular Modeling

High resolution liquid-state NMR emerged as a structure determination technique for biological macromolecules in 1985. From the beginning, molecular modeling has had a central place in the derivation of NMR solution structures. There are several reasons for this. First, the energy parameters, typically derived from a molecular dynamics or molecular mechanics force field, play a central role in calculating and refining the structure. This is because experimental data are scarce, being available for only a fraction of the atoms (mostly the hydrogens). An additional difficulty is that most of the data describe relative positions of atoms and do not directly correspond to the global structure of the molecule. Second, models are not built manually but are automatically calculated by appropriate algorithms.

In this way the conformational space consistent with the data is sampled randomly to test whether the data determine the structure uniquely. Consequently, a lot of effort has gone into the development of algorithms to fit the experimental data. The methods used for NMR structure calculations are usually adapted from algorithms originally developed for different purposes in molecular modeling. Third, the wealth of dynamic information obtained by NMR and the difficulties in interpreting it in structural terms have led to a close interaction with MD simulation.

EXPERIMENTAL DATA

A. Deriving Conformational Restraints from NMR Data

The principal sources of structural data [10] are the nuclear Overhauser effect (NOE), which gives information on the spatial proximity of protons (up to a distance of about 4 A); coupling constants, which give information on dihedral angles; and residual dipolar couplings [11,12], which give information on the relative orientation of a bond vector to the molecule (e.g., to the chemical anisotropy tensor or an alignment tensor. With residual dipolar couplings one can, for example, define the relative orientation of domains. Because of the increasing number of experimental terms, we can get an increasingly complete description of the molecule in solution. The NOE is, however, still the richest source of structural information and at the same time the most problematic to analyze. Therefore, in this chapter we mostly deal with the treatment of NOEs in determining NMR solution structures. The other energy terms are included in structure refinements in a very similar manner.

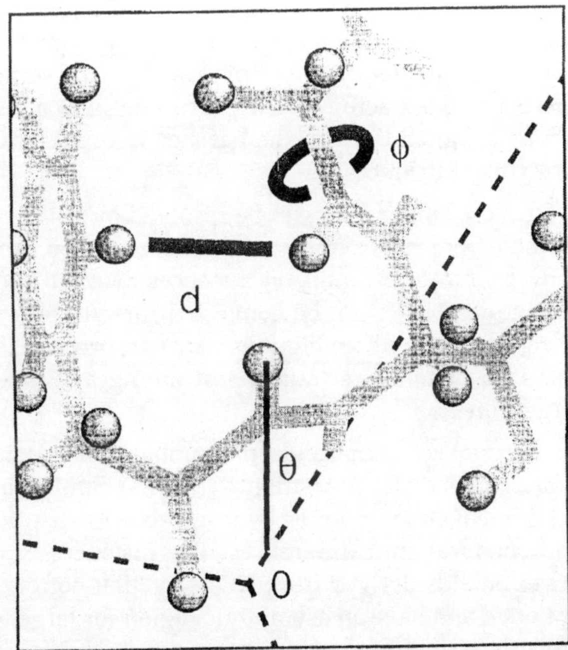

Fig. 2.8. The principal sources of structural data are the NOEs, which give information on the spatial proximity *d* of protons; coupling constants, which give information on dihedral angles ϕ; and residual dipolar couplings, which give information on the relative orientation θ of a bond vector with respect to the molecule (to the magnetic anisotropy tensor or an alignment tensor). Protons are shown as spheres. The dashed line indicates a coordinate system rigidly attached to the molecule.

The first step for any structure elucidation is the assignment of the frequencies (chemical shifts) of the protons and other NMR-active nuclei (^{13}C, ^{15}N). Although the frequencies of the nuclei in the magnetic field depend on the local electronic environment produced by the three-dimensional structure, a direct correlation to structure is very complicated. The application of chemical shift in structure calculation has been limited to final structure refinements, using empirical relations for proton and ^{13}C chemical shifts and ab initio calculation for ^{13}C chemical shifts of certain residues. In addition, hydrogen bonding can be deduced from NMR data by analyzing the exchange of labile protons. Only the hydrogen bond donor can be determined in this way.

The hydrogen bond acceptor is difficult to observe experimentally, and it has only recently been realized that scalar ("through-bond") couplings can be measured across hydrogen bonds. Most often, the hydrogen bond acceptor is inferred from a preliminary structure. Alternatively, it can be treated like an ambiguous NOE.

B. Distance Restraints

In an isolated two-spin system, the NOE (or, more accurately, the slope of its buildup) depends simply on d^{-6}, where d is the distance between two protons. The difficulties in the interpretation of the NOE originate in deviations from this simple distance dependence of the NOE buildup (due to spin diffusion caused by other nearby protons, and internal dynamics) and from possible ambiguities in its assignment to a specific proton pair.

Molecular modeling methods to deal with these difficulties are discussed further below. Usually, simplified representations of the data are used to obtain preliminary structures. Thus, lower and upper bounds on the interproton distances are estimated from the NOE intensity, using appropriate reference distances for calibration. The bounds should include the estimates of the cumulative error due to all sources such as peak integration errors, spin diffusion, and internal dynamics. The dispersion of proton chemical shifts is usually incomplete in the one-dimensional spectrum of a macromolecule, resulting in many degenerate resonances. As a result, few NOEs can be assigned only on the basis of resonance assignments and without any knowledge of the structure of the molecule.

Unless ambiguities can be resolved by using additional information, such as the peak shape or data from heteronuclear experiments, the remaining NOEs are ambiguous and cannot be converted into restraints on distances between proton pairs. Nevertheless, the information from ambiguous NOEs can be converted directly into structural restraints. The structure calculation or refinement with ambiguous data can proceed in a way directly analogous to refinement with standard distance restraints, restraining a "d^{-6}-summed distance" $\bar{D}$ by means of a distance target function.

By analogy with standard unambiguous distance restraints between atom pairs, we call these "ambiguous distance restraints" (ADRs). Similar methods can be applied to ambiguities in other experimental data, such as hydrogen bonds, disulfide bridges, and paramagnetic shift broadening and chemical shift differences. The distances in the structure are restrained to the upper and lower bounds derived from NOEs by "flat-bottom" potentials. The potential should be gradient-bounded and have an asymptotic region for large violations that is linear. Then, for large restraint violations, the force approaches a maximum value or can even be decreased, depending on the parameters.

This makes the optimization numerically more stable and seems to improve convergence by transiently allowing larger violations during the calculation, thus allowing the structure to gradually escape deep local minima. The limitation of the gradient of the potential is particularly important for calculations with ADRs and for data sets that potentially contain noise peaks, since it facilitates the appearance of violations due to incorrect restraints. A standard harmonic potential would put a high penalty on large violations and would introduce larger distortions into the structure.

C. The Hybrid Energy Approach

Even if the set of data from NMR experiments is as complete as possible, it is insufficient to define the positions of all the atoms in the molecule, simply because most of the data are measured for protons only. The positions of the other atoms have to be inferred, using values of bond lengths, bond angles, planarity, and van der Waals radii that are known a priori.

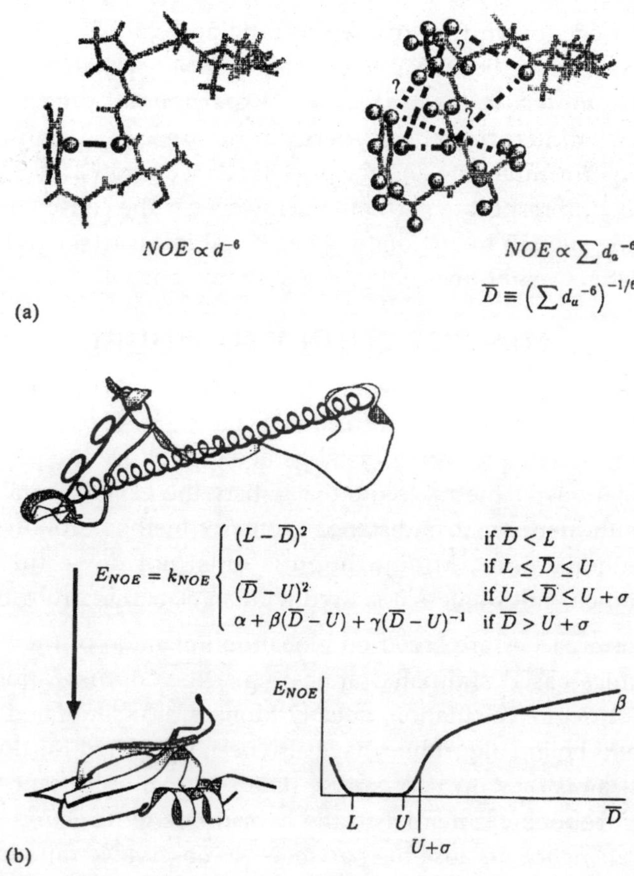

Fig. 2.9. Use of unambiguous or ambiguous distance restraints in an optimization calculation.

(*a*) The distance $\overline{D}$ that is restrained can be a distance measured between two protons in the molecule or a "$(\Sigma\, d^{-6})^{-1/6}$ summed distance" with contributions from many proton pairs, where the sum runs over all contributions to a cross-peak that are possible due to chemical shift degeneracy. The question marks indicate ambiguities in the assignment of the NOE. For clarity, a situation with only two assignment possibilities is shown. There can be many more possibilities with experimental data.

(*b*) The restraining potential is gradient bounded to avoid large forces for large violations. k_{NOE} is the energy constant, and *U* and L are upper and lower bounds derived from the size of the NOE. The parameter σ determines the distance at which the potential switches from harmonic to asymptotic behavior, β is the asymptotic slope of the potential, and the coefficients α and β are determined such that the potential is continuous and differentiable at *U* + σ. If *D* is between *L* and *U*, the energy and gradient are zero.

A molecular dynamics force field is a convenient compilation of these data. The data may be used in a much simplified form (*e.g.*, in the case of metric matrix distance geometry, all data are converted into lower and upper bounds on interatomic distances, which all have the same weight). Similar to the use of energy parameters in X-ray crystallography, the parameters need not reflect the dynamic behavior of the molecule. The force constants are chosen to avoid distortions of the molecule when experimental restraints are applied. Thus, the force constants on bond angle and planarity are a factor of 10-100 higher than in standard molecular dynamics force fields.

Likewise, a detailed description of electrostatic and van der Waals interactions is not necessary and may not even be beneficial in calculating NMR structures. The problem of finding conformations of the molecule that satisfy the experimental data is then that of finding conformations that minimize a hybrid energy function E_{hybrid}. which contains different contributions from experimental data and the force field (see below). These contributions need to be properly weighted with respect to each other. However, if the chosen experimental upper and lower bounds are wide enough to avoid any geometrical inconsistencies between the force field and the data, this relative weight does not play a predominant role.

MINIMIZATION PROCEDURES

Finding the minimum of the hybrid energy function is very complex. Similar to the protein folding problem, the number of degrees of freedom is far too large to allow a complete systematic search in all variables. Systematic search methods need to reduce the problem to a few degrees of freedom. Conformations of the molecule that satisfy the experimental bounds are therefore usually calculated with metric matrix distance geometry methods followed by optimization or by optimization methods alone. Minimization is often not powerful enough for structure calculations of macromole-cules unless it is used with an elaborate protocol.

More powerful approaches are based on global optimization of the hybrid energy function by molecular dynamics based simulated annealing. Other optimization methods have been suggested for NMR structure calculation, notably Monte Carlo simulated annealing and genetic algorithms. Branch-and-bound algorithms have also been suggested for docking rigid monomers with ambiguous restraints or with very sparse data sets. An important feature of the latter is the addition of a hydrophobic potential to the hybrid energy function, which serves to pack secondary structure elements. Because the parameter-to-observable ratio is rather low, structures are calculated repeatedly with the same restraints.

The aim is a random sampling of the conformational space consistent with the restraints. In metric matrix distance geometry, randomness is achieved by the random selection of distance estimates within the bounds. In optimization calculations, one achieves random searching by either selecting a starting conformation very far from the folded structure (*e.g.*, an extended strand or by choosing starting conformations that are random (either in torsion angles or in Cartesian coordinates).

A. Metric Matrix Distance Geometry

A distance geometry calculation consists of two major parts. In the first, the distances are checked for consistency, using a set of inequalities that distances have to satisfy (this part is called "bound smoothing"); in the second, distances are chosen randomly within these bounds, and the so-called metric matrix (M_{ij}) is calculated. "Embedding" then converts this matrix to three-dimensional coordinates, using methods akin to principal component analysis.

There are many extensive reviews on metric matrix distance geometry, some of which provide illustrative examples. In total, we can distinguish five steps in a distance geometry calculation:

1. Bound smoothing
2. Distance selection and metrization
3. Construction of the metric matrix
4. Embedding
5. Refinement (optimization)

Bound smoothing serves two purposes: to check consistency of the distances and to transfer information between atoms. Distances have to satisfy the triangle inequalities in a metric space of any dimension (the sum of two sides of a triangle has to be larger than the third. To ensure consistency of the distances in three-dimensional space, more inequalities would be necessary (the triangle, tetrangle, pentangle, and hexangle inequalities). Only the tetrangle inequality is of practical use, and it is usually not employed because of high computational costs.

This inequality transfers information from one diagonal of a tetrangle to the other; in two dimensions this is the parallelogram equation $|a + b| = |c + d|$. The most important consequence of bound smoothing is the transfer of information from those atoms for which NMR data are available to those that cannot be observed directly in NMR experiments. Within the original experimental bounds, the minimal distance intervals are identified for which all N^3 triangle inequalities can be satisfied. A distance chosen outside these intervals would violate at least one triangle inequality. For example, an NOE between protons p_i, and p_j and the covalent bond between p_j and carbon C_j imposes upper and lower bounds on the distance between p_i and C_j, although this distance is not observable experimentally nor is it part of E_{chem}. The second step concerns *distance selection and metrization.*

Bound smoothing only reduces the possible intervals for interatomic distances from the original bounds. However, the embedding algorithm demands a specific distance for every atom pair in the molecule. These distances are chosen randomly within the interval, from either a uniform or an estimated distribution, to generate a trial distance matrix. Uniform distance distributions seem to provide better sampling for very sparse data sets. Note that although the bounds on the distances satisfy the triangle inequalities, particular choices of distances between these bounds will in general violate them. Therefore, if all distances are chosen within their bounds independently of each other (the method that is used in most applications of distance geometry for NMR structure determination), the final distance matrix will contain many violations of the triangle inequalities.

The main consequence is a very limited sampling of the conformational space of the embedded structures for very sparse data sets despite the intrinsic randomness of the technique. In spite of these limitations, the algorithm is remarkably stable in its simplest form. *Metrization* guarantees that all distances satisfy the triangle inequalities by repeating a bound-smoothing step after each distance choice. The order of distance choice becomes important; optimally, the

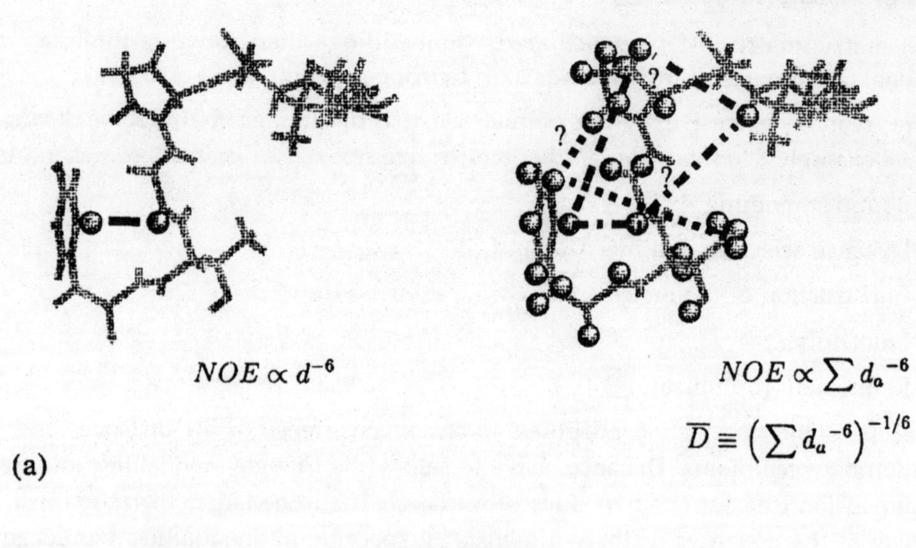

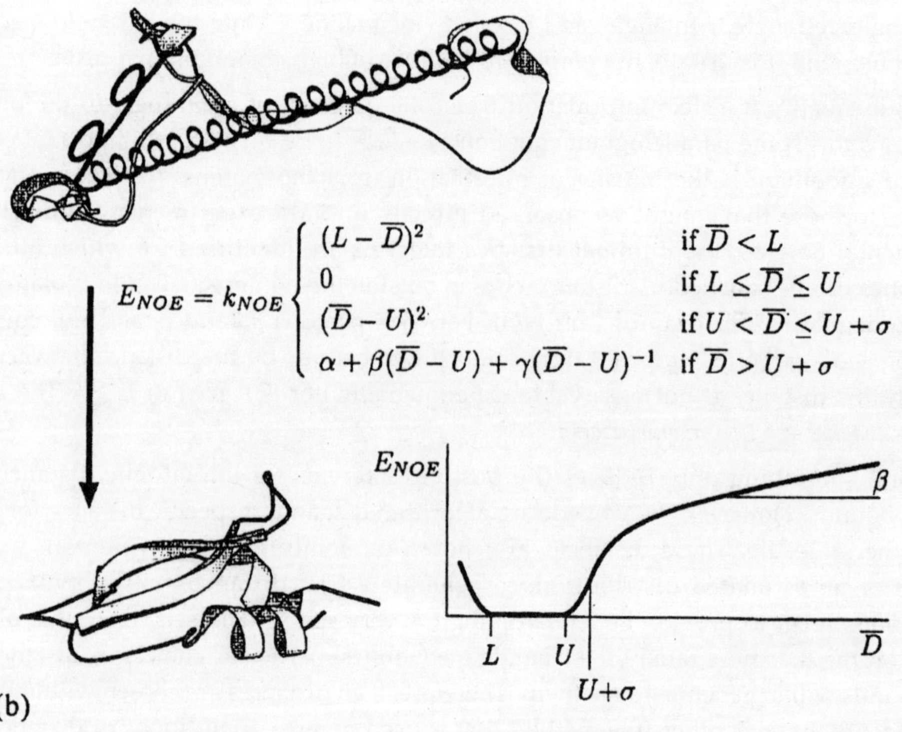

Fig. 2.10. Flow of a distance geometry calculation. On the left is shown the development of the data; on the right, the operations. d_{ij} is the distance between atoms i and j; L_{ij} and U_{ij} are lower and upper bounds on the distance; L'_{ij} and U'_{ij} are the smoothed bounds after application of the triangle inequality; d_{i0} is the distance between atom i and the geometric center; N is the number of atoms; (M_{ij}) is the metric matrix; $\vec{r}_i$ is the positional vector of atom i; $\vec{e}_i$ is the first eigenvector of (M_{ij}) with eigen value λ_i; x_i, y_i, and z_i, are the x-, y-, and z-coordinates of atom i. *(1-5 correspond to the numbered list on p....)*

distances are chosen in a completely random sequence. Metrization is a very computer-intensive operation. Computer time can be saved by using a partially random sequence and terminating the process after $4N$ distances (a three-dimensional object is completely specified by $4N - 10$ distances). Metrization leads to a much better sampling of conformational space and dramatically improves the local quality of the structures when few long-range connectivities are present.

The better sampling of space comes at a certain price: The embedded structures may show errors in the topology that are not seen without metrization. This may be due to the enforced propagation of an error in a distance choice to many other distances through the triangle inequality. The *metric matrix* is the matrix of all scalar products of position vectors of the atoms when the geometric center is placed in the origin. By application of the law of cosines, this matrix can be obtained from distance information only. Because it is invariant against rotation but not translation, the distances to the geometric center have to be calculated from the interatomic distances. The matrix allows the calculation of coordinates from distances in a single step, provided that all $N_{\text{atom}}(N_{\text{atom}} - 1)/2$ interatomic distances are known. *Embedding* is the calculation of coordinates from the metric matrix by methods akin to principal component analysis. The eigenvectors of the metric matrix contain the principal coordinates of the atoms. If the distances correspond to a three-dimensional object, only three eigenvalues of the matrix are nonzero, and the first eigenvector contains all x-coordinates, the second all x-coordinates, and the third all z-coordinates. If the distances are not consistent with a three-dimensional object (the usual situation with sparse NMR data, when the majority of distances come from the random number generator), there will be more than three positive eigenvalues.

The eigenvector expansion is then truncated after the first three eigenvalues; this corresponds to a projection of a higher dimensional object into three-dimensional space. *Refinement* of the embedded structures is always necessary to remove distortions in the structure. One shortcoming of the embedding algorithm is that data cannot be weighted according to their certainty in any way. During the projection, bond lengths are distorted in the same way as long-range distances guessed by the random number generator within possibly very wide bounds. Also, chirality information is completely absent during bound smoothing and embedding.

The first step in the refinement is the selection of the correct enantiomer, which may be achieved on the basis of the chirality of Cα atoms, secondary structure elements, or partial refinement of both enantiomers and choice of the enantiomer with lower energy. If the distances satisfy the triangle inequalities, they are embeddable in some dimension. One possible solution is therefore to try to start refinement in four dimensions and use the allowed deviation into the fourth dimension as an additional annealing parameter.

The advantages of refinement in higher dimensions are similar to those of soft atoms discussed below. A time-saving variant of the distance geometry procedure described above is substructure embedding. Here, about a third of the atoms are chosen *after* the bound smoothing step and embedded. This procedure was originally used to improve the performance of the distance geometry algorithm by adding the distances from the embedded and partially refined structures back to the distance list. The substructures can be refined directly with simulated annealing by filling in the missing atoms approximately in their correct position.

B. Molecular Dynamics Simulated Annealing

In Cartesian coordinates, molecular dynamics-based simulated annealing (MDSA) refinement consists of the numerical solution of Newton's equations of motion. The specific advantage of molecular dynamics over energy minimization is the larger radius of convergence due to possible uphill motions over large energy barriers. Together with variation of temperature or energy scales, very powerful minimization strategies can be implemented. Scaling the temperature, the overall weight on E_{hybrid} or all masses m_i are formally equivalent. The independent scaling of each contribution E_l by its weight factor w_t gives rise to a large number of possible simulated annealing schemes. We call annealing schemes that vary the w_t independently "generalized annealing schemes."

The initial velocities are usually assigned from a Maxwell distribution at the desired starting temperature, and the temperature is controlled (*e.g.*, by coupling to a heat bath. For the use of MD as an optimization technique, it is convenient to use uniform masses $m_i \equiv m$ for all i. This, in combination with uniform energy constants in the force field, allows the use of larger time steps in the molecular dynamics, because differences in vibrational frequencies are avoided (the time step is determined by the highest vibrational frequency). Recently, MD constrained to torsion angle space [torsion angle dynamics (TAD)] was introduced to refinement calculations.

Earlier versions of the equations of motion for molecular dynamics in torsion angle space were very inefficient to solve owing to the need for a matrix inversion at every time step. Newer algorithms break down the necessary operations into a series of multiplications of small matrices and are therefore much more efficient.

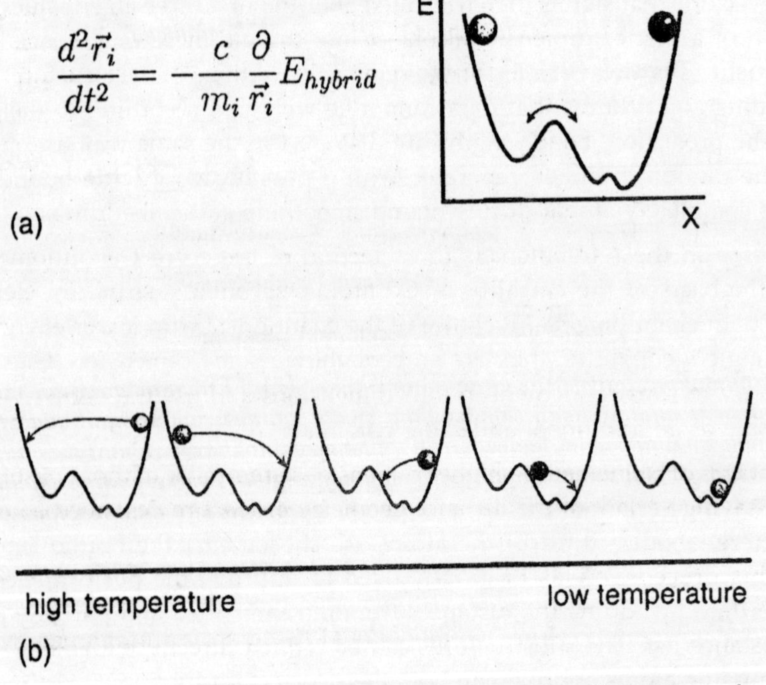

Fig. 2.11. (*a*) Solving Newton's equations of motion at constant energy allows the molecule to overcome energy barriers in E_{hybrid}. The quantities $\vec{r}_i$ and m_i are the coordinate vectors and masses, respectively, of atom *i*, and E_{hybrid} is the target function of the minimization problem, containing different contributions from experimental data and from a priori knowledge (*i.e.*, the force field), (*b*) With temperature variation, powerful minimization schemes can be implemented, allowing for large energy barriers to be crossed at high temperatures, ultimately leading to the identification of the "global" minimum.

The application of TAD in standard MD calculations may require the development of dedicated force fields to emulate the missing flexibility by a reparametrization of the non-bonded potential. This is not necessary for its application in NMR structure calculation, because the energy parameters developed for this purpose already assume in most cases a rigid covalent geometry, either by employing high force constants or by using only torsion angles as degrees of freedom. The advantage of TAD is that the geometry of the molecule does not have to be maintained by high force constants, which lead to high vibrational frequencies. Therefore, longer time steps at higher temperatures can be used with TAD, and the refinement protocols are numerically more stable.

C. Folding Random Structures by Simulated Annealing

Various simulated annealing protocols have been suggested to fold random structures with experimental restraints. The choice of starting structure determines the optimal protocol. The most obvious choices are random distributions of dihedral angles (as indicated in Fig. 2.9). The minimization procedure has to try to avoid entanglement of the chain while properly relaxing large forces in the starting conformation, which could arise from overlapping atoms or distance restraints violated by a large amount.

This is achieved by a combination of soft non-bonded interactions, a violation-tolerant form of the distance restraint potential, and high temperature dynamics. To achieve convergence with an annealing protocol using Cartesian dynamics, multistage generalized annealing protocols were introduced. The first stage is a high temperature search where the molecule adopts approximately the correct fold. In this stage, the non-bonded interactions are reduced to allow the chain to intersect itself, and the representation of the non-bonded interactions may be further simplified by computing them for only a fraction of the atoms.

The protocol is also adaptable to ambiguous restraint lists by a specifically reduced weight w_{ambig} on the ADRs, which is varied independently of w_{ambig}. A detailed description can be found elsewhere. With mostly unambiguous data, this protocol has been successfully used for proteins with up to 160 residues. Although virtually all structures converge to the correct fold for small proteins, we observe that approximately one-third of the structures are misfolded for larger proteins, or for low data density, or many ambiguities.

We have also used this protocol for most structure calculations with the automated NOE assignment method ARIA discussed in the next section. Calculations starting from random Cartesian coordinates and using standard Newton dynamics illustrate the flexibility of the generalized annealing approach. The extremely bad geometry of the initial structures requires that the weights on the covalent geometry terms start with very low values, which are then slowly increased during the calculation.

All torsion angle terms (dihedral angles, planarity, and chirality) are removed from E_{hybrid} because of the difficulty in calculating them for random Cartesian structures. Enantiomer selection and regularization are necessary with this protocol much as they are with MMDG embedded structures. The principal advantage of the use of random Cartesian coordinates over that of random dihedral coordinates is that the former give better sampling for highly ambiguous data. The initial structure does not bias toward intraresidue or sequential assignments of ambiguous NOEs.

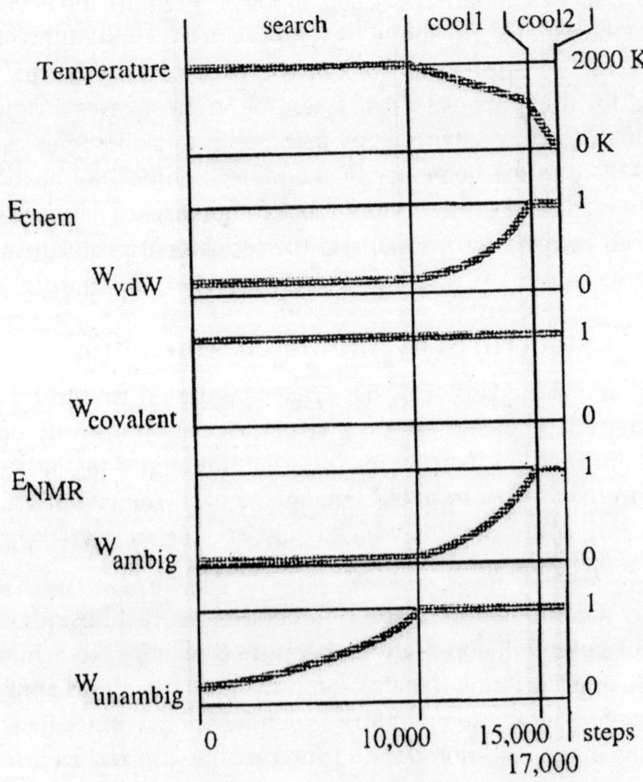

$$
\begin{aligned}
E_{hybrid} &= \sum_l w_l E_l \\
&= w_{bond}E_{bond} + w_{angle}E_{angle} + w_{improper}E_{improper} \\
&\quad + w_{nonbonded}E_{nonbonded} \\
&\quad + w_{unambig}E_{unambig} + w_{ambig}E_{ambig} + \ldots
\end{aligned}
$$

Fig. 2.12. Schematic representation of a Cartesian dynamics protocol starting from random torsion angles. The weights w for non-bonded (*i.e.*, van der Waals) interactions, unambiguous distance restraints, and ambiguous distance restraints are varied independently. The covalent interactions are maintained with full weight, $w_{covalent}$, for the entire protocol. Weights for other experimental terms may be varied in an analogous way. Coupling constant restraints and anisotropy restraints are usually used only in a refinement stage.

A TAD protocol may have a three-stage organization similar to that of the Cartesian MDSA protocol, with two TAD stages (one high temperature, one cooling) and a final Cartesian cooling stage. The starting temperatures can be set to much higher values (up to 50,000 K).

Weights on experimental and non-bonded terms differ in the different stages, with higher weights on the experimental terms in the high temperature stage, but the principal parameter that is varied during simulated annealing is the temperature. TAD protocols used with the program DYANA are even simpler, with only temperature variation in the simulated annealing stage, which is followed by conjugate gradient minimization. In general, TAD shows better convergence than Cartesian dynamics. For nucleic acid structures, for example, the convergence rate can be very low both for MMDG and for Cartesian dynamics owing to the low restraint density. The sampling of conformational space by TAD for very sparse data sets should be comparable to Cartesian dynamics protocols and better than for MMDG without metrization. Depending on the implementation, ambiguous distance restraints can be used throughout the protocol as with Cartesian dynamics. With its implementation in several NMR structure determination programs, including X-plor, CNS, and DYANA, the field seems to converge toward this calculation method.

AUTOMATED INTERPRETATION OF NOE SPECTRA

The methods discussed in this section extend the original concept of deriving structures from experimental NMR data in two ways. First, during the structure calculation, part of the assignment problem is solved automatically. This allows specification of the NOE data in a form closer to the raw data, which makes the refinement similar to X-ray refinement. Second, the quality of the data is assessed.

A. Recognition of Incorrect Restraints: The Structural Consistency Hypothesis

Structure calculation algorithms in general assume that the experimental list of restraints is completely free of errors. This is usually true only in the final stages of a structure calculation, when all errors (*e.g.*, in the assignment of chemical shifts or NOEs) have been identified, often in a laborious iterative process. Many effects can produce inconsistent or incorrect restraints, *e.g.*, artifact peaks, imprecise peak positions, and insufficient error bounds to correct for spin diffusion. Restraints due to artifacts may, by chance, be completely consistent with the correct structure of the molecule.

However, the majority of incorrect restraints will be inconsistent with the correct structural data (*i.e.*, the correct restraints and information from the force field). Inconsistencies in the data produce distortions in the structure and violations in some restraints. Structural consistency is often taken as the final criterion to identify problematic restraints. It is, for example, the central idea in the "bound-smoothing" part of distance geometry algorithms, and it is intimately related to the way distance data are usually specified: The error bounds are set wide enough that all data are geometrically consistent. The problem in using violations to identify incorrect restraints is twofold. First, one has to distinguish between violations that appear because of insufficient convergence power of the structure calculation algorithm and violations due to incorrect restraints.

Violations caused by incorrect restraints will be consistent (*i.e.*, they will be present in the majority of structures), whereas insufficient convergence will produce violations that are randomly distributed. This reasoning has been formalized in the "self-correcting distance geometry" method, which calculates structures iteratively and modifies the list of restraints after each iteration. Consistent violations are identified by calculating the fraction of structures in which a particular restraint is violated by more than a threshold (*e.g.*, 0.5 Å). If this fraction exceeds a certain value (*e.g.*, 0.5), the restraint is removed from the list for the calculation in the next iteration. Second, it is possible that an incorrect restraint produces a systematic violation of another restraint. Currently, this can be ruled out only by manually checking the results, where more information on the data can be used to evaluate the restraint (*e.g.*, by inspecting the peak shape).

B. Automated Assignment of Ambiguities in the NOE Data

Assigning ambiguous NOEs is one of the major bottlenecks in NMR solution structure determination, comparable to map fitting in X-ray crystallography. In principle, ambiguous NOEs need not be explicitly assigned if they are used as ADRs, because the assignment is done implicitly in the structure calculation. This is because the summed distance $\bar{D}$ is strongly weighted toward the shortest of the contributing distances. If ADRs are used in the refinement of an already reasonably well determined structure, this weighting is expected to, in most cases, favor the really dominating contribution to the ambiguous NOE.

The implicit assignments, achieved through weighting with the distances in the structure, will be mostly correct, and the path to the final structure satisfying all data will be relatively smooth. The case is very different when ADRs are used for calculating structures ab initio, starting from random structures. Obviously, most of the initial implicit assignments from the random structures are incorrect, and the path toward the final structure is much more difficult owing to additional local minima in the energy. During the calculation, the in-terproton distances, and with them the weighting on different assignment possibilities, need to change. However, convergence can be achieved by treating the amiguous NOEs appropriately in generalized simulated annealing protocols.

C. Iterative Explicit NOE Assignment

The main difficulties with a fully automated method lie in defining rules for explicit assignment based on an ensemble of structures with possibly incorrect features and providing mechanisms for correcting incorrect assignments. Two fully automated iterative assignment methods have been proposed, one based on ADRs (ARIA: ambiguous restraints for iterative assignment) and the other on self-correcting distance geometry. Methods to assign ambiguous NOEs follow a sequence of steps. Structures are calculated and NOEs are assigned in an iterative way. In the first iteration, the assignments have to be based on the frequencies alone.

In the following iterations, the assignments of ambiguous NOEs are derived from the structures by comparing interproton distances corresponding to each assignment possibility. In the "traditional" approach, one possibility would be chosen by hand, and a peak would not be used if an unambiguous assignment were impossible. In contrast, the automated methods use ambiguous peaks during the structure calculation. The key difference between these methods is in how ambiguous peaks are converted into distance restraints.

The program ARIA generates one ADR for each ambiguous peak, whereas the program NOAH creates an unambiguous restraint for each assignment possibility. In general, the advantages of using an automated method may be comparable to those of SA refinement in X-ray crystallography, where many of the operations necessary to refine a structure can be done automatically and the remaining manual interventions are easier because the SA refinement usually results in a more easily interpreted electron density map. Automated methods are usually used in combination with manual assignment. However, fully automated assignment of the NOEs is possible.

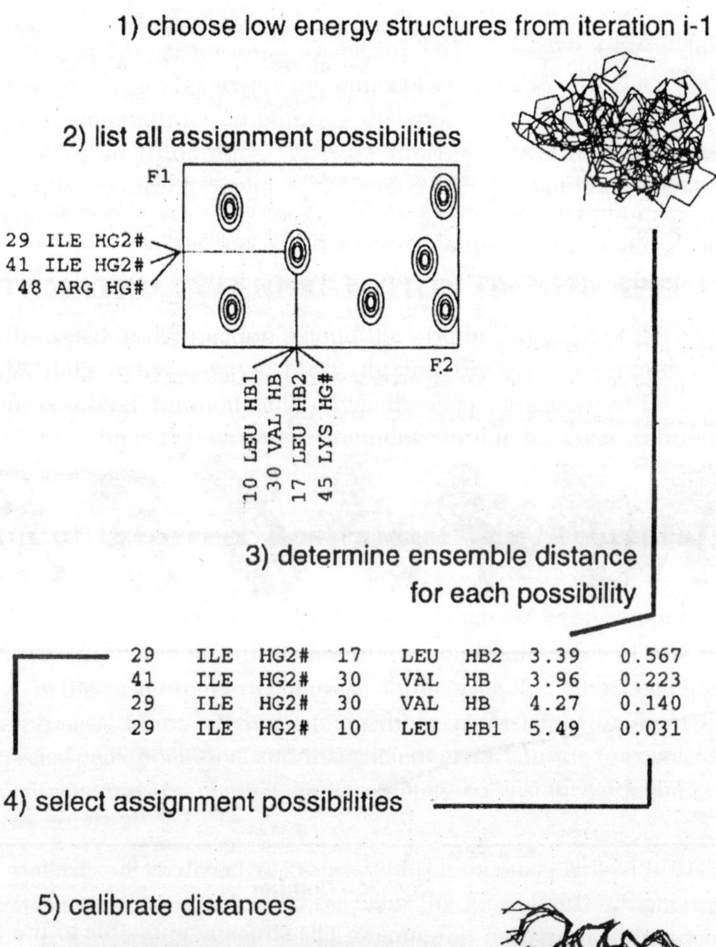

Fig. 2.13. Steps in automated assignment. (1) Select the S_{conv} lowest energy structures from iteration $i - 1$ that are used to interpret the spectra. (2) For each peak, list all possible assignments compatible with the resonances within a frequency range. (3) Extract a distance for each assignment possibility from the ensemble of structures. (4) Use the distances to assign ambiguous NOEs. (5) Calibrate the peak volumes to obtain distance restraints. (6) Calculate structures based on the new restraints.

D. Symmetrical Oligomers

Symmetrical oligomers present a special difficulty for NMR spectroscopy, because all symmetry-related hydrogens will have equivalent magnetic environments and therefore will be degenerate in chemical shift. Only one monomer is "seen" in the spectra. In principle, every single NOE peak in the spectrum is therefore ambiguous. This ambiguity that arises from the symmetry can be treated with the same concept as ambiguities due to limited spectral resolution with ambiguous distance restraints. Structure determinations of symmetrical oligomers were reviewed in detail recently [70].

The same principal ideas are incorporated in the calculation protocols for symmetrical oligomers as with asymmetrical systems, *i.e.,* the weight is reduced specifically for ADRs or all distance restraints. In addition, the symmetry of the system restricts conformational space and is maintained during the calculation by additional restraints. An attractive potential between the monomers can be used in the beginning of the protocol to prevent them from drifting apart.

The special difficulties with symmetrical oligomer calculations arise for two reasons: First, all NOEs are ambiguous a priori; second, the assignments of neighboring residues are strongly correlated. A minimization method such as simulated annealing that moves single atoms (or rigid parts of amino acids) may not be optimal for moving larger parts of the structure coherently if a whole set of NOEs needs to be implicitly reassigned. As a result, the structure calculation has a lower convergence rate than for asymmetrical systems. A combination of annealing calculations with other optimization approaches (*e.g.*, a recent branch-and-bound algorithm may be a more efficient approach. However, the present approaches based on annealing alone have been successful in several cases for quite complex systems (up to a tetramer and a hexamer.

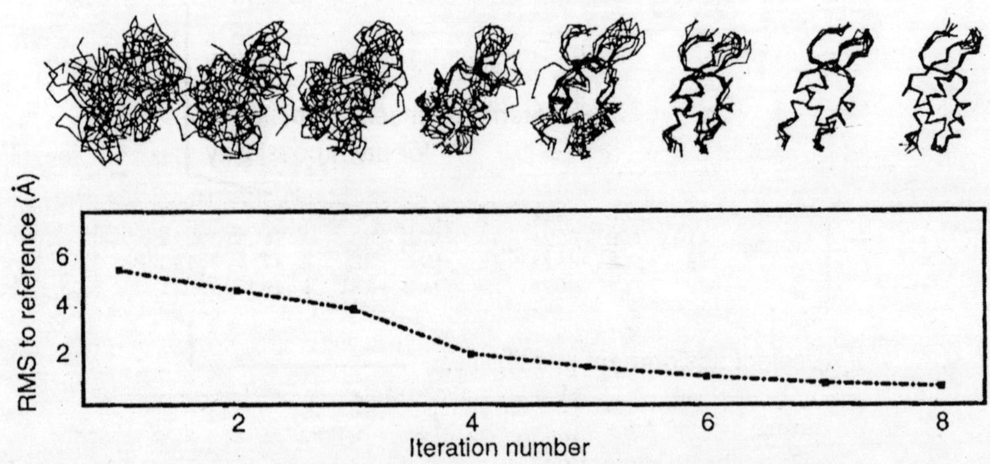

Fig. 2.14. Example of a fully automated assignment. The structure ensemble of the seven lowest energy structures at each iteration is shown. These structures are used for the violation analysis and for a partial assignment of ambiguous NOEs. In the first iteration the structures are calculated with all restraints, where each restraint has all assignment possibilities. In each subsequent iteration, consistently violated restraints are removed, and assignment possibilities are selected with increasingly tight criteria such that at the end of the eight iterations most NOEs are unambiguously assigned.

TREATMENT OF SPIN DIFFUSION

Depending on experimental parameters, NOE intensities, will be affected by spin diffusion. Magnetization can be transferred between two protons via third protons such that the NOE between the two protons is increased and may be observed even when the distance between the two protons is above the usual experimental limit. This is a consequence of the d^{-6} distance dependence of the NOE. Depending on the conformation, it can be more efficient to move magnetization over intermediate protons than directly. From a given structure, the NOE effect can be calculated more realistically by complete relaxation matrix analysis. Instead of considering only the distance between two protons, the complete network of interactions is considered.

Approximately, the relaxation matrix (R_{ij}) depends on all interproton distances and on parameters describing overall and local motion. The long mixing times often necessary in heteronuclear NOE experiments would make spin diffusion estimates especially valuable. To calculate heteronuclear NOEs realistically, the transfer efficiencies between protons and heteronuclei should be incorporated into the equations.

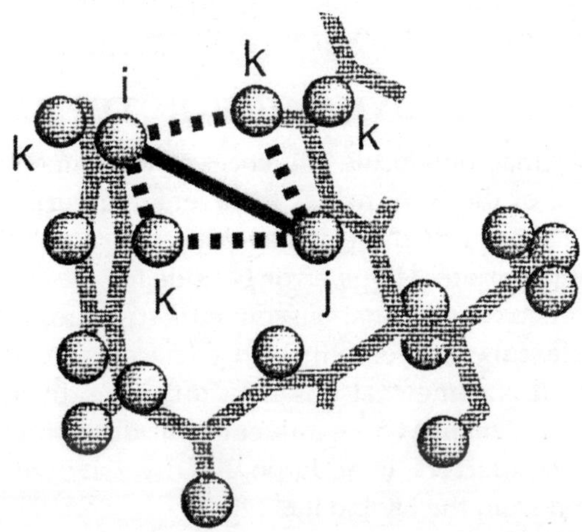

Fig. 2.15. Effects of spin diffusion. The NOE between two protons (indicated by the solid line) may be altered by the presence of alternative pathways for the magnetization (dashed lines). The size of the NOE can be calculated for a structure from the experimental mixing time, τ_m, and the complete relaxation matrix, (R_{ij}), which is a function of all interproton distances d_{ij} and functions describing the motion of the protons. 7 is the gyromagnetic ratio of the proton, τ is the Planck constant, *i* is the rotational correlation time, and ω is the Larmor frequency of the proton in the magnetic field. The expression for (R_{ij}) is an approximation assuming an internally rigid molecule.

One approach to include spin diffusion corrections in a structure calculation is a direct refinement against NOE intensities, analogous to X-ray crystal structure refinement. In this approach, forces are calculated directly from the difference between the experimental NOE intensities and those calculated from the structure via the relaxation matrix. This necessitates, however, an expensive evaluation of derivatives of the simulated NOE spectra with respect to coordinates at every minimization step. Approximations and faster methods to evaluate the gradients make this direct approach more feasible.

Various pseudo-energy functions have been proposed. The simplest form (harmonic in the difference between experimental and calculated NOE) places a predominant weight on the largest intensities (shortest distances), which are most often due to intraresidue interactions and will therefore contribute little to determining the conformation of the molecule. By using pseudo-energy functions depending on the sixth root of the difference between calculated and experimental NOEs, the weight is distributed more equally. Other approaches use complete relaxation matrix analysis to obtain spin diffusion corrected distances from the NOE intensities which are then used in conventional distance-restrained optimization. This is more efficient in the use of CPU time, because the gradients do not have to be evaluated and a full relaxation matrix calculation is necessary only a few times in a refinement. These methods invert the calculation of the NOE intensities to calculate the relaxation matrix (R_{ij}) from a complete spectrum (NOE_{ij}), including the diagonal peaks. Since this is impossible to obtain experimentally for macromolecules, approximate iterative schemes are used.

One method uses preliminary structures to calculate NOE intensities that are merged with the incomplete experimental data to obtain a complete spectrum. A next generation of structures is then calculated with the distances, and an improved estimate of the relaxation matrix can be obtained. Another approach "shortcuts" the structure calculation and uses properties of the relaxation matrix itself (*e.g.*, the relation of the diagonal elements to the off-diagonal elements) to iteratively correct the relaxation matrix. Integration errors in the data have to be properly taken into account, otherwise they can lead to incorrect distance estimates.

INFLUENCE OF INTERNAL DYNAMICS ON THE EXPERIMENTAL DATA

Internal dynamics of the macromolecule influences all experimental data that can be measured by NMR. The d^{-6} weighting of the NOE makes the averaging very nonlinear, and the measured distance may appear much shorter than the average distance. The "distance geometry approach" to the problem is to use appropriately large error bounds for the distances and a rough estimate of dynamics from the diversity of the final ensemble of structures. Although this approach has given qualitatively satisfactory agreement with dynamics measurements and theoretical calculations in some cases, it is somewhat unsatisfactory. The diversity reflects the distribution of experimental data. Internal dynamics does influence this distribution, but experimental artifacts and overlap are also important factors. In addition, the diversity will depend on exactly how the distance bounds are derived from the NOE data.

Furthermore, local dynamics can result in locally conflicting data, while multiple conformations appear in the calculated structures predominantly in regions with little data. The measured NOE is an average over time and a large ensemble of structures, whereas in a standard structure calculation the lower and upper bounds refer to instantaneous distances. Methods have been proposed to account for the averaging in the interproton distance by fitting either a dynamics trajectory to the measured distance, by means of time-averaged distance restraints, or an ensemble of structures. Formally, an ensemble-averaged distance restraint is equivalent to an ambiguous distance restraint. The difference is just a scale factor. Therefore, we can understand an ensemble-averaged NOE as an NOE that is ambiguous between different conformers in the ensemble. The most serious problem with ensemble average approaches is that they introduce many more parameters into the calculation, making the parameter-to-observable ratio worse. The effective number of parameters has to be restrained. This can be achieved by using only a few conformers in the ensemble and by determining the optimum number of conformers by cross-validation.

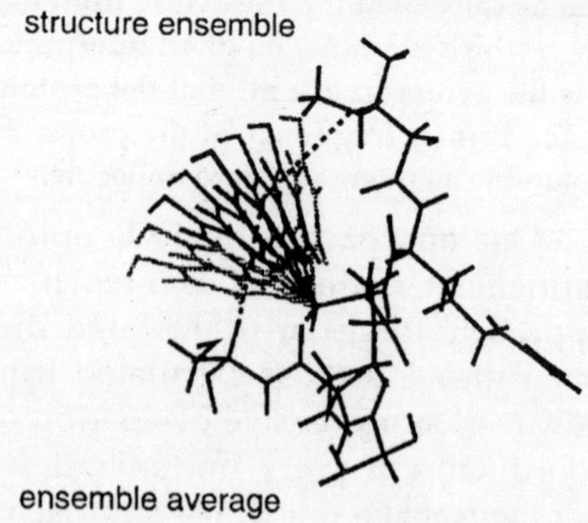

Fig. 2.16. Treating internal dynamics during the refinement process. Due to dynamics and the d^{-6} weighting of the NOE, the measured distance may appear much shorter than the average distance. This can be accounted for by using ensemble refinement techniques. In contrast to standard refinement, an average distance is calculated over an ensemble of C structures (ensemble refinement) or a trajectory (time-averaged refinement). The time-averaged distance is defined with an exponential window over the trajectory. *T* is the total length over the trajectory, *t* is the time, and τ is a "relaxation time" characterizing the width of the exponential window.

A more indirect way of restraining the effective number of parameters is to restrict the conformational space that the molecule can search during refinement. For example, with a full molecular dynamics force field and low temperatures, only a small fraction of the conformational space is accessible. A more direct way to restrict the number of parameters would be to use motional models. Normal modes have been used, for example, to model NMR order parameters obtained from relaxation studies. Another principal difficulty is that the precise effect of local dynamics on the NOE intensity cannot be determined from the data.

The dynamic correction factor describes the ratio of the effects of distance and angular fluctuations. Theoretical studies based on NOE intensities extracted from molecular dynamics trajectories are helpful to understand the detailed relationship between NMR parameters and local dynamics and may lead to structure-dependent corrections. In an implicit way, an estimate of the dynamic correction factor has been used in an ensemble relaxation matrix refinement by including order parameters for proton-proton vectors derived from molecular dynamics calculations. One remaining challenge is to incorporate data describing the local dynamics of the molecule directly into the refinement, in such a way that an order parameter calculated from the calculated ensemble is similar to the measured order parameter.

STRUCTURE QUALITY AND ENERGY PARAMETERS

The well-known difficulties in calculating three-dimensional structures of macromolecules from NMR data mentioned above (sparseness of the data, imprecision of the restraints due to spin diffusion and internal dynamics) also make the validation of the structures a challenging task. The quality of the data and the energy parameters used in the refinement can be expected to influence the quality of structures. Several principles can be used to validate NMR structures. First, the structure should explain the data.

Apart from the energy or target function value returned by the refinement program, this check can be performed with some independent programs (*e.g.*, AQUA/PROCHECK-NMR [90], MOLMOL. The analysis of the deviations from the restraints used in calculating the structures is very useful in the process of assigning the NOE peaks and refining the restraint list. As indicators of the quality of the final structure they are less powerful, because violations have been checked and probably removed. A recent statistical survey of the quality of NMR structures found weak correlations between deviations from NMR restraints and other indicators of structure quality. A similar problem arises with present cross-validated measures of fit, because they also are applied to the final clean list of restraints. Residual dipolar couplings offer an entirely different and, owing to their long-range nature, very powerful way of validating structures against experimental data. Similar to cross-validation, a set of residual dipolar couplings can be excluded from the refinement, and the deviations from this set are evaluated in the refined structures. Second, the structures should satisfy the a priori information used in the refinement in the form of the energy parameters. Programs like PROCHECK-NMR check for deviation from expected geometries and close non-bonded contacts. Finally, structural properties that depend directly neither on the data nor on the energy parameters can be checked by comparing the structures to statistics derived from a database of solved protein structures. PROCHECK-NMR and WHAT IF use, *e.g.*, statistics on backbone and side chain dihedral angles and on hydrogen bonds. PROSA uses potentials of mean force derived from distributions of amino acid-amino acid distances.

RECENT APPLICATIONS

Molecular modeling is an indispensable tool in the determination of macromolecular structures from NMR data and in the interpretation of the data. Thus, state-of-the-art molecular dynamics simulations can reproduce relaxation data well and supply a model of the motion in atomic detail. Qualitative aspects of correlated backbone motions can be understood from NMR structure ensembles. Additional data, in particular residual dipolar couplings, improve the precision and accuracy of NMR structures qualitatively. Standard calculation methods developed for small proteins are sufficiently powerful to solve protein structures and complexes in the 30 kDa range and beyond and protein-nucleic acid complexes.

Torsion angle dynamics offers increased convergence, in particular for nucleic acids, which are more difficult to calculate because of the sparseness of NMR data. Examples of structures for which automated assignment methods were used from the start are still rare. However, automated methods are being used increasingly as a powerful tool in structure determination in combination with manual assignment.

MODELS OF VOLTAGE GATING 3

In this chapter we will learn how to model current flowing through ion channels and how to include the feedback regulation of the channel behavior by voltage. But let us be very clear about what we are modeling in this chapter. Only in the most idealized sense are we modeling the behavior of excitable cells like neurons. In fact, we will be creating *point models* of patches of membrane, with only implicit consideration through the use of ion concentrations of the idea that the membrane encloses a compartment. We do not consider at all that a neuron is a highly complicated cell that integrates signals within *dendrites* and a *soma* before producing *action potentials* in an excitable *axon*. Electrophysiology is the study of ionic currents and electrical activity in cells and tissues. Because this field has its roots in classical physics, traditionally it has been the most quantitative field in cell physiology.

The groundbreaking work of the physiologists Hodgkin and Huxley and others in elucidating the mechanism of action potentials in the squid giant axon before and after the Second World War was the first major breakthrough of dynamical modeling in physiology. In the latter half of the twentieth century, the introduction of the patch-clamp technique established firmly that ionic currents are carried by proteins that act as gated ionic pores. More recently, genetic engineering techniques have been employed to clone, modify, and characterize the gating mechanisms of many types of channels (Hille 2001). In this chapter we focus on voltage gated ionic currents.

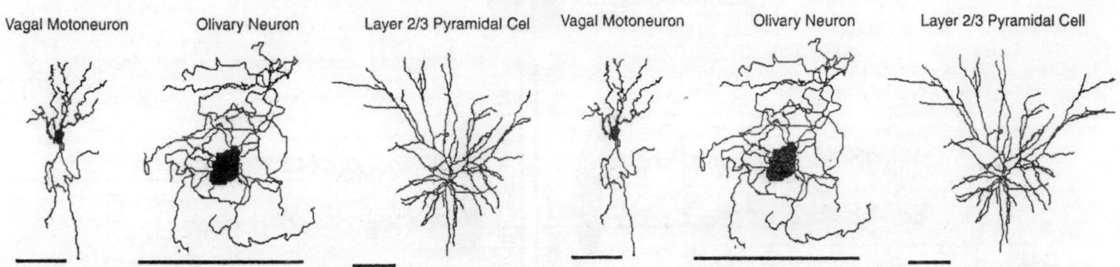

Fig. 3.1. Examples of the diverse shapes of mammalian neurons. Reprinted from Koch and Segev (2000).

We begin by reviewing the basic concepts of electrical behavior in cells. Next, we describe classical activation and inactivation kinetics and how the voltage clamp technique can be used to study these currents. We use the Morris-Lecar model for action potentials in the giant barnacle muscle to illustrate how voltage gated channels can interact to produce oscillations and action potentials.

The Morris-Lecar model is nonlinear but involves only two variables. With only two variables, we can analyze the dynamics of the equations for this model using phase plane techniques. For completeness, we close with brief introductions to the Hodgkin-Huxley model of the squid giant axon and FitzHugh-Nagumo models.

Table 2.1. Important Definitions in Electrophysiology

Definition	*Abbreviation*	*Value*
Avogadro's number	N	$6.02 . 10^{23}$/mol
Faraday's constant	F	$9.648 . 10^{4}$C/mol
elementary charge	e	$1.602 . 10^{-19}$ C
gas constant	R	8.315 J/(mol . K)
joule	J	1 V . C
volt	V	1 J/C
ampere	A	1 C/s

BASIS OF THE IONIC BATTERY

The electrical behavior of cells is based upon the transfer and storage of charge. We are used to thinking about electricity as the movement of electrons, but current can be carried by any charged particle, including ions such as K^+, Na^+, and Ca^{2+} in solution. As we will see, the ability of cells to generate electrical signals is entirely dependent on the evolution of ion-specific pumps and pores that allow the transfer of charge up and down gradients. Ion pumps use energy in the form of ATP to transport ions against a concentration gradient.

Before we begin our discussion of the ionic battery, it is useful to recall several important definitions from elementary physics listed in Table 3.1. An ion's *valence* is the number of charges, plus or minus, that it carries. A given number of *divalent* Ca^{2+} ions would carry twice the amount of charge as the same number of *univalent* K^+ ions.

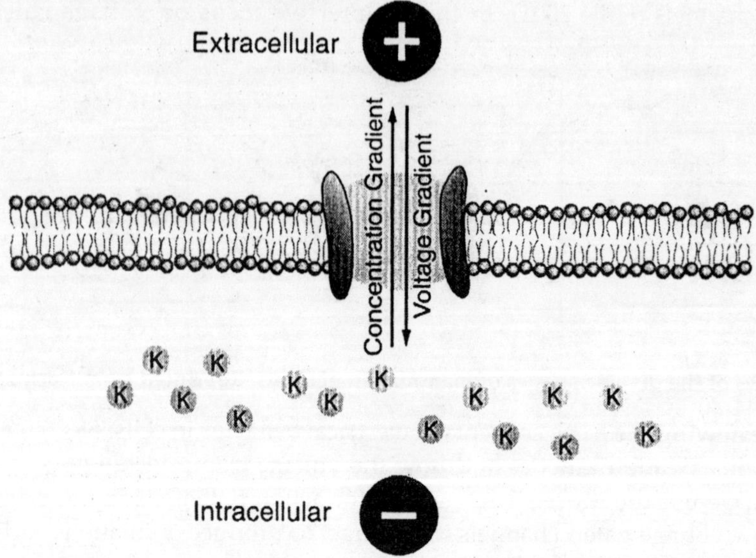

Fig. 3.2. The basis of the ionic battery.

Using these relationships, we see that the transfer of 1 mole of K^+ ions in a period of one second would carry a current equal to Faraday's constant. It requires one joule of energy (*i.e.* ATP consumed by ion pumps) to separate one coulomb of charge across one volt of potential (which is the definition of volt).

The Nernst Potential: Charge Balances Concentration

Biological fluids such as cytoplasm and extracellular fluid contain numerous ions. Consider the case where the two ions K^+ and any monovalent anion A^- are in solution such that the concentration is different across the impermeable membrane but the two ions are equal in concentration on the same side of the membrane. As shown in Fig. 3.3A, before we make any changes, there is no potential difference across the membrane because the charge between the K^+ ions and the A^- ions is balanced on each side due to the equivalent concentrations. As shown in Fig. 3.3B, if we insert a non-selective pore into the membrane, concentration and charge equilibrate such that there are equal concentrations of each ion on both sides of the membrane, and the voltage across the membrane is again zero.

It is when we insert into the membrane an ion-selective pore that allows only the passage of K^+ that the phenomenon shown in Fig. 3.3C occurs. Because $[K^+]$ is greater on one side of the membrane, K^+ ions diffuse through the K^+ pore down the concentration gradient. Because the membrane is not permeable to the anion A^-, each K^+ ion that passes down the concentration gradient carries a positive charge that is not balanced by an accompanying A^-. Because the transfer of these charges establishes an electrical potential gradient, K^+ ions continue to move from high concentration to low concentration until the growing force due to the electrical potential difference is balanced by the (opposite) force generated by the concentration difference.

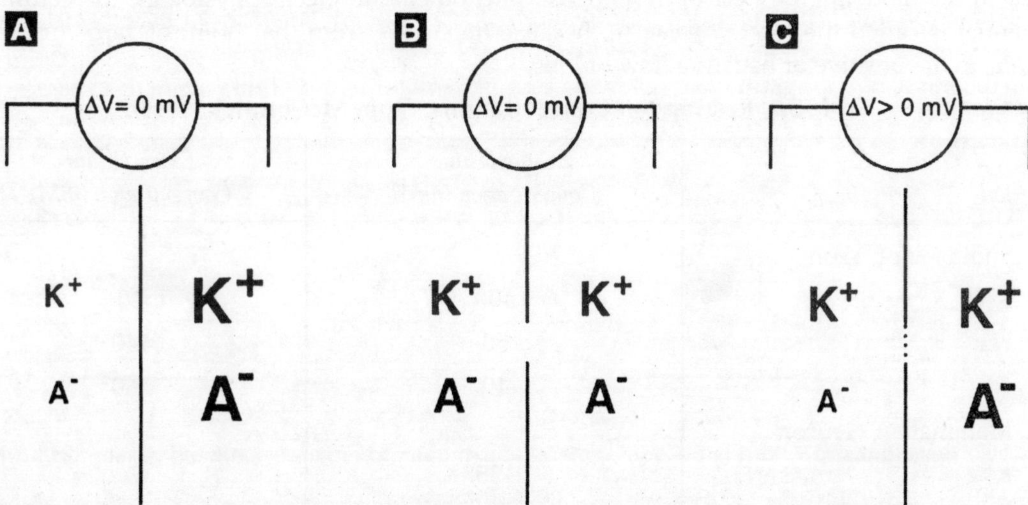

Fig. 3.3. (A) Concentration and charge are balanced on each side of the membrane, so there is no ΔV across the membrane. (B) Due to the nonselective pore, charge and concentration are balanced everywhere, and so there is no ΔV across the membrane. (C) A K^+ selective pore allows K^+ but not A^- to pass through the membrane. K^+moves to equilibrate concentration until counterbalanced by the accumulating negative charge, because A^- cannot move. The "voltmeter" is seeing excess + charge on the left side and excess-charge on the right side.

The *equilibrium potential,* where the electrical and osmotic forces are balanced, is given by the *Nernst equation.* The Nernst equation is derived from the expression for the change in Gibbs free energy when one mole of an ion of valence z is moved across a membrane:

$$\Delta G = -RT \ln \frac{[\text{ion}]_{\text{out}}}{[\text{ion}]_{\text{in}}} + \Delta VFz. \qquad \text{...(3.1)}$$

At equilibrium, ΔG is zero. Rearranging gives us the Nernst potential:

$$\Delta G = V_{\text{nernst}} = \frac{RT}{zF} \ln \frac{[\text{ion}]_{\text{out}}}{[\text{ion}]_{\text{in}}}$$

$$= 2.303 \frac{RT}{zF} \log_{10} \frac{[\text{ion}]_{\text{out}}}{[\text{ion}]_{\text{in}}}$$

$$= \frac{61.5}{z} \log_{10} \frac{[\text{ion}]_{\text{out}}}{[\text{ion}]_{\text{in}}} \text{ (at 37°C)} \qquad \text{...(3.2)}$$

where R and F are given in Table 2.1, T is temperature (in kelvin), and z is the valence of the ion as previously defined. At body temperature, RT/F is approximately 60 mV. Therefore, a 10-fold difference in the concentration of a monovalent ion like K^+ would result in approximately –60 mV of potential difference across a membrane ($[K^+]$ is greater inside a neuron). Because the Nernst potential represents the equilibrium of the thermodynamic system, the potential difference evolves to that given by the Nernst equation regardless of the initial starting potential. This tendency for the system to move toward the equilibrium potential is the basis of the *ionic battery* used in the modeling of electrophysiological phenomena. In electrophysiology, the equilibrium potential is called the *reversal potential,* because departure from that point of zero current flux results in the positive or negative flow of ions.

Table 3.2. Resting Ion Concentrations. From McCormick (1999).

Ion	*Cytoplasmic Concentration (mM)*	*Extracellular Concentration (mM)*
Squid Giant Axon		
K^+	400	20
Na^+	50	440
Cl^-	40	560
Mammalian Neuron		
K^+	135	3
Na^+	18	145
Cl^-	7	120

The Resting Membrane Potential

The Nernst potential is the equilibrium potential for one permeant ion. In reality, no channel is perfectly selective for a given ion, and there are various channels selective for various ions in a given cell as well. The Goldman-Hodgkin-Katz (GHK) equation is related to the Nernst equation, but considers the case where there are multiple conductances. The GHK equation determines the resting membrane potential of a cell from a weighted sum of the various conductances:

$$V_{GHK} = \frac{RT}{F}\ln\frac{P_K[K^+]_{out} + P_{Na}[Na^+]_{out} + P_{Cl}[Cl^-]_{in}}{P_K[K^+]_{in} + P_{Na}[Na^+]_{in} + P_{Cl}[Cl^-]_{out}} \quad ...(3.3)$$

where P_i is the relative permeability for ion i, which must be determined experimentally. While it looks like a straightforward extension of the Nernst equation, the GHK equation requires assumptions about both the interaction of ions and their ability to diffuse within channels.

We are not going to deal with these complicating details here, and in addition, we will assume perfect selectivity for our ion channels. If we define the *conductance*, $g = 1/R$ as the reciprocal of the resistance, we can use a similar weighted-sum formalism for calculating the membrane potential of a cell:

$$V_m = \frac{\Sigma_i(V_i \cdot g_i)}{\Sigma_i g_i}, \quad ...(3.4)$$

where V_i is the Nernst-equation-derived reversal potential for ion i calculated using (3.2). For example, the membrane potential for a cell containing Na^+, K^+, and Cl^- ions would be

$$V_m = \frac{(V_{Na} \cdot g_{na}) + (V_K \cdot g_K) + (V_{Cl} \cdot g_{Cl})}{g_{Na} + g_K + g_{Cl}} \quad ...(3.5)$$

This is the expression for the resting membrane potential that we will use throughout the remainder of the book. Because the resting membrane potential is the weighted average of Nernst potentials for the various ions, the ion with the greatest permeability contributes the most (see Exercise 1c).

The Nernst potential represents an equilibrium between electrical and osmotic forces on an ion across a membrane: It is what the membrane potential would be if a particular ion were at equilibrium across a membrane. Each ion has its own characteristic Nernst potential given by equation (3.2), which is determined by its concentration gradient across the membrane. The actual membrane potential is some "average" of the Nernst potentials of the several ions that are able to cross the membrane; each ion contributing to the actual membrane potential according to the permeability (or conductance) of the membrane for that particular ion. The resting potential across a membrane given by (3.5) is a steady state in which the currents carried by different ions across the membrane cancel each other out. The GHK potential is a steady state solution of a different kinetic equation for total current, based on different assumptions about how ions cross the membrane.

THE MEMBRANE MODEL

We know from Ohm's law that current flows down a voltage gradient in proportion to the resistance in the circuit. Current is therefore expressed as

$$I = \frac{V}{R} = gV \quad ...(3.6)$$

To a first approximation, our task in modeling electophysiological phenomena is to describe how the conductance of the membrane to various ions changes with time and then to keep track of the changes in current and voltage that result.

The conceptual idea behind contemporary electrophysiological models originates in the work of K.S. Cole, who pioneered the notion that cell membranes could be likened to an electronic circuit. Cole's basic circuit elements are (1) the phospholipid bilayer, which acts as a capacitor in that it accumulates ionic charge as the electrical potential across the membrane changes; (2) the ionic permeabilities of the membrane, which act as resistors in an electronic circuit; and (3) the electrochemical driving forces, which act as batteries driving the ionic currents. These ionic and capacitive currents are arranged in a parallel circuit, as shown in Fig. 3.4. This analogy to electrical circuits is now widely relied upon for developing models of electrical activity in membranes.

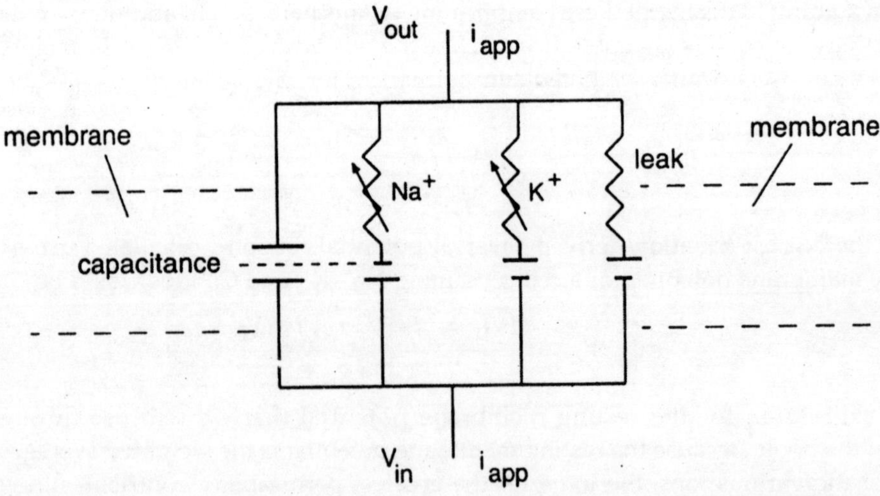

Fig. 3.4. The equivalent electrical circuit for an electrically active membrane. The capacitance is due to the phospholipid bilayer separating the ions on the inside and the outside of the cell. The three ionic currents, one for Na$^+$, one for K$^+$, and one for a non-specific leak, are indicated by resistances. The conductances of the Na$^+$ and K$^+$ currents are voltage dependent, as indicated by the variable resistances. The driving force for the ions is indicated by the symbol for the electromotive force, which is given in the model by the difference between the membrane potential $V = V_{in} - V_{out}$ and the reversal potential.

Equations for Membrane Electrical Behavior

Given the several conductances and their reversal potentials, we calculate using (3.5) what the membrane potential will be after the system has stabilized. This equations for the resting membrane potential tell us nothing about how the system evolves to the steady state. Because we are interested in the time course of the membrane voltage, we have to study the dynamics of the various currents that flow in and out of the cell. We can approximate the current flow through a single K$^+$ channel using Ohm's law and an assumption that the reversal potential stays constant:

$$I_K = -g_K(V - V_K). \qquad ...(3.7)$$

Here g_K is the conductance of the K^+ channel and the leading negative sign is necessary because of our definitions for "in" and "out." V_K is the K^+ reversal potential determined by the Nernst equation, and $V - V_K$ represents the driving force across the membrane provided by the ionic battery. We assume that the reversal potential for a given ion remains constant, which is equivalent to assuming that restorative mechanisms such as ionic pumps can keep pace with electrical activity on a time scale that prevents the ionic battery from running down. This is a reasonable assumption for a large cell, which would have a small surface area to volume ratio. In a small cell, with a large surface to volume ratio, the ion transfer necessary to change the membrane potential might have a large effect on the intracellular ionic concentration and thus the strength of the ionic battery.

Of course, numerous ions are responsible for the electrical behavior in a cell, and the total current is the sum of the individual ionic currents.

$$I_{ion}\, \Sigma I_i = \Sigma -g_i(V - V_i) \;=\; -g_K(V - V_K) - g_{Na}(V - V_{Na}) - \ldots \quad \text{...(3.8)}$$

To translate the electric circuit diagram into ODEs, we use the traditional interpretation of each circuit element along with Kirchoff's law. Assuming that the membrane acts as a capacitor, the capacitive current across the membrane can be written

$$I_{cap} = C\frac{dV}{dt}, \quad \text{...(3.9)}$$

where C is the capacitance of the membrane and V is the membrane potential, defined as the electrical potential difference between the inside and outside of the cell. To establish the differential equation satisfied by the voltage V, Kirchoff's law of charge conservation is applied to the circuit in Fig. 3.4. Kirchoff's law dictates that capacitive current must balance with the ionic current and any currents that might be applied, say, through experimental manipulation. This implies that

$$I_{cap} = I_{ion} + I_{app}, \quad \text{...(3.10)}$$

where the sum is over all the ionic currents. Using the expressions in (3.8) – (3.10) this can be rewritten

$$C\frac{dV}{dt} = -\sum_i g_i(V - V_i) + I_{app}. \quad \text{...(3.11)}$$

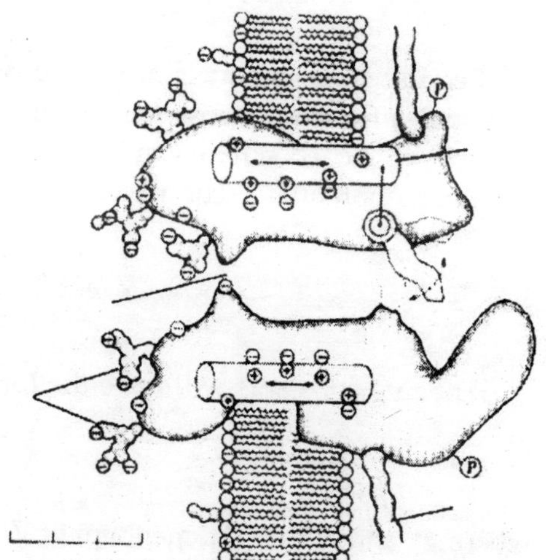

Fig. 3.5. Mechanistic cartoon of a gated ionic channel showing an aqueous pore that is selective to particular types of ions. The portion of the transmembrane protein that forms the "gate" is sensitive to membrane potential, allowing the pore to be in an open or closed state.

To solve this differential equation for voltage, we must know how the gated conductances g_i depend on V (and possibly time). In general, the g_i will not be linear functions of V, and therefore the problem is to find the time and possible voltage dependence of the various conductances.

ACTIVATION AND INACTIVATION GATES

Channels can be thought to *have gates* that regulate the permeability of the pore to ions, as illustrated schematically in Fig. 3.5. These gates can be controlled by membrane potential, producing *voltage gated* channels; by chemical ligands, producing *ligand gated* channels; or by a combination of factors. In a series of experiments, Alan Hodkgin, Andrew Huxley, and others established experimentally the voltage dependence of ion conductances in the electrically excitable membrane of the squid giant axon.

Models of Voltage-Dependent Gating

The mathematical description of voltage-dependent activation and inactivation gates is based on the mechanism

$$C \underset{k^-}{\overset{k^+}{\rightleftharpoons}} O, \qquad ...(3.12)$$

What distinguishes a voltage-dependent gating mechanism from a passive mechanism is the voltage dependence of the rate constants. Recall from (3.1) – (3.5) that the fraction of open channels *fo* satisfies the differential equation

$$\frac{dfo}{dt} = \frac{-(f_O - f_\infty)}{\tau}, \qquad ...(3.13)$$

where

$$f_\infty = \frac{k^+}{k^+ + k^-} \text{ and } \tau = \frac{1}{k^+ + k^-}. \qquad ...(3.14)$$

Because ionic channels are composed of proteins with charged amino acid side chains, the potential difference across the membrane can influence the rate at which the transitions from the open to closed state occur. According to the Arrhenius expression for the rate constants, the membrane potential V contributes to the energy barrier for these transitions:

$$k^+ \propto \exp\left(\frac{-\Delta V^+}{RT}\right) \text{ and } k^+ \propto \exp\left(\frac{-\Delta V^-}{RT}\right). \qquad ..(3.15)$$

The rate constants will have the form

$$k^+ = k_o^+ \exp(-\alpha V) \text{ and } k^- = k_o^- \exp(-\beta V), \qquad ...(3.16)$$

where k_o^+ and k_o^- are independent of V. Substituting the relationships in (3.16) into the expressions for f_∞ and τ and rearranging, we obtain

$$f_\infty = \frac{1}{1 + k_o^-/k_o^+ \exp((\alpha - \beta)V)}, \qquad ...(3.17)$$

$$\tau = \frac{1}{k_o^+ \exp(-\alpha V)} \cdot \frac{1}{1 + k_o^-/k_o^+ \exp((\alpha - \beta)V)}. \qquad ...(3.18)$$

We can define

$$S_o = \frac{1}{\beta - \alpha} \qquad ...(3.19)$$

and

$$V_o = \frac{\ln(k_o^-/k_o^+)}{\beta - \alpha} \qquad ...(3.20)$$

If we substitute S_o and V_o into (3.17) and (3.18), we have

$$f_\infty = \frac{1}{1 + \exp(-(V - V_o)/S_o)} \qquad ...(3.21)$$

$$\tau = \frac{\exp(\alpha V)}{k_o^+} \cdot \frac{1}{1+\exp(-(V-V_o)/S_o)}. \quad ...(3.22)$$

Finally, both of these expressions can be rewritten in terms of hyperbolic functions :

$$f_\infty = 0.5(1 + \tanh((V - V_0)/2S_0)), \quad ...(3.23)$$

$$\tau = \frac{\exp(V(\alpha+\beta)/2)}{2\sqrt{k_o^+ k_o^-}\cosh((V-V_o)/2S_o)}. \quad ...(3.24)$$

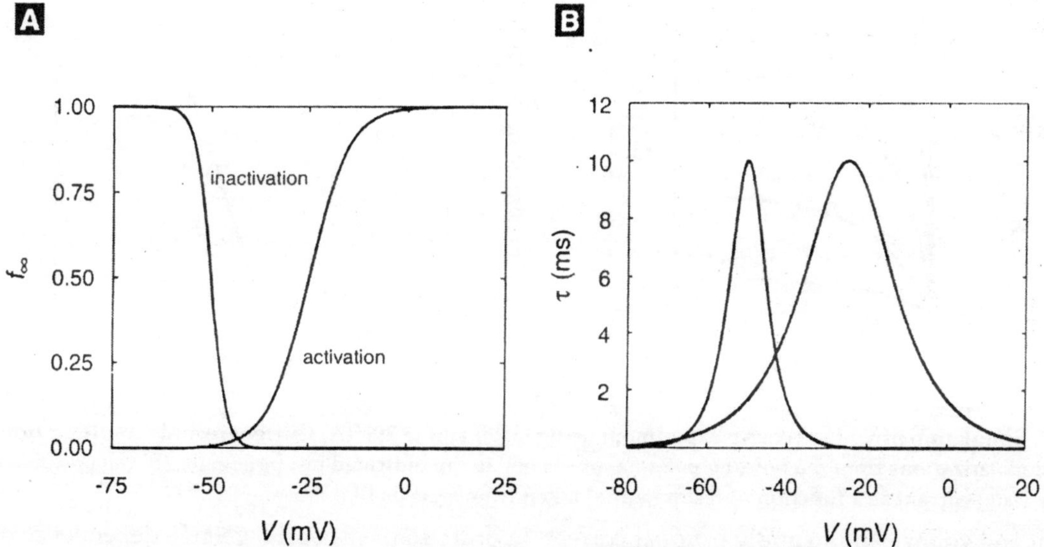

Fig. 3.6. (A) Equilibrium open fractions (f_∞) for an inactivation gate (V_0 = –50 mV and S_0 = –2 mV) and activation gate (V_0 = –25 mV and S_0 = 5 mV) as a function of voltage. (B) The characteristic relaxation times τ for the activation and inactivation gates in (A) as a function of voltage, which are peaked around the values of V_0 and have a width determined by S_0.

Recall that f_∞ gives the fraction of channels open at equilibrium at the membrane potential V. Thus for a fixed value of V it gives the open fraction after transient changes in f_0 have damped out with a characteristic time τ.

An activation gate tends to open and an inactivation gate tends to close when the membrane is depolarized. Whether a gate activates or inactivates with depolarization is determined by the sign of S_0: a positive sign implies activation and a negative sign inactivation. This is illustrated in Fig. 3.6A, where the dependence of f_∞ on V has been plotted for an activation gate with V_0 = –25 mV and S_0 = 5 mV and an inactivation gate with V_0 = –50 mV and S_0 = –2 mV.

Notice that the magnitude of S_0 determines the steepness of the dependence of f_∞ on V, whereas the value of V_0 determines the voltage at which half of the channels are open. The dependence of τ on V for these activation and inactivation gates is illustrated in Fig. 3.6B (assuming that $\alpha = -\beta$ and $2\sqrt{k_0^+ k_0^-} = 0.2\ \text{ms}^{-1}$). When $\alpha = -\beta$ and $\phi = 1/\left(2\sqrt{k_0^+ k_0^-}\right)$,

$$\tau = \frac{\phi}{\cosh((V-V_0)/2S_0)}. \quad ...(3.25)$$

The Voltage Clamp

In order to measure the voltage across a cell membrane or the current flowing through a membrane, microelectrodes are inserted into cells. These electrodes can be used both to measure

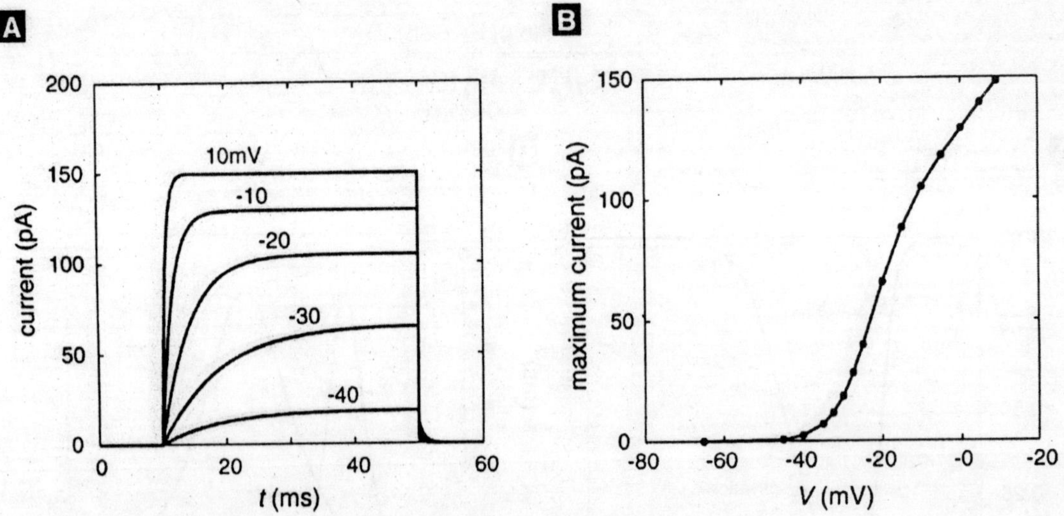

Fig. 3.7. Simulation of voltage clamp experiment using (3.28) and (3.29). (A) Current records resulting from 40 ms depolarizations from the holding potential of –60 mV to the indicated test potentials. (B) the maximum (steady state) current as a function of test potential taken from records like those in (A).

current and voltage and to apply external current. In order to measure the voltage dependence of the activation and inactivation of ion conductances, a technique called the *voltage clamp* is used. This is an electronic feedback device that adjusts the applied current I_{app} to match and counter the membrane currents such that the membrane voltage is held constant. To see what this accomplishes, consider a membrane with a single gated ionic current. If we assume that the total conductance is the result of the activation of many channels, the conductance g that we have used above can be defined as the product of the maximum possible conductance $\bar{g}$ and the fraction of open channels f_o that we have already encountered.

$$g = f_o\bar{g}. \quad \text{...(3.26)}$$

Table 3.3. Consistent Electrical Units

Name (Symbol)	*Units*	*Abbreviation*
voltage (V)	10^{-3} volt	mV
time (t)	10^{-3} second	ms
conductance (g)	10^{-9} siemens	nS
capacitance (C)	10^{-12} farad	pF
current (I)	10^{-12} ampere	pA

We can include this new relationship in the differential equation for membrane potential that we have already seen:

$$C\frac{dV}{dt} = -fo\bar{g}(V - V_{\text{rev}}) + I_{\text{app}}, \quad ...(3.27)$$

where V_{rev} is the reversal potential given by the Nernst equation. If we can apply a current that is equal and opposite to the current flowing through the membrane,

$$I_{\text{app}} = \bar{g}\xi(V - V_{\text{rev}}), \quad ...(3.28)$$

then the right-hand side of (3.27) is zero and the voltage must be constant. Because V is constant, $\xi = f_o$ and the time dependence of the applied current comes only from the dependence of f_o on t as determined by the gating equation covered in Chapter 1:

$$\frac{d\xi}{dt} = \frac{-(f_O - f_\infty)}{\tau}. \quad ...(3.29)$$

Thus the time dependence of the applied current provides a direct measurement of the gated current at a fixed voltage. Note that throughout the remainder of the text we will drop the overbar with the understanding that conductances g_i refer to maximum conductances to be scaled by gating variables.

To carry out a voltage clamp measurement like this it is necessary to block all but a single type of current. While this is not always possible, specific toxins and pharmacological agents have proven useful. For example, tetrodotoxin (TTX) from the puffer fish selectively blocks voltage gated Na^+ currents.

It is not difficult to simulate a voltage clamp measurement using (3.28) – (3.29). However, to carry out either an experimental measurement or a simulation, a consistent set of electrical units must be used. As we have seen, the standard unit for membrane potential is millivolts (mV), and because the characteristic times for voltage-dependent gates t are in milliseconds (ms), this is taken as the standard unit of time. Currents are typically expressed in $\mu A/cm^2$ and capacitances as $\mu F/cm^2$. For a typical cell of area 10^{-6} cm^2, this translates to a whole-cell current of picoamperes (1 pA = 10^{-12} A) and a whole cell capacitance of picofarads (1 pF = 10^{-12} F). Cellular dimensions are usually reported in micrometers ("microns"), and there are 10^{-8} square centimeters per square micron. Because most biological channels have a conductance g on the order of 1 to 150 pS, whole-cell conductances of nanosiemens are

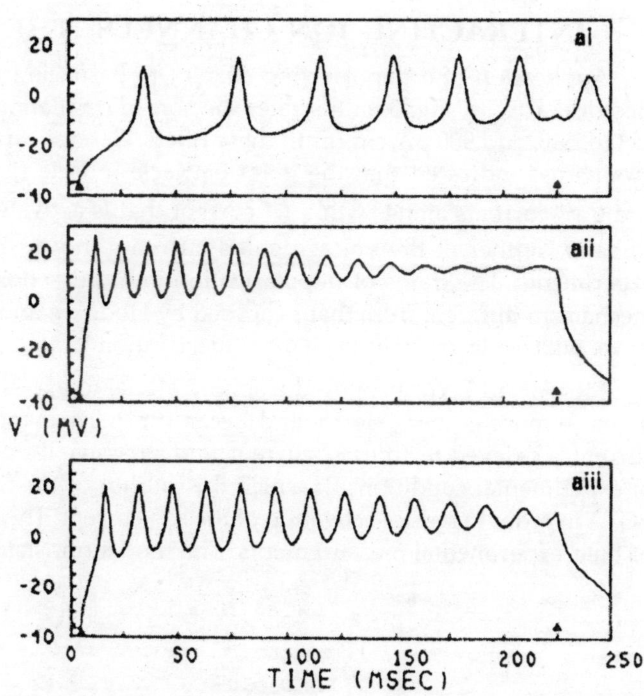

Fig. 3.8. Depolarization-induced electrical activity in giant barnacle muscle fibers; the arrows indicated the start and end of the depolarizing currents.

usually expressed in mS/cm^2, because the units of $V - V_{rev}$ are mV. This standard set of units is summarized in Table 2.3. An alternative consistent set of units uses current in femtoamperes (1 fA = 10^{-15} A), conductance in pS, and capacitance in fF.

To simulate a voltage clamp experiment we have solved (3.29) and plotted the resulting current given by (3.28). The simulation shown in Fig. 3.7A represents a typical set of experiments in which the membrane potential is clamped at a holding potential (−60 mV in Fig. 3.7), then changed to various test potentials for a fixed interval (40 ms), and finally returned to the holding potential. The value of the holding potential generally is chosen so that there is little or no current through the channel. This greatly simplifies the interpretation of the current at the test voltages. Fig. 3.7A shows the current that develops during this protocol for 5 test voltages V_{test}. The increase in current when the potential is clamped at the test values is governed by the exponential increase in f_O with characteristic time $\tau(V_{test})$. When the potential is clamped again at the holding potential, the resulting current is called the *tail current.* Its decline is also exponential, but because $V = -60$ mV during this period, the characteristic time is now $\tau(-60$ mV).

Fig. 3.7B gives a plot of the steady-state current as a function of the test voltage. According to (3.28) it can be expressed as $I = gf_\infty(V - V_{rev})$. Thus for an activating current like that in the simulations, when V is large enough, $f_\infty \approx 1$ and the current is a linear function of V. The curvature in Fig. 3.7B at lower voltages is caused by the shape of the activation function f_∞. In the jargon of circuit theory, currents like this are said to *rectify.* The delay in the onset of the maximum current, which is determined by the value of τ, has led to channels like the one simulated in Fig. 3.7 being referred to as *delayed rectifiers.*

INTERACTING ION CHANNELS: THE MORRIS-LECAR MODEL

Application of a depolarizing current to barnacle muscle fibers produces a broad range of electrical activity. Fig. 3.8 illustrates the sort of oscillations that are induced by current injections of 180, 540, and 900 μA/cm^{-2} into these fibers. Careful experimental work by a number of research groups has indicated that the giant barnacle muscle fiber contains primarily voltage gated K^+ and Ca^{2+} currents along with a K^+ current that is activated by intracellular Ca^{2+}, a so-called K^+_{Ca} current. Neither of the voltage gated currents shows significant inactivation in voltage clamp experiments. The trains of depolarization-induced action potentials in Fig. 3.8 must occur via a mechanism different from that proposed by Hodgkin and Huxley for the squid giant axon, which, as we will see later, include channel inactivation.

Morris and Lecar proposed a simple model to explain the observed electrical behavior of the barnacle muscle fiber (Morris and Lecar 1981). Their model involves only a fast activating Ca^{2+} current, a delayed rectifier K^+ current, and a passive leak. They tested the model against a number of experimental conditions in which the interior of the fiber was perfused with the Ca^{2+} chelator EGTA in order to reduce activation of the K_{Ca} current. Their simulations provide a good explanation of their experimental measurements. The model translates into two equations:

$$C\frac{dV}{dt} = -g_{Ca}m_\infty(V - V_{Ca}) - g_K w(V - V_K) - g_L(V - V_L) + I_{app}, \quad ...(3.30)$$

$$\frac{dw}{dt} = \frac{\phi(w_\infty - w)}{\tau}. \quad ...(3.31)$$

Here m_∞ is the fraction of voltage-dependent Ca^{2+}channels open, and this is a function of voltage but not time. Furthermore, w is the fraction of open channels for the delayed rectifier K^+ channels, and the conductances g_L, g_{Ca}, and g_K are for the leak, Ca^{2+}, and K^+ currents, respectively. We use w

rather than the previously used f_o for the fraction of open channels for historical reasons. The functions

$$m_\infty = 0.5[1 + \tanh((V - v_1)/v_2)], \quad ...(3.32)$$

$$w_\infty = 0.5[1 + \tanh((V - v_3)/v_4)], \quad ... (3.33)$$

$$\tau = 1/\cosh((V - v_3)/(2 \cdot v_4)), \quad ...(3.34)$$

Table 3.4. Morris-Lecar Oscillator Parameters (Type II)

Parameter	*Value*
C	20μF/cm²
V_K	–84 mV
g_K	8 mS/cm²
V_{Ca}	120 mV
g_{Ca}	4.4 mS/cm²
V_{leak}	–60 mV
g_{leak}	2 mS/cm²
$v1$	–1.2 mV
$v2$	18 mV
$v3$	2 mV
$v4$	30 mV
ϕ	0.04/ms

are the equilibrium open fractions for the Ca^{2+} current and the K^+ current, and the activation time constant for the delayed rectifier. Representative parameters are given in Table 3.4. Again note that m is not a dynamic variable. The reason for this is that we have assumed that the time constant for m is short enough that m is always in steady state, $m = m_\infty$. The idea of fast and slow processes is arguably one of the most important concepts in modeling.

We have solved the Morris-Lecar equations for four values of the applied current I_{app} and plotted the time series for V in Fig. 3.9A. Current in the Morris-Lecar model is specified in μA/cm², however to simplify notation somewhat we will assume that the cell has a total surface area of 10^{-6} cm², so that μA/cm² corresponds to 1 pA of total current. In the absence of applied current the equations have a stable steady state near –60 mV. Although increasing I_{app} to 60 pA produces a brief transient action potential, the effect of the depolarization simply produces a steady state near –35 mV. Depolarization with a current of 150 pA, on the other hand, produces a steady train of action potentials reminiscent of those observed experimentally in Fig. 3.8. In the presence of depolarizing currents much greater than this, the simulated barnacle cell can no longer sustain continuous spiking, as shown at I_{app} = 300 pA.

Phase Plane Analysis

The mechanistic features underlying continuous spiking and action potentials can be understood easily using phase plane analysis. Because of the ease of representation in two dimensions, two variable models such as the Morris-Lecar model are particularly amenable to this technique. Phase plane analysis is a powerful way to determine how the behavior of a system will change with changes in the various parameters in a system. Several types of plots are utilized as part of what is generically called phase plane analysis:

- a *phase portrait* consists of the variables describing a system plotted against each other rather than as a function of time to produce a *trajectory* in *phase space.* A phase portrait tells us how the variables interact for a given set of parameters.
- A *vector field* shows us the direction in which a system will evolve from any location in phase space.
- *Nullclines* are plotted in phase space, and show us the values of a pair of variables at which one of the variables does not change. In other words, for a coupled system of equations $X(x, y, t)$ and $Y(x, y, t)$ nullclines are the solutions to the equations

$$\frac{dX}{dt} = 0, \frac{dY}{dt} = 0 \qquad ...(3.35)$$

Note that there is a nullcline for each variable. The points of intersection of two nullclines are particularly interesting points in phase space that we will discuss below.

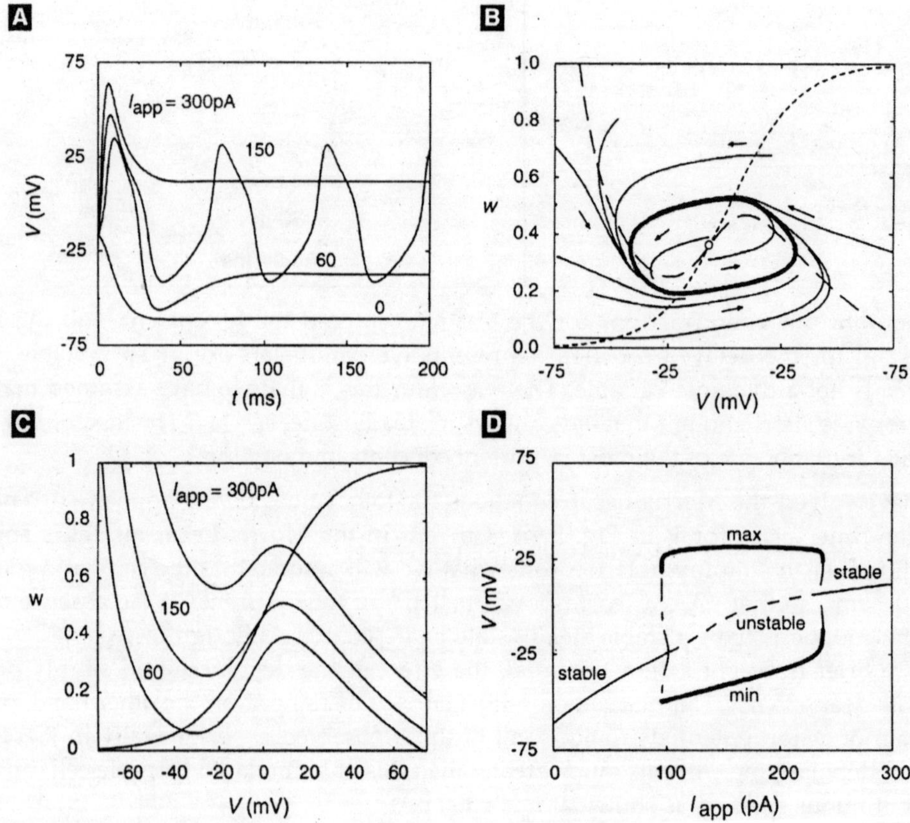

Fig. 3.9. (A) Voltage simulation for the Morris-Lecar equations using the indicated applied currents and parameters as in Table 3.4. Oscillations occur at I_{app} = 150 pA. (B) The phase plane for the Morris-Lecar model for I_{app} = 150 pA. The heavy line is the limit cycle corresponding to the oscillation in (A), and the lighter lines are short trajectories that circulate in the counterclockwise direction toward the stable limit cycle. The short arrows indicate the vector field. The *V*-nullcline is the long-dashed line and the *w*-nullcline is the short-dashed line. (C) Nullclines for several values of I_{app}. (D) A bifurcation diagram that catalogues the dynamical states of the Morris-Lecar model as a function of I_{app} with the other parameters fixed. The maximum and minimum of *V* on the limit cycle are represented by the heavy lines. Compare the values for I_{app} = 150 pA (long-dashed line) with the voltage record in (A).

Phase plane analysis plots for the Morris-Lecar model are shown in Fig. 3.9B and Fig. 3.9C. Fig. 3.9B shows trajectories, nullclines, and the vector field for I_{app} = 150 pA, which leads to the pattern of repetitive spiking. The phase portrait consists of a number of representative trajectories together with a unique, closed trajectory indicated by the heavy line. This trajectory is called a *stable limit cycle* because it is the cyclic curve to which all the neighboring trajectories converge no matter where in phase space they originate. The trajectories circulate around the steady state in a counterclockwise direction as indicated by the velocity vector field, which is shown as small arrows.

The nullclines are indicated by the lighter segmented lines. The V-nullcline, which has the inverted "N" shape, is given by the long-dashed line. It is the solution to

$$0 = -g_{Ca}m_{\infty}(V - V_{Ca}) - g_K w(V - V_K) - g_L(V - V_L) + I_{app} \quad ...(3.36)$$

for each value of W.

The w-nullcline is the solution to

$$0 = \frac{\phi(w_{\infty} - w)}{\tau} \quad ...(3.37)$$

for each value of V.

Note that the limit cycle in Fig. 3.9B circulates around the intersection of the two nullclines. We saw in Fig. 3.9A that different values of I_{app} lead to different behaviors and that I_{app} = 150 pA was the only value of the four tested that results in oscillations. Fig. 3.9C shows how changing the value of I_{app} affects the nullclines of the system. We see that increasing I_{app} from 60 pA to 150 pA and then to 300 pA raises the V nullcline but leaves the w nullcline unaltered. If we examine Fig. 3.9A and Fig. 2.9C carefully, we notice something interesting:

As we saw above, the limit cycle obtained when I_{app} = 150 pA circulates around the intersection of the nullclines at that parameter value. At I_{app} = 300 pA the system evolves to a steady state that corresponds to the intersection of the nullclines obtained for that I_{app}. The same is true for I_{app} = 60 pA. We know that a point on a nullcline corresponds to a point at which the variable of interest is not changing, and so it makes sense that the intersection of nullclines represents a combination of variables for which the system as a whole does not change. The intersection of the nullclines for I_{app} = 60 pA and I_{app} = 300 pA are examples of *fixed points,* and represent *stable steady states.* What is different about the intersection of nullclines obtained when I_{app} =150 pA?

Stability Analysis

Fixed points can be either stable or unstable and the intersection of nullclines obtained when I_{app} = 150 pA is an unstable fixed point. Notice that the two trajectories that start nearest to the intersection of the nullclines in Fig. 3.9B diverge away from the intersection and toward the limit cycle. The trajectory of the system is thus driven away from the intersection of the nullclines while at the same time being constrained to orbit around it. The existence of a stable limit cycle should be no more surprising than the existence of stable steady states. It corresponds to a closed trajectory to which all neighboring trajectories converge.

The tools provided for phase plane analysis and stability analysis in some ODE packages make it easy to determine steady states, limit cycles, and stability, and therefore how a system like the Morris-Lecar model will behave. Appendix A reviews how the *eigenvalues* determine the global stability properties of a linear system. Eigenvalues describe the solution of the linear system:

Negative real eigenvalues indicate a stable solution, positive real, eigenvalues indicate an unstable solution, and complex eigenvalues indicate the presence of oscillations in the solution. Stability analysis for nonlinear systems is not quite so straightforward, in that we can only determine the stability in such a system in a very small region around the a fixed point. In essence we linearize the system, either analytically (with a *Taylor series* or numerically (with a software package). We then use the same tools that we used for the linear system in the small linearized region around the fixed point in the nonlinear system. This book emphasizes the use of computational tools rather than mathematical analysis.

An intuitive understanding of the concept of stability can be obtained without mathematical rigor. However, it must be stressed that there are myriad subtleties that can be appreciated only with an understanding of the underlying mathematics. The topic of stability analysis of linear and nonlinear equations is covered in more depth in Appendix A, and the student should be familiar with the analytical techniques discussed there in order to appreciate the output of computational tools. Other excellent sources covering this material on a reasonably introductory level are available, and a particularly detailed analysis of the firing properties of the Morris-Lecar system has been published. The simulations in Fig. 3.9A show that a stable limit cycle occurs for the Morris-Lecar model only for certain values of I_{app}, such as I_{app} = 150 pA. We can determine the stability of the fixed points corresponding to different values of I_{app} using a numerical package.

The intersection of the nullclines for I_{app} = 150 pA indicated by the open circle in Figure 3.9B is an unstable steady state at $V = -.460$ and $w = 0.459$, a fixed point with eigenvalues $\lambda\pm = 0.264$ and 0.033 (values are rounded). It is unstable because the eigenvalues are positive real numbers, and yet these parameters result in stable oscillations. This is an excellent example of how the local stability of a fixed point does not tell the whole story for a nonlinear system. Unlike the case with linear systems, here positive eigenvalues can be associated with stable oscillations in

Table 3.5. Fixed Points and their Eigenvalues

I_{app}	V	w	*Eigenvalue*	*Eigenvalue*
0 pA	–60.855 V	0.015	–0.037	–0.096
60 pA	–37.755V	0.070	–0.055 + *i*0.063	–0.055 – *i*0.063
110 pA	–19.219V	0.196	0.055 + *i*0.045	0.055 – *i*0.045
150 pA	–0.460 V	0.459	0.264	0.033
180 pA	6.656 V	0.577	0.025 + *i*0. 139	0.025 –*i*0.139
300 pA	14.302V	0.694	–0.137 + *i*0.1 17	–0.137 –*i*0.1 17

phase space around the unstable fixed point. Using a systematic stepping procedure, we can test the stability of the system over a wide range of parameter values.

Table 3.5 shows the eigenvalues for the values of I_{app} that we have examined plus some additional intermediate values. As I_{app} is increased from $I_{app} = 0$, the eigenvalues are seen to change from negative (stable node) to complex with negative real parts (stable focus), to complex with positive real parts (unstable focus) as oscillations emerge. The oscillations at I_{app} = 150 correspond to eigenvalues that are both real and positive (unstable node). As I_{app} is increased further, the eigenvalues become complex again with positive real parts, and eventually become complex with negative real parts as oscillations cease. Fixed points at which the qualitative character of the solution changes with a change in a parameter are called *bifurcation* points.

By making even smaller changes in I_{app} and testing the stability of the steady state, it is possible to locate the specific values at which the stability of the steady state changes. This procedure could become very laborious, and so numerical algorithms have been developed that allow investigation of bifurcations to be done automatically. In the bifurcation diagram shown in Fig. 3.9D, the characteristic values of the membrane potential are plotted on the ordinate as a function of I_{app}. The thin full and dashed lines are the steady-state values of V for each value of I_{app}, with the full lines representing stable steady states and the dashed lines unstable states. The points $I_{app} \approx 94$ pA and 212 pA are bifurcation points where the stability of the steady state changes. Near these points two new dynamical features appear: a stable limit cycle and an unstable limit cycle. This type of bifurcation is called a *subcritical Hopf bifurcation,* as discussed in Appendix A. Note that while the detail is not apparent in Fig. 3.9D, the unstable limit cycle turns back and then coalesces at a turning point bifurcation with the stable limit cycle at a value of I_{app} that is smaller than the bifurcation point, and therefore there is a small region of bistability. The bifurcation diagram in Fig. 3.9D also records the maximum and minimum values of V on the limit cycles with heavy full lines. For example, at I_{app} = 150 pA the points on the heavy line correspond to the maximum and minimum of the spikes in Fig. 3.9A.

Why Do Oscillations Occur?

If the following three conditions on the Morris-Lecar equations hold, then oscillations will occur:

- the V nullcline has the inverted "N" shape like that in Fig. 3.9B;
- a single intersection of the V– and w-nullclines occurs between the maximum and minimum of the "N;"
- the rate of change of V is much greater than w.

All three conditions are met for the parameter values in Table 3.4, giving rise to oscillations in Fig. 3.9A. The importance of the slow change in w, *i.e.,* the "delay" of the delayed rectifier, can

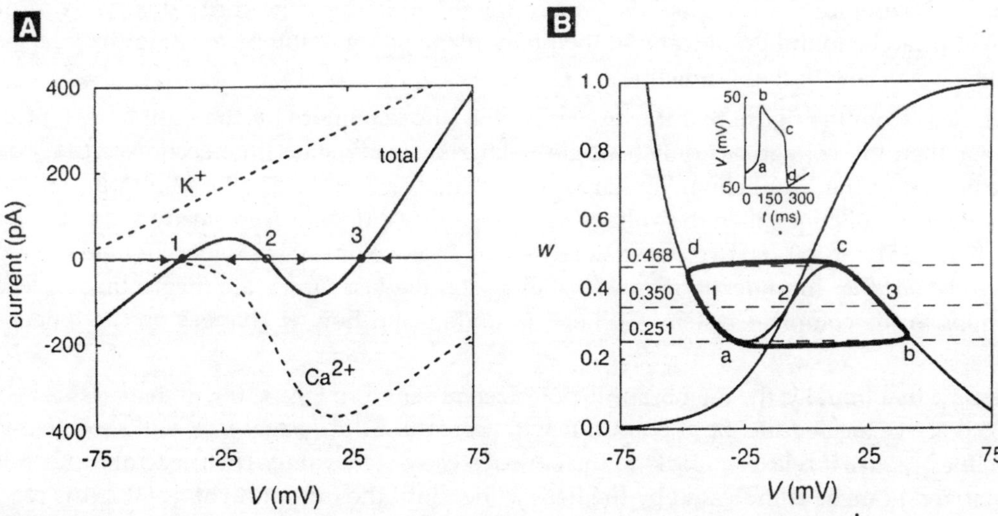

Fig. 3.10. (A) The K^+, Ca^{2+}, and total current ($I_{Ca} + I_K + I_{leak} - I_{app}$) when w = 0.35. States 1 and 3 are stable steady states, and state 2 is unstable as indicated by the velocity vectors. (B) The phase plane for the Morris-Lecar model for I_{app} = 150 pA, except that $\tau(V)$ has been increased by a factor often. The values w = 0.468 and 0.251 correspond to the maximum and minimum of the V-nullcline. The points 1, 2, and 3 at w = 0.350 are the steady states in (A). Inset shows the voltage record for a single spike.

be seen by examining the trajectories in Fig. 3.9B. If the rate of change of w were large with respect to V, then the trajectories would not depolarize and hyperpolarize rapidly as they do in Fig. 3.9A. This is tested by decreasing the value of the parameter ϕ, thereby increasing the value of the characteristic time for relaxation of w.

By altering the time scale we see why oscillations occur when the rate of change of V is very much faster than w and the nullclines have the shape in Fig. 3.9. In this case, we can treat changes in V under the assumption that w is constant, and we need only to consider the voltage equation with w fixed:

$$C\frac{dV}{dt} = -g_{Ca}m_{\infty}(V - V_{Ca}) - g_K w(V - V_K) - g_L (V - V_L) + I_{app}. \quad \text{...(3.38)}$$

Because only the voltage is changing on this time scale, we can examine its dynamical behavior using the one-dimensional phase portrait, rather than a phase plane. This is shown in Fig. 3.10A, where the total current, which is proportional to the rate of change of V, is plotted along with the Ca^{2+} and K^+ currents for $w = 0.35$ and $I_{app} = 150$ pA.

The total current vanishes at three points, which are the steady states for the voltage when w is fixed. States 1 and 3 are stable, and state 2 is unstable, as indicated by the velocity vectors in the figure. Note that at state 1 the membrane is polarized and the outward K^+ current dominates the Ca^{2+} inward current, whereas the opposite holds true at state 3. As we shall see, the oscillations can be thought of as transitions back and forth between the polarized and depolarized states, driven by slow changes in activation of the delayed rectifier current. Fig. 3.10A also explains why the V-nullcline in the Morris-Lecar model has the inverted "N" shape: The K^+ and leak currents exceed the inward Ca^{2+} current at polarized voltages between states 1 and 2, whereas the Ca^{2+} current exceeds the other currents between states 2 and 3. This voltage-dependent competition between inward and outward currents leads to a maximum and minimum in total current and therefore the inverted "N" shape for the nullcline. The three steady states of the voltage also can be found graphically in the phase plane of Fig. 3.10B by locating the intersection of the line $w = 0.35$ with the V-nullcline.

It is clear from the figure that if w exceeds 0.468 (the maximum on the right branch of the V-nullcline), then the voltage has only a single polarized steady state (intersection with w) on the far left branch of the V-nullcline. Similarly, if w is smaller than 0.251 (the minimum on the left branch of the V-nullcline), then the voltage has only a single depolarized steady state on the right branch. For $0.251 \le w \le 0.468$ two stable steady states and one unstable state occur, and the voltage is said to be *bistable.* To understand how bistability on the fast time scale (neglecting w) leads to oscillations in the complete system, we need to understand how w changes on the longer time scale.

Assume that initially, the membrane is polarized at state 1 in Fig. 3.10B, with $w = 0.35$. Because $w = 0.35$ is above the w-nullcline at point 1, w will decrease. As w decreases, V will stay close to the V-nullcline because it relaxes rapidly to the closest steady-state value. The trajectory thus follows the polarized branch, as indicated by the heavy line, until the minimum at $w = 0.251$ is reached. Beyond the minimum (near a), stable polarized states no longer exist, and V rapidly relaxes to the only remaining steady state (near b), which is on the depolarized, far right branch of the V-nullcline. During the depolarization, however, the w-nullcline is crossed. Thus on the depolarized branch, w increases and tracks the V-nullcline upward until the maximum at $w = 0.468$ is reached (near c) and the membrane rapidly repolarizes to the polarized branch (near d). The abrupt transitions from the polarized to depolarized branch and back again have led to the name *relaxation oscillator* for systems of equations that have well-separated time scales.

For the parameters in Fig. 3.9, the Morris-Lecar model is not a relaxation oscillator. However, when the characteristic time for w is increased by a factor of 10 (by decreasing ϕ), the limit cycle (heavy line in Fig. 3.10B) closely approximates that for a relaxation oscillator. If the characteristic time were increased sufficiently, or the characteristic time for V were decreased sufficiently by altering the capacitance, then the trajectory would coincide with the bistable portions of the V-nullclines, and the rapid excursions of the voltage would occur precisely at w = 0.251 and 0.468. The inset in Fig. 3.10B shows a single voltage spike that illustrates the rapid upstroke and downstroke for the limit cycle. The inset also illustrates that the shape of the depolarized and polarized portions of the spike reflects the shape of the two branches of the V-nullcline.

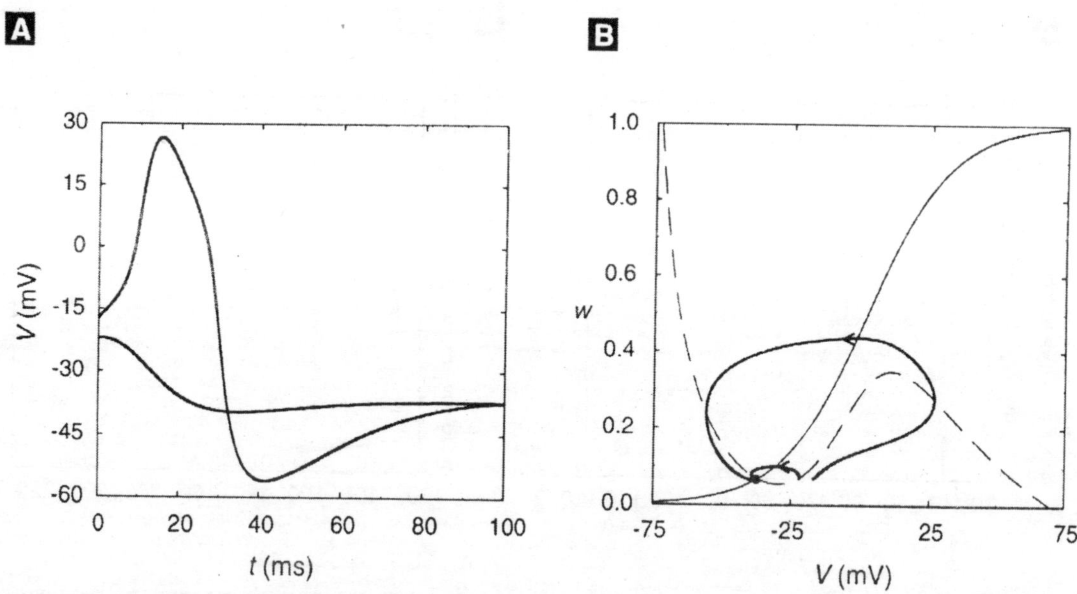

Fig. 3.11. Excitability in the Morris-Lecar model for I_{app} = 60 pA. (A) An initial deviation of the voltage to –22 mV relaxes rapidly to the steady-state voltage, whereas a deviation to –17 mV produces an action potential. (B) The trajectories in (A) represented in the phase plane. When V changes much faster than w, the location of the V-nullcline (long-dashed line) sets the threshold for action potential spikes.

Excitability and Action Potentials

Another dynamical feature of the Morris-Lecar model is *excitability*. A useful working definition of excitability is that a system is excitable when small perturbations return to the steady state, but larger (*i.e.*, above a threshold) perturbations cause large transient deviations away from the steady state. An example of this is shown in Fig. 3.11 A, where the time course of the voltage with an applied current of 60 pA and initial conditions of $w(0) = w^{ss} = 0.070$ and either $V(0) = -22$ or –17 mV are used. For sub-threshold $V(0)$ such as –22 mV, the voltage increases only slightly before it decreases monotonically to its steady state-value of about –37 mV. For suprathreshold deviations, like that for $V(0) = -17$ mV, the voltage increases dramatically, producing an action potential spike before returning to steady-state.

The explanation for excitability can be understood most easily in the phase plane. In Fig. 3.11B we have plotted the trajectories for the two initial conditions in Fig. 3.11A along with the two nullclines. The trajectories for subthreshold initial conditions like $V(0) \leq -22$ mV start at points in phase space *above* the *V*-nullcline and *below* the w nullcline. This implies that the initial velocity vector, and therefore the initial trajectory, points in the direction of smaller potentials and larger values of the activation of the delayed rectifier K^+ current. Thus the voltage begins to decrease and continues to do so because the K^+ current activates as w increases. This contrasts with the trajectory that starts at $V(0) = -17$ mV, for which the initial velocity vector points in the direction of increasing voltage. Even though w is increasing in this region, which is also below the w-nullcline, the rate of increase of voltage exceeds that of w, and the trajectory moves to higher voltages until it crosses the V-nullcline and begins to decrease.

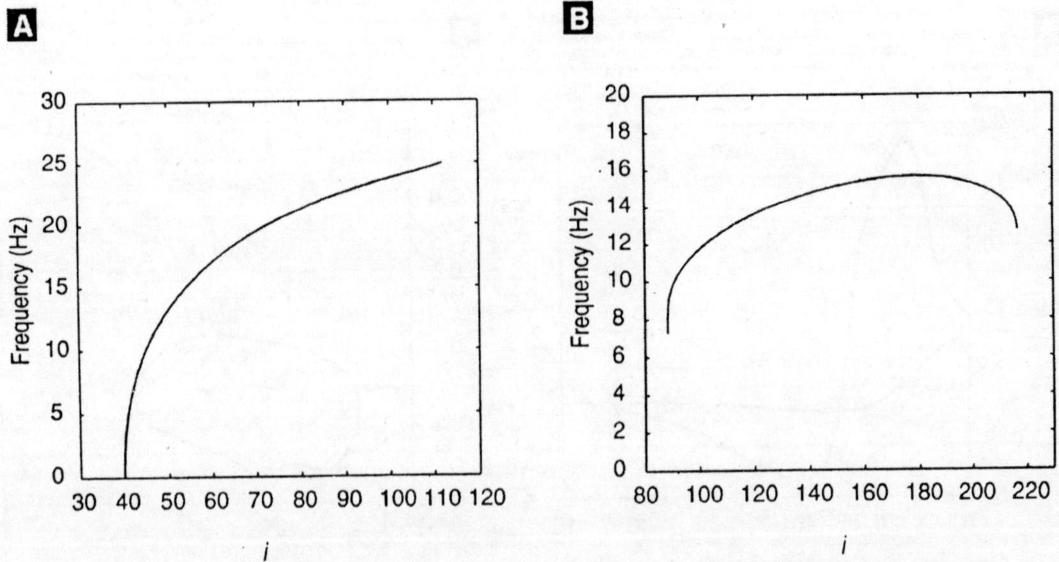

Fig. 3.12. Examples of voltage-frequency plots for the Morris-Lecar model with parameters that result in (A) Type I dynamics and (B) Type II dynamics. Note that the axes are somewhat different. Plotted according to Rinzel and Ermentrout (1998).

The threshold value of $V(0)$ above which action potentials occur depends on the shape of the nullclines and the rate of activation of w (altered by ϕ). However, it is close to the point where a line drawn parallel to the V axis at w^{ss} crosses the V-nullcline. It is not hard to check that it will approach this point as w is made to change more slowly than V.

Type I and Type II Spiking

Many investigators are now interested in applying biophysical models of neurons and other spike-generating mechanisms to the study of information transfer using information-theoretic measures. It is particularly important, therefore, that the dynamical behavior of a model be characterized over the space of parameters and that the characteristics of the spike generating model match the intended use. Both the Hodgkin-Huxley model, described below, and the Morris-Lecar model produce trains of action potentials with commonly used parameter sets and when sufficiently depolarized. Muscle fiber and axons innervating muscle fiber are examples of situations where a strong, consistent signal is required. In our investigations of the Morris-Lecar model, we have seen that depolarization beyond threshold results in oscillations that begin at a characteristic nonzero frequency. This behavior has been defined as a Type II oscillator. Many neurons exhibit firing properties that are fundamentally different from those of the Type II oscillator.

Models of these cells can produce arbitrarily low frequencies of oscillations. These models are classified as Type I oscillators. The Morris-Lecar model can exhibit Type I or Type II behavior, depending on the parameters that are chosen, and the Hodgkin-Huxley model also can be modified to show this behavior. See Table 3.6 for the Morris-Lecar Type I parameters that differ from Type II parameters. Examples of voltage-frequency plots for the Morris-Lecar model in these two regimes are shown in Fig. 3.12. The Morris-Lecar Type II oscillator might not be appropriate for the study of subtle aspects of information transfer or intra-neuronal coupling. We have seen that Type II spiking results from a subcritical Hopf bifurcation as input current is increased. Type I spiking results from a saddle-node bifurcation. We also note that there are other levels of complexity for spiking and oscillating models. Models of bursting cells, which exhibit trains of oscillations separated by periods of quiescence, require additional slow variables.

THE HODGKIN-HUXLEY MODEL

Ordinarily, a discussion of membrane electrophysiology might begin with the famous model of the squid giant axon developed by Hodgkin and Huxley (the HH model). Our goal here has been to facilitate a clear mathematical understanding of oscillatory behavior in systems of ion channels. While it remains a seminal accomplishment in the history of physiology, the HH model is complicated and not amenable to the phase plane methods of analysis that we have used to understand dynamical electrical behavior in cells. The student should be familiar with the HH model, for historical reasons and because it is still widely used. We cannot cover the work of Hodgkin and Huxley in the detail it deserves, but the concepts learned in the study of the Morris-Lecar oscillator will aid the student in understanding the HH model. We recommend that the student consult Bertil Hille's *Ionic Channels of Excitable Membranes* (Hille 2001), which contains one of the best treatments of the HH model and the history behind it. "Hille" is also the best consolidated source of information about the biophysics of ion channels, and it should be familiar to any person attempting to model electrical behavior.

Table 3.6. Morris-Lecar Type I Oscillator Parameters

Parameter	*Value*
g_{Ca}	4 mS/cm^2
$v3$	12 mV
$v4$	17.4 mV
ϕ	0.066/ms

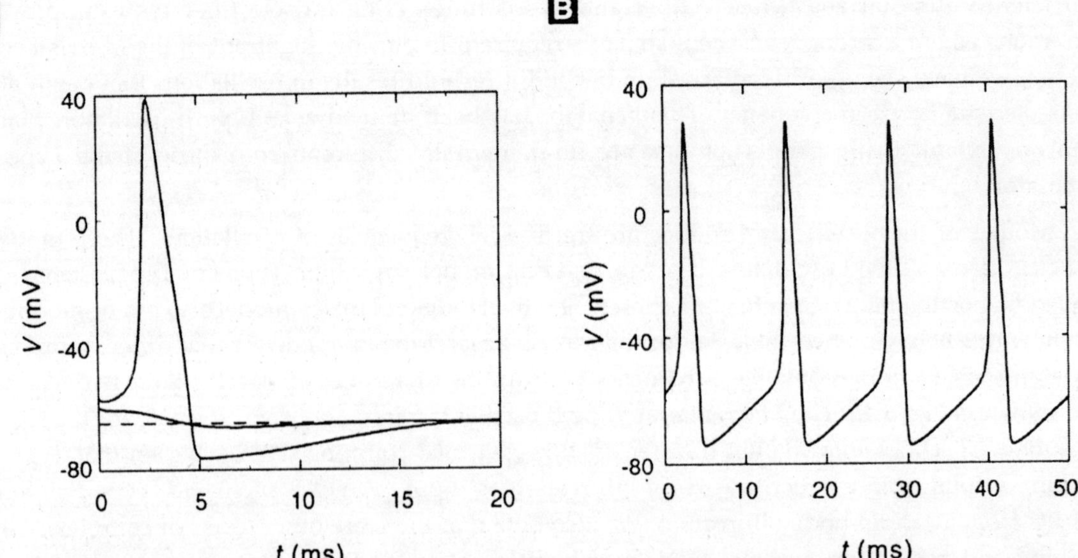

Fig. 3.13. (A) The solution of the Hodgkin-Huxley equations for three different initial values of the membrane potential and no applied current. When the initial value exceeds ca. –59 mV, an action potential is produced. (B) Continuous spiking occurs under the same conditions with an applied current I_{app} = 15.

The HH model is empirical. Many voltage clamp experiments were performed by Hodgkin and Huxley, and their data were fit to expressions that they incorporated into the model without consideration of an underlying mechanism for the channel gates. One of the remarkable aspects of Hodgkin and Huxley's work is that their model was developed without a molecular understanding of the mechanism. In fact, it required almost thirty years of intensive research after their work to formulate a realistic cartoon of the mechanisms underlying the ionic currents in cells.

Although Hodgkin and Huxley justified these expressions on empirical grounds, it is possible to derive the gating expression used in the Hodgkin-Huxley model using mechanistic models. From their voltage clamp and other measurements, Hodgkin and Huxley deduced that the sodium conductance involved two voltage-dependent "gates," an *activation* gate and an *inactivation,* gate and that the potassium conductance had a single activation gate. Note that the presence of an inactivation gate differs from the Morris-Lecar model. To account for these facts they represented the ionic conductances in the following form:

$$g_{Na} = \bar{g}_{Na} m^3 h, \qquad \text{...(3.39)}$$

$$g_K = \bar{g}_K n^4 \qquad \text{...(3.40)}$$

where the terms $\bar{g}$ represent maximal conductances, and m and n are the activation gating variables and h the inactivation. The exponents on m and n were chosen for the best fit to experimental data. These gating variables were postulated to satisfy linear differential equations where the variables "relax" to voltage-dependent values, e.g., m_∞, that vary between zero and one with voltage-dependent time constants, *e.g.*, τ_m.

Putting the ODEs for the gating variables together with (3.11) gives the primary equations for the Hodgkin-Huxley model:

$$CdV/dt = \bar{g}_{Na}m^3h(V - V_{Na}) - \bar{g}_K n^4(V - V_K),$$

$$- \bar{g}_{\text{leak}}(V - V_{\text{leak}}) + I_{\text{app}}, \quad ...(3.41)$$

$$dm/dt = -(m - m_\infty)/\tau_m, \quad ...(3.42)$$

$$dh/dt = -(h - h_\infty)/\tau_h, \quad ...(3.43)$$

$$dn/dt = -(n - n_\infty)/\tau_n. \quad ...(3.44)$$

Hodgkin and Huxley added the third conductance $\bar{g}_{\text{leak}}$ to their voltage equation to account for a small voltage-independent conductance that they attributed to a "leak" in the membrane, possibly through their microelectrode. The nonlinear terms in (3.41) are obvious in the activation and inactivation gates. Not so obvious are the nonlinearides in (3.42)-(3.44). However, all the voltage-dependent terms in those equations are nonlinear functions of V as well.

The student should solve the HH equations to provide a basis for comparison and further exploration. The equations for the Hodgkin-Huxley spike generator are readily available from other sources, and it is not difficult to run simulations with a package for solving ODEs. One must be careful to choose a suitable numerical method of solution due to the nonlinear equations. Fig. 3.13 shows several simulations that can be made with the HH equations. The *Runge-Kutta* method has been used with a time step of 0.05, which is appropriate because reducing the time step to 0.01 gives no noticeable change in the results.

Fig. 3.13A shows calculations with $I_{\text{app}} = 0$, but with $V(0) = -65$, -60, and -57 mV. The steady state for the voltage with these parameters is clearly -65 mV; however, if the initial value of V exceeds about -59 mV, then the equations produce an action potential spike, and we have another example of a Type II oscillator. The Hodgkin-Huxley equations can also produce repetitive firing of action potentials. This is illustrated in Fig. 3.13B. The parameters in that simulation are identical to those in Fig. 3.13A, except that $I_{\text{app}} = 15$.

FITZHUGH-NAGUMO CLASS MODELS

We have seen that a fast variable with a cubic nullcline for voltage and a slower variable with a monotonically increasing nullcline for channel opening are sufficient to produce oscillations. In some cases, and particularly for mathematical analysis, it may be beneficial to completely abstract the mathematical equations for an oscillator from the underlying physical processes in order to understand general properties of these systems. Because two-variable systems are useful for the phase plane methods we have discussed, we can contrive such a system.

The two differential equations will be functions of both state variables:

$$\frac{dv}{dt} = f(v, w),$$

$$\frac{dv}{dt} = g(v, w). \quad ...(3.45)$$

By definition, the nullcline is obtained by setting the right-hand side of the equation to zero and plotting the function in the (v, w) phase plane. We can create the behavior we desire by specifying the shapes of $f(v, w)$ and $g(v, w)$. The slow variable, w, is easiest, because we want $g(v, w)$ to be similar to a line describing the increase of v with w. Note that v here is the dependent variable for the equation for a line:

$$g(v, w) : v = \gamma w, \quad \text{...(3.46)}$$

and therefore

$$\frac{dw}{dt} = v - \gamma w, \quad \text{...(3.47)}$$

where γ is the slope parameter.

The fast variable v will have a form similar to (3.47) with w as the dependent variable:

$$f(v, w): w = f(v), \quad \text{...(3.48)}$$

and therefore

$$\frac{dv}{dt} = f(v) - w. \quad \text{...(3.49)}$$

Because we know that w needs to be a cubic function of v, let us try

$$f(v) = Bv(v - \beta)(\delta - v), \quad \text{...(3.50)}$$

where A is a parameter to scale the amplitude of the curve. The equations for the complete system are therefore

$$\frac{dv}{dt} = Av(v - \beta)(\delta - v) - Cw,$$

$$\frac{dw}{dt} = \varepsilon(v - \gamma w). \quad \text{...(3.51)}$$

The parameter ε has been added to more easily control the speed of one variable relative to the other, and the parameter C affects the coupling strength. For an appropriate choice of all of the parameters, we would expect that this system of abstracted differential equations would produce oscillations.

This dynamical paradigm appeared first in cellular neuroscience in simplifications of the Hodgkin-Huxley equations by FitzHugh (1961), and later as part of an independent development by Nagumo et al. (1962). In fact, the various parameters we have contrived have physical interpretations in the context of a simplification of the Hodgkin-Huxley circuit diagram we discussed earlier. This type of system, involving a linear nullcline for the slow variable and a cubic nullcline that has the inverted "N" shape for the fast variable, are given the generic name *FitzHugh-Nagumo* (or FH-N) models.

CONCLUDING REMARKS

Voltage gated ion channels and the currents that flow through them underlie much of the electrical behavior of cells. We derived a mechanism-based model for ion channels and found that we could produce oscillatory behavior in electrical membranes with only two variables. Therefore, we could analyze the underlying dynamics with phase plane techniques and other methods of dynamical systems analysis. The dynamical features of "slow" variables coupled to "fast" variables with either "N"-shaped or inverted "N"-shaped nullclines are characteristic of FitzHugh-Nagumo type oscillators common to many biological mechanisms at the cellular level. In addition to producing oscillations in the barnacle muscle, the same dynamical structures will appear in mechanistic models of insulin secretion and Ca^{2+} oscillations.

CALCIUM WAVES AND SPARKS 4

In this chapter we shall discuss a variety of intracellular Ca^{2+} wave phenomena, but always from the perspective that the distance scales of interest are large enough that Ca^{2+} transport is well-modeled by conservation equations based on a continuum description of matter. Although recent experimental and theoretical work suggests that the macroscopic behavior of propagating Ca^{2+} waves (*e.g.*, wave speed) may depend in subtle ways on the density and distribution of intracellular Ca^{2+} release channels, we postpone consideration of intracellular heterogeneities such as clusters of Ca^{2+} release channels until later in the chapter. This makes sense because both the mathematics and simulation methods used to study nonlinear wave propagation in *homogeneous* media are simpler than the *heterogeneous* case. This simplicity should facilitate the development of intuition regarding nonlinear wave propagation. Throughout the chapter a recurring theme will be the manner in which Ca^{2+} buffers, through their important association with free Ca^{2+}, can influence wave phenomena dependent on diffusion. The chapter concludes with calculations of localized Ca^{2+} elevations due to intracellular Ca^{2+} release, *i.e.*, Ca^{2+} "puffs" or "sparks," elementary events that sum to produce Ca^{2+} waves.

MICROFLUOROMETRIC MEASUREMENTS

Described several examples of global or cell-wide Ca^{2+} excitability and oscillations. Experimental observations of such intracellular Ca^{2+} dynamics are often made using microfluorimetry, an experimental technique that involves loading Ca^{2+} indicator dyes into cells and instrumentation that optically excites these indicators and measures emission. These Ca^{2+} indicator dyes are themselves Ca^{2+} buffers (often highly mobile) that can potentially affect intracellular Ca^{2+} signaling. For example, the differential fraction of free to bound cytosolic Ca^{2+}, denoted by f_i ($[Ca^{2+}]_i$), will be determined by both exogenous as well as endogenous Ca^{2+} buffers.

Although a measured fluorescence signal is only indirectly related to the dynamics of intracellular Ca^{2+}, it is relatively straightforward to determine the free Ca^{2+} concentration during a cell-wide Ca^{2+} response using the time course of measured fluorescence (Grynkiewicz *et al.* 1985). If the Ca^{2+} and indicator dye concentrations are homogeneous throughout the cell, the equilibrium relation

$$[CaB] = \frac{[Ca^{2+}]_i \, [B]_T}{K + [Ca^{2+}]_i} \qquad ...(4.1)$$

is valid as long as $[Ca^{2+}]_i$ changes slowly compared to the equilibration time of the buffers. Because this equilibration time is on the order of milliseconds, this condition is usually satisfied for global Ca^{2+} responses, which occur with a time scale of seconds or tens of seconds.

If the equilibrium relation accurately describes the relationship between the concentration of bound indicator dye ([CaB]) and free Ca^{2+} ($[Ca^{2+}]_i$), then it is a simple matter to "backcalculate" the free Ca^{2+} concentration as a function of time. For a single excitation wavelength measurement (*e.g.*, using a nonratiometric dye such as fluo-3 at low concentration), we can idealize the indicator fluorescence as the sum of two components,

$$F = \eta_B[B] + \eta_{CaB}[CaB], \qquad ...(4.2)$$

where η_B and η_{CaB} are proportionality constants for free and bound dye, respectively. When $\eta_B < \eta_{caB}$, the maximum and minimum observable fluorescences are given by $F_{min} = \lim_{[Ca^{2+}]_i \to 0} F = \eta_B[B]_T$ and $F_{max} = \lim_{[Ca^{2+}]_i \to \infty} F = \eta_{CaB}[B]_T$. Using the equilibrium relation it can be shown (see Exercise 1) that

$$[Ca^{2+}]_i = K\frac{[CaB]}{[B]} = K\frac{F - F_{min}}{F_{max} - F}. \qquad ...(4.3)$$

If only the Ca^{2+}-bound indicator fluorescences strongly, then $F \approx \eta_{caB}[CaB]$, and a slightly simpler expression results from substituting $F_{min} = 0$.

It should be noted that the validity of relies on the stability of instrument sensitivity, optical path length, and dye concentration between measurements of F, F_{min}, and F_{max}. Because determining F_{min} and F_{max} usually involves titrating the indicator released from lysed cells, this is difficult to achieve in practice.

In whole-cell Ca^{2+} measurements, fluorescence intensities can be measured at two excitation wavelengths (λ and λ') using indicator dyes such as fura-2. Such *mtiometric* measurements can be related to the underlying free Ca^{2+} signal by supplementing with

$$F' = \eta'_B[B] + \eta'_{CaB}[CaB], \qquad ...(4.4)$$

where the primes indicate the second excitation wavelength. Using the first equality of the fluorescence ratio $R = F/F'$ can be inverted to give

$$[Ca^{2+}]_i = K\frac{\eta_B - \eta'_B R}{\eta'_{CaB} R - \eta_{CaB}} = K\left(\frac{R - R_{min}}{R_{max} - R}\right)\left(\frac{\eta'_B}{\eta'_{CaB}}\right), \qquad ...(4.5)$$

where for the second equality we use $R_{min} = \lim_{[Ca^{2+}]_i \to 0} F/F' = \eta_B/\eta'_B$ and $R_{max} = \lim_{[Ca^{2+}]_i \to \infty} F/F' = \eta_{CaB}$. If λ' is chosen to be a wavelength at which the calibration spectra at different Ca^{2+} concentrations cross one another, then $\eta'_B \approx \eta'_{CaB}$, and the last factor in (4.5) is eliminated. An advantage of the ratiometric method is its insensitivity to changes in dye concentration and instrument sensitivity between measurements.

A MODEL OF THE FERTILIZATION CALCIUM WAVE

When mature *Xenopus laevis* oocytes (eggs) are loaded with an indicator dye (*e.g.*, Ca^{2+}-green dextran) and stimulated by the fusion of sperm, a propagating wave of intracellular Ca^{2+} release can be observed by backcalculating the free Ca^{2+} concentration ($[Ca^{2+}]_i(x, t)$ from a time-dependent fluorescence signal ($F(x, t)$). *This fertilization* Ca^{2+} *wave* is an important step in early development. It triggers the fusion of cortical granules (vesicles) with the plasma membrane, a process that initiates the raising of the viteline envelope and a long-lasting block to polyspermy. The cell divisions that initiate development *of Xenopus* begin only after the fertilization Ca^{2+} wave has propagated throughout the entire cell. The eggs of many species, from starfish to mammals, exhibit propagating Ca^{2+} waves upon fertilization.

Because fertilization Ca^{2+} waves such as those shown in Fig. 4.1 still occur when the extracellular medium is nominally Ca^{2+} free, the phenomenon appears to be largely independent of Ca^{2+} influx. On the other hand, the fertilization Ca^{2+} wave absolutely requires functional IP_3 receptors. When IP_3-mediated Ca^{2+} release is blocked by any one of several experimental manipulations (*e.g.*, upon introduction of heparin into the cytosol), the fertilization Ca^{2+} wave is not observed.

We expect Ca^{2+} diffusion to play an important role in the fertilization Ca^{2+} wave because *Xenopus laevis* eggs are large (approximately 1.2 mm in diameter). Indeed, Fig. 4.1 indicates that the Ca^{2+} concentration within the egg depends very much on both spatial position and time. For this reason a whole cell model of this phenomenon would be deficient, and instead researchers mathematically describe the fertilization Ca^{2+} wave in the *Xenopus* egg and other cell types using a combination of spatial modeling and whole-cell modeling approaches.

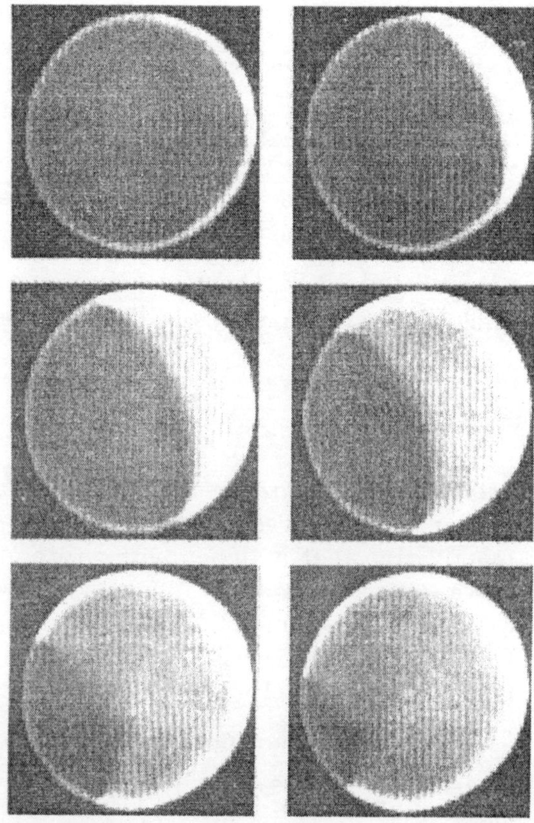

Fig. 4.1. Microfluorimetric images of sperm-induced fertilization calcium waves in mature *Xenopus laevis* eggs loaded with the indicator dye Ca^{2+}-green dextran. The wave takes approximately 5 minutes to cross the length of the egg (1.2 mm).

Here we introduce such a spatial whole-cell model of the fertilization Ca^{2+} wave by recalling that the Li-Rinzel reduction of the DeYoung-Keizer model of the IP_3R can be combined with sigmoidal SERCA pump kinetics and a passive ER leak to create a whole-cell model of Ca^{2+} handling in pituitary gonadotrophs. Using a variation of this model to represent Ca^{2+}-induced Ca^{2+} release (CICR) and reuptake by the endoplasmic reticulum (ER), we follow and account for Ca^{2+} diffusion in both the cytosol and ER by writing the following reaction-diffusion system:

$$\frac{\partial[Ca^{2+}]_i}{\partial t} = f_i[D_i\nabla^2[Ca^{2+}]_i + j_{IP_3R} + j_{LEAK} - j_{SERCA}], \quad ...(4.6)$$

$$\frac{\partial w}{\partial t} = [w_\infty([Ca^{2+}]_i, [IP_3]) - w]/\tau, \quad ...(4.7)$$

$$\frac{\partial[Ca^{2+}]_{ER}}{\partial t} = f_{ER}[D_{ER}\nabla^2[Ca^{2+}]_{ER} - (\bar{V}_i/\bar{V}_{ER})(j_{IP_3R} - j_{LEAK} + j_{SERCA})], \quad ...(4.8)$$

where $j_{LEAK} = v_{LEAK}$ $[Ca^{2+}]_{ER}$ $-$ $([Ca^{2+}]_i)$, $j_{IP_3R} = v_{IP_3R} m_\infty^3 w^3$ $([Ca^{2+}]_{ER} - [Ca^{2+}]_i)$, j_{SERCA} $= v_{SERCA}[Ca^{2+}]_i^2/([Ca^{2+}]_i^2 + K_{SERCA}^2)$, and $\bar{V}$'s are the volumes as before. In these equations, w is the fraction of IP_3Rs *not* inactivated and the open probability of the IP_3Rs is given by $P_O = m_\infty^3 w^3$ where the fraction of activated IP_3Rs (m_∞^3) is assumed to be an instantaneous function of Ca^{2+} and IP_3 concentrations,

$$m_\infty^3 = \left(\frac{[IP_3]}{[IP_3 + d_1]}\right)^3 \left(\frac{[Ca^{2+}]_i}{[Ca^{2+}]_i + d_5}\right)^3. \qquad ...(4.9)$$

Although the Li-Rinzel reduction of the DeYoung-Kiezer model gives a time constant of IP_3R inactivation that is dependent on Ca^{2+} and IP_3 concentration, $\tau([Ca^{2+}]_i, [IP_3])$, for simplicity we will assume that τ is constant (2 sec). A diagram of the fertilization Ca^{2+} wave model components and fluxes is presented in Fig. 4.2.

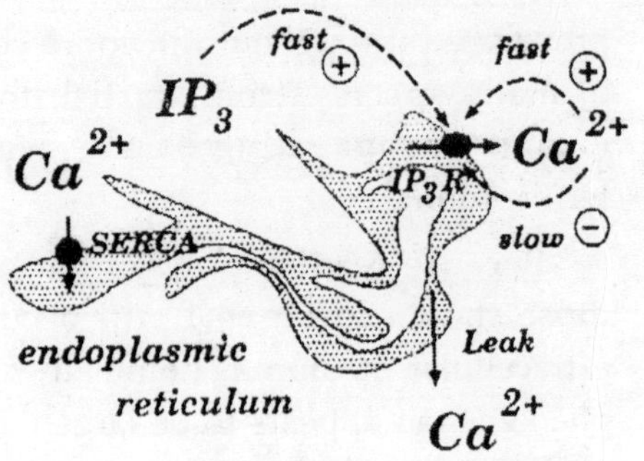

Fig. 4.2. Schematic diagram of the *Xenupus* egg fertilization Ca^{2+} wave model. Ca^{2+} enters the cytosol from the ER via a passive leak and the IP_3R, which is activated by both Ca^{2+} and IP_3 on a fast time scale and inhibited by Ca^{2+} on a slow time scale, all at the cytoplasmic face. The ER is refilled by a SERCA-type Ca^{2+}-ATPase pump. Diffusion of Ca^{2+} in both the cytosol and ER is accounted for using effective diffusion coefficients (constant) that account for interactions with Ca^{2+} buffers (not shown) and the volume fractions of both compartments.

Note that the maximum conductance of IP_3-mediated Ca^{2+} release (v_{IP_3R}), passive leak rate (v_{LEAK}), and maximum rate of ATP-dependent reuptake (v_{SERCA}) are constants. Thus, the model assumes homogeneous Ca^{2+} release and reuptake dynamics, that is, a uniform and high density of inositol 1,4,5-trisphosphate (IP_3) receptors and sarco-endoplasmic reticulum Ca^{2+}-ATPases. Also note that (4.7) includes no diffusion term for the simple reason that IP_3Rs (and thus the gating variable representing their inactivation) has a fixed spatial location and does not diffuse. Throughout this chapter we will assume that $[IP_3]$ is uniform and constant.

INCLUDING CALCIUM BUFFERS IN SPATIAL MODELS

In this spatial whole-cell model given by (4.6)-(4.8), the quantities f_i and f_{ER} multiply both reaction and diffusion terms. As discussed these factors scale the Ca^{2+} release and reuptake rates to account for the proportion of Ca^{2+} bound to buffer in both cytosolic and ER compartments.

Note that the coefficients of the Laplacians in (4.6) and (4.8) imply constant *effective diffusion coefficients* for the cytosol and ER given by $D_i^{eff} = f_i D_i$ and $D_{ER}^{eff} = f_{ER} D_{ER}$, both of which are expected to be much smaller than the free Ca^{2+} diffusion coefficient. This assumption is here made for convenience and is strictly true only for low-affinity Ca^{2+} buffers.

To see that this is the case, consider how one Ca^{2+} buffer (perhaps representing a Ca^{2+} indicator dye) would be added to the fertilization Ca^{2+} wave model. The association and dissociation of Ca^{2+} with the indicator dye would contribute bimolecular reaction terms in the cytosolic compartment, giving

$$\frac{\partial[\text{Ca}^{2+}]_i}{\partial t} = D_C\nabla^2[\text{Ca}^{2+}]_i + R + j_T, \qquad ...(4.10)$$

$$\frac{\partial[\text{B}]}{\partial t} = D_B\nabla^2[B] + R, \qquad ...(4.11)$$

$$\frac{\partial[\text{CaB}]}{\partial t} = D_{CB}\nabla^2[\text{CaB}] - R, \qquad ...(4.12)$$

where D_C, D_B, and D_{CB} are diffusion constants for free Ca^{2+}, free buffer, and bound buffer, respectively,

$$R = k^+[\text{B}][\text{Ca}^{2+}]_i + k^-[\text{CaB}],$$

and $j_T = j_{\text{IP}_3\text{R}} + j_{\text{LEAK}} - j_{\text{SERCA}}$ is the sum of all Ca^{2+} fluxes into and out of the ER.

Notice that if the diffusion of buffer doesn't depend on whether or not Ca^{2+} is bound ($D_B \approx D_{CB}$), then (4.11) and (4.12) can be summed to give

$$\frac{\partial[\text{B}]}{\partial t} = D_B\,\nabla^2\,[\text{B}]_\text{T}$$

where $[\text{B}]_\text{T} = [\text{B}] + [\text{CaB}]$ is the total buffer concentration profile (free plus bound). This equation implies that if $[\text{B}]_\text{T}$ is initially uniform (not a function of position), it will remain uniform for all time. We can thus eliminate (4.11) and obtain a reduced system given by (4.10) and (4.12) with

$$R = -k^+\,([\text{B}]_\text{T} - [\text{CaB}])\,[\text{Ca}^{2+}]_i + k^-[\text{CaB}].$$

THE EFFECTIVE DIFFUSION COEFFICIENT

If the buffer reactions are rapid with respect to the diffusion, it is possible to simplify our model further using the rapid buffer approximation. Assuming local equilibrium (*i.e.*, chemical equilibrium at every point in space), we can write an equilibrium expression for total cell Ca^{2+} concentration:

$$[\text{Ca}^{2+}]_\text{T} = [\text{Ca}^{2+}]_i + [\text{CaB}] = [\text{Ca}^{2+}]_i + \frac{[\text{Ca}^{2+}]_i\,[\text{B}]_\text{T}}{[\text{Ca}^{2+}]_i + \text{K}}.$$

Next, we eliminate reactions terms (and the fast time scale) by adding (4.10) and (4.12), to give

$$\frac{\partial[\text{Ca}^{2+}]_\text{T}}{\partial t} = \frac{\partial[\text{Ca}^{2+}]_i}{\partial t} + \frac{\partial[\text{CaB}]}{\partial t} = D_C\nabla^2[\text{Ca}^{2+}]_i + D_B\nabla^2[\text{CaB}] + j_T. \qquad ...(4.13)$$

The reader can confirm that substituting $[Ca^{2+}]_T$ and [CaB] in terms of $[Ca^{2+}]_i$ and taking derivatives gives the so-called rapid buffer approximation,

$$\frac{\partial[\text{Ca}^{2+}]_i}{\partial t} = \beta\left((D_C + D_B\gamma)\nabla^2[\text{Ca}^{2+}]_i - \frac{2\gamma D_B}{K + [\text{Ca}^{2+}]_i}(\nabla^2[\text{Ca}^{2+}]_i)^2 + j_T\right), \qquad ...(4.14)$$

where $\gamma = K[\text{B}]_\text{T}/(\text{K} + [\text{Ca}^{2+}]_i)^2$ and $\beta = 1/(1 + \gamma)$. Although this equation may appear ominous, note that using the rapid buffer approximation we have been able to eliminate the two extra equations (for [B] and [CaB]) needed to explicitly account for an indicator dye in the cytosolic compartment.

Another valuable feature of (4.14) is that it provides some insight into the effect of Ca^{2+} buffers on Ca^{2+} transport in cells by allowing us to identify the Ca^{2+}-dependent effective diffusion coefficient,

$$D^{\text{eff}} = \beta(D_C + \gamma D_B)$$

Furthermore, if this buffer has low affinity so that $[Ca^{2+}]_i \ll K$ and $r \approx [B]_T/K$, then as promised at the beginning of this section this diffusion coefficient is approximately constant:

$$D^{\text{eff}} \approx \frac{K}{K+[B]_T}\left(D_C + \frac{[B]_T}{K} D_B\right).$$

For simplicity the remainder of this chapter assumes a rapid and low-affinity buffer.

SIMULATION OF A FERTILIZATION CALCIUM WAVE

Now that we have justified the reduced effective diffusion coefficients used in the fertilization Ca^{2+} wave model, let us further simplify (4.6)-(4.8) by assuming that ER depletion is minimal, *i.e.*, $[Ca^{2+}]_{ER} = c_{ER}$ is approximately constant. We will also assume that the dynamics of $[Ca^{2+}]_i$ are much slower than the gating variable for Ca^{2+} inactivation of the IP_3R, so that w is well approximated by $w_\infty([Ca^{2+}]_i)$. With these assumptions, (4.6) and (4.7) reduce to the single-variable reaction-diffusion equati

$$\frac{\partial c}{\partial t} = D\nabla^2 c + f(c), \qquad ...(4.15)$$

where we have written $c = [Ca^{2+}]_i$, $D = f_i D_i$, a

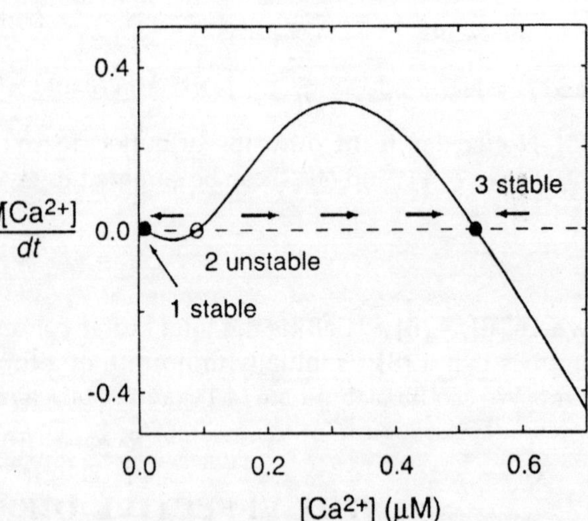

Fig. 4.3. Using parameters as in Exercise 4, the cubic rate function $dc/dt = f(c)$ of the fertilization Ca^{2+} wave model, (4.15), leads to a bistable phase portrait (*x*-axis). Equilibrium point 1 represents a stable resting state at basal $[Ca^{2+}]_i$ (ER replete) while equilibrium 3 is a stable resting point at high $[Ca^{2+}]_i$ (ER empty). Equilibrium 2 lies between points 1 and 3, is unstable, and corresponds to a threshold for Ca^{2+}-induced Ca^{2+} release.

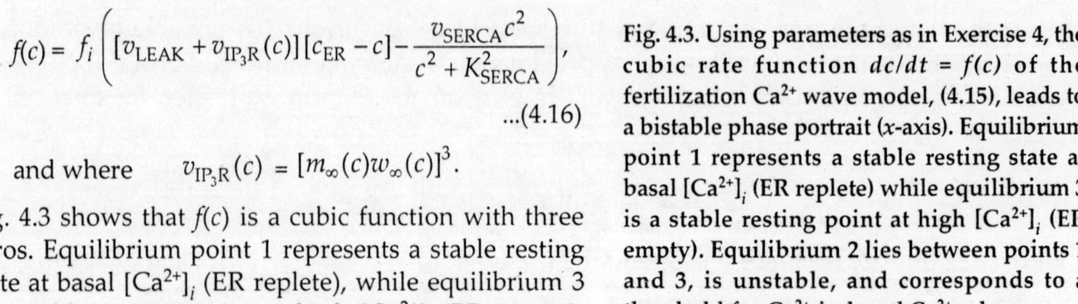

$$f(c) = f_i\left([v_{\text{LEAK}} + v_{\text{IP}_3\text{R}}(c)][c_{\text{ER}} - c] - \frac{v_{\text{SERCA}}c^2}{c^2 + K_{\text{SERCA}}^2}\right), \qquad ...(4.16)$$

and where $\quad v_{\text{IP}_3\text{R}}(c) = [m_\infty(c) w_\infty(c)]^3.$

Fig. 4.3 shows that $f(c)$ is a cubic function with three zeros. Equilibrium point 1 represents a stable resting state at basal $[Ca^{2+}]_i$ (ER replete), while equilibrium 3 is a stable resting point at high $[Ca^{2+}]_i$ (ER empty). Equilibrium 2 lies between points 1 and 3, is unstable, and corresponds to a threshold for Ca^{2+}-induced Ca^{2+} release.

SIMULATION OF A TRAVELING FRONT

We follow to simulate our minimal model of a fertilization Ca^{2+} wave given by (4.15). Assuming a one-dimensional geometry ($0 \le x \le L$), we discretize space so that $C_j(t)$ is an approximation to $c(x_j, t)$, where $x_j = j\Delta x, 0 \le j \le j$, and $\Delta x = L/J$. In this way we arrive at the system of ODEs

$$\begin{aligned} \frac{dc_0}{dt} &= D\frac{c_1 - c_0}{(\Delta x)^2} + f(c_0), \\ \frac{dc_j}{dt} &= D\frac{c_{j+1} - 2c_j + c_{j-1}}{(\Delta x)^2} + f(c_j), \\ \frac{dc_J}{dt} &= D\frac{c_{J-1} - c_J}{(\Delta x)^2} + f(c_J), \end{aligned} \quad ...(4.17)$$

where the equations for c_0 and c_J are found using a discretized version of no-flux boundary conditions $\partial c/\partial x|_{x=0,L} = 0$ to specify values at the "ghost points" $c_{-1} = c_0$ and $c_{J+1} = c_J$.

Fig. 4.4 shows a simulation of a fertilization Ca^{2+} wave calculated using (4.18). Our assumed one dimensional geometry suggests that we interpret the calculation as a propagating planar signal ($[Ca^{2+}]_i$ not a function of y or z, but only x). Initial conditions are chosen so that elevated $[Ca^{2+}]_i$ in a small region triggers a wave traveling leftwards at a velocity of 3.45 μm/s. The basal $[Ca^{2+}]_i$ before the front passes is given by equilibrium point 1 (0.01 μM), and after the front passes elevated $[Ca^{2+}]_i$ is given by equilibrium point 3 (0.53 μM). Because $[Ca^{2+}]_i$ doesn't return to basal values after the wave passes, this type of traveling wave is referred to as a *propagating front.*

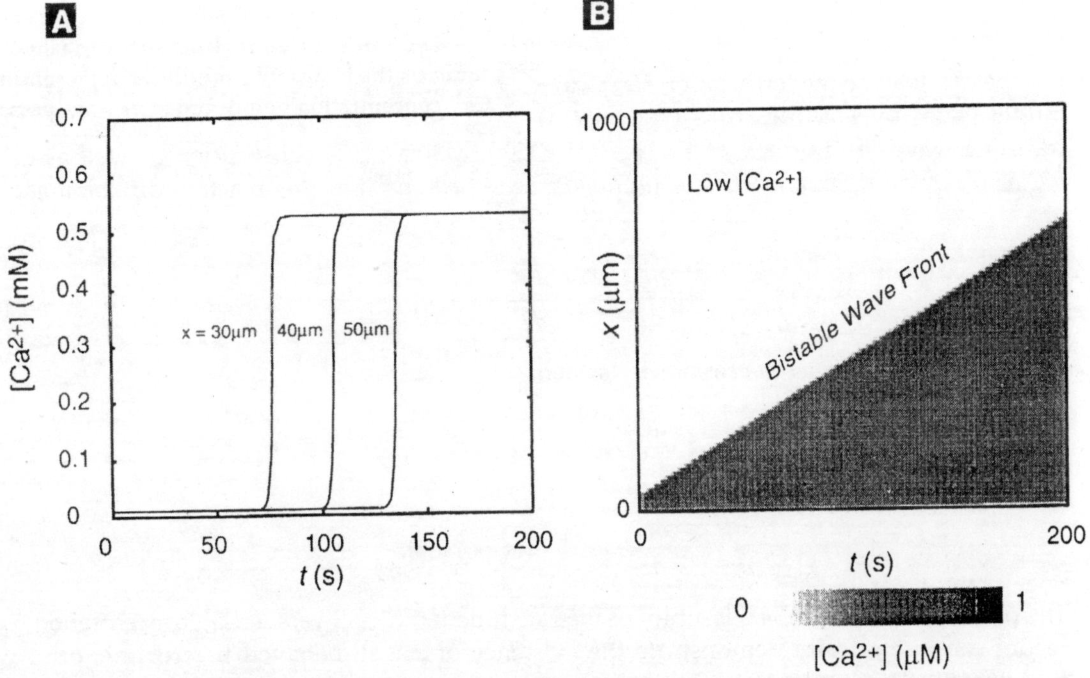

Fig. 4.4. Simulation of a fertilization Ca^{2+} wave following (4.15) results in a traveling front. (A) Time courses $c(x_i, t)$ (the Ca^{2+} profile) at three spatial positions x_i = 30, 40, 50 μm. (B) An array plot of $c(x, t)$. The slope of the front gives a velocity of 3.45 *μm/s*.

In the fertilization Ca^{2+} wave model, we could say that the physiological state of the cell cytoplasm is bistable, that is, if the spatial component of the model is neglected (for example, by setting $D = 0$ in (4.15)), the scalar ODE ($dc/dt - f(c)$) that remains exhibits bistability. Alternatively, imagine walling off (or lifting out) a small region of the spatial model and investigating its dynamics. If the region is small enough, the time scale for diffusion in the isolated region ($T = L^2/D$) will be much faster than the dynamics associated with the reaction terms $f(c)$. Thus, the Ca^{2+} profile will be approximately uniform ($\partial c/\partial x = 0$), and the region will behave as a *compartmental model* in which diffusion no longer plays a role. The propagating front observed in the spatial model, Figure 4.4, is possible precisely because any such isolated compartmental model would exhibit bistability. Indeed, if the physiological state of the cytosol were not bistable, the basal and elevated $[Ca^{2+}]_i$ at the wave front and back could not persist. We will see below that spatial phenomena observed in reaction-diffusion models can be categorized by the qualitative dynamics of the ODEs obtained by assuming that all profiles are uniform.

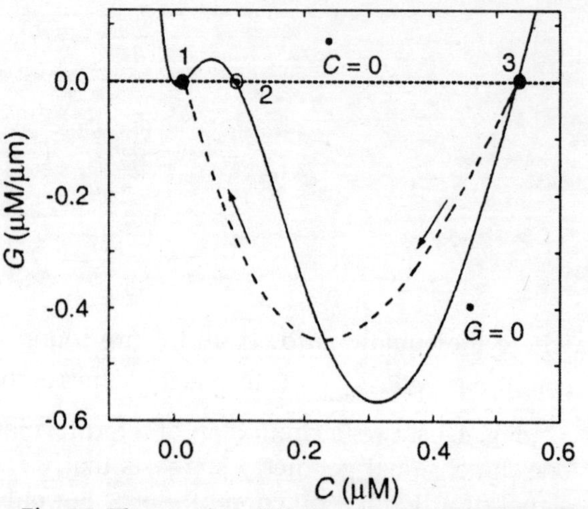

Fig. 4.5. The reaction-diffusion equation model of the fertilization Ca^{2+} wave, (4.18), can be transformed into traveling wave coordinates ($z = x - vt$) resulting in a first-order system of ODEs, (4.19)-(4.20). $C(z)$ is Ca^{2+} concentration and $G(z) - dC/dz$ is the rate function. Nullclines for C and G are shown (dotted and solid lines, respectively). When wave speed is v - 4.96 μm/s a heteroclinic orbit (dashed line) connects the two stable equilibria, representing the Ca^{2+} concentration before and after wave passage.

We can analyze the fertilization Ca^{2+} wave model (4.15) by looking for a (rightward) traveling wave solution $c(x, t) = C(x - vt) = C(z)$. Making this substitution as well as $\partial C/\partial x = (dC/dz)(\partial z/\partial x) = dC/dz$ and $\partial C/\partial t = (dC/dz)(\partial z/\partial t) = -vdC/dz$ into this reaction-diffusion equation gives

$$-v\frac{dC}{dz} = D\frac{d^2C}{dz^2} + f(C), \quad ...(4.18)$$

which can be written as the first-order system

$$\frac{dC}{dz} = G, \quad ...(4.19)$$

$$\frac{dG}{dz} = -\frac{1}{D}[vG + f(C)]. \quad ...(4.20)$$

The phase portrait in Fig. 4.5 is a plot of the rate function $C' = G$ versus Ca^{2+} concentration C. The reader can numerically demonstrate the existence of a well-behaved heteroclinic orbit when $v = 4.96$ μm/s. This heteroclinic orbit does not exist for smaller or large values of v and corresponds to the traveling front presented in Fig. 4.4.

Intuitively, we expect the speed of traveling front solutions such as those presented in Fig. 4.4 and Fig. 4.5 to depend on the diffusion coefficient D. Indeed, when the rate function $f(c)$ is a cubic polynomial as the wave speed is proportional to the square root of the diffusion coefficient. This is clear from the discussion. Although in the fertilization Ca^{2+} wave model $f(c)$ is not a polynomial, we can demonstrate the same principle by noting that the substitution $D^* = \alpha^2 D$ in (4.18) can be offset by scaling the wave speed ($v^* = \alpha v$). To see this, we define a scaled wave coordinate $z^* = x - v^*t$ that implies $-v^* \, dC/dz^* = -v dC/dz$ or $dC/dz^* = \alpha^{-1} dC/dz$. Thus, any solution of (4.18) is also a solution of

$$v^* \frac{dC}{dz^*} = D^* \frac{d^2C}{dz^{*2}} + f(C).$$

In Exercise 8 the reader can repeat the simulations of Fig. 4.4 and/or Fig. 4.5 to numerically confirm that the traveling front velocity is proportional to the square root of the diffusion coefficient. Interestingly, exogenous buffers have been shown to alter the speed of Ca^{2+} waves in mature *Xenopus* oocytes, an effect that may be due to changes in the effective diffusion coefficient for Ca^{2+}.

The astute reader may have noticed that we have reported two different wave velocities for two different calculations of the same traveling front. Recall that Fig. 4.4 used the method of lines to simulate the fertilization Ca^{2+} wave model (4.15) and that we observed a speed of 3.45 μm/s. On the other hand, Fig. 4.5 shows that the first-order system of ODEs resulting from a transformation into traveling wave coordinates, (4.19)-(4.20), has a heteroclinic orbit that connects the two stable equilibria for wave velocity of $v = 4.96$ μm/s. The suppression of the wave speed in the calculation using the method of lines is due to *discretization error.*

CALCIUM WAVES IN THE IMMATURE XENOPUS OOCYCTE

Although we have begun our discussion of models of intracellular Ca^{2+} waves by focusing on traveling fronts in *mature Xenopus laevis* eggs, IP_3-dependent Ca^{2+} responses occur in many cell types and take various forms. It is instructive to note here that Ca^{2+} wave phenomena in the *immature X. laevis* oocyte are qualitatively distinct from the traveling fronts discussed above. Although there are many differences between immature and mature *Xenopus* oocytes/eggs, from a dynamical point of view these different Ca^{2+} wave phenomena are the result of different underlying dynamics of Ca^{2+} handling that could be distinguished even in the absence of diffusion. While our model of the fertilization Ca^{2+} wave demonstrated how bistable cellular dynamics can lead to a traveling front, the ER of the immature *Xenopus* oocyte is excitable (and may even be oscillatory) as opposed to bistable.

Fig. 4.6 presents phase planes for the reaction terms of the fertilization Ca^{2+} wave model (4.6)-(4.7), which illustrate the qualitative difference between excitable and bistable dynamics. To make this comparison, we relax our assumption that the dynamics of $[Ca^{2+}]_i$ are much slower than the gating variable for Ca^{2+} inactivation of the IP_3R (so that w is once again given by (4.7) rather than $w_\infty([Ca^{2+}]_i)$). The model now has two dynamic variables ($[Ca^{2+}]_i$ and w), the nullclines of which are shown in panel A of Fig. 4.6. The intersections of these nullclines imply three equilibria (as before, points 1 and 3 are stable, while 2 is unstable). Conversely, panel B of Fig. 4.6 presents the (w, $[Ca^{2+}]_i$) phase portrait with adjusted IP_3R parameter values (Ca^{2+}-dependence of inactivation). The reaction terms of the model now represent excitable cytoplasm. A small perturbation in Ca^{2+} concentration results in a long excursion of the trajectory representing Ca^{2+} release from the ER, IP_3R inactivation, and ATP-dependent Ca^{2+} reuptake.

SIMULATION OF A TRAVELING PULSE

A simulation of Ca^{2+} wave phenomena mediated by ER Ca^{2+} excitability can be implemented in a manner similar to (4.18). To do this we discretize the

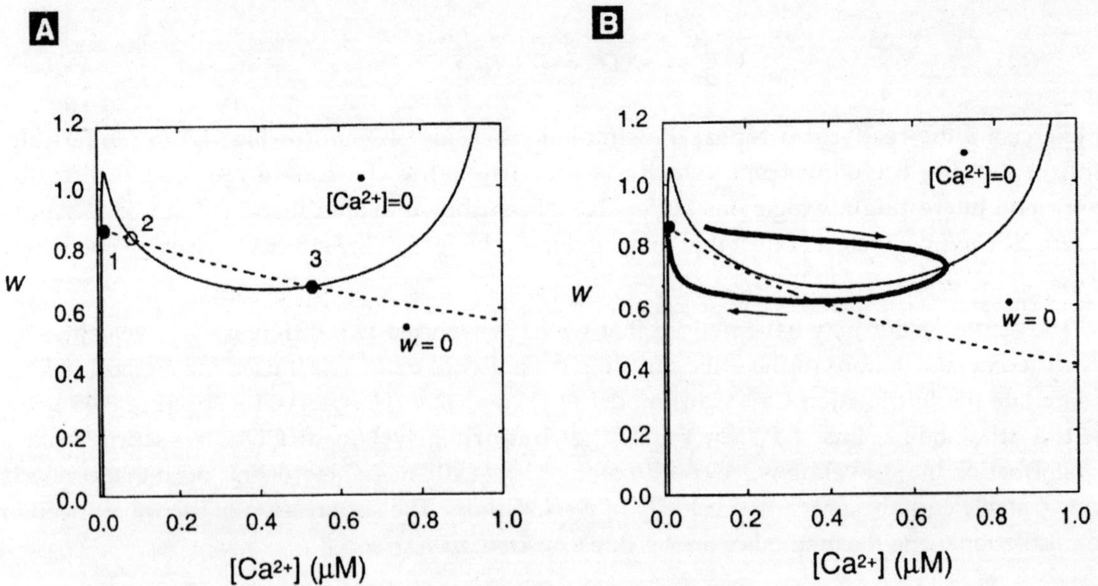

Fig. 4.6. The (w, $[Ca^{2+}]_i$) phase plane for the reaction terms of the fertilization Ca^{2+} wave model (4.6)-(4.7) can exhibit bistability (A) or excitability (B) with a change of IP_3R parameters. Parameters as in Exercise 4 except $d_2 = 2$ μM in the bistable case, while $d_2 = 1$ μM in the excitable case. In the Li-Rinzel reduction of the DeYoung-Keizer IP_3R model, this change in d_2 increases equilibrium IP_3R inactivation (note the downward/leftward shift in w nullcline) and eliminates the high $[Ca^{2+}]_i$ equilibrium present in (A).

gating variable $w_j \approx w(x_j, t)$ at $J + 1$ mesh points ($0 \le j \le J$) representing the fraction on noninactivated IP_3Rs at each location in space ($x_j = j\Delta x$). Of course, there will be no Laplacian term in the equations for w_j because the IP_3Rs (and thus their state) do not diffuse. The model now takes the form

$$\frac{dc_0}{dt} = D\frac{c_1 - c_0}{(\Delta x)^2} + g(c_0, w_0), \qquad \frac{dw_0}{dt} = [w_\infty(c_0) - w_0]/\tau,$$

$$\frac{dc_j}{dt} = D\frac{c_{j+1} - 2c_j + c_{j-1}}{(\Delta x)^2} + g(c_j, w_j), \qquad \frac{dw_j}{dt} = [w_\infty(c_j) - w_j]/\tau, \qquad \text{...(4.21)}$$

$$\frac{dc_J}{dt} = D\frac{c_{J-1} - c_J}{(\Delta x)^2} + g(c_J, w_J), \qquad \frac{dw_J}{dt} = [w_\infty(c_J) - w_J]/\tau.$$

Fig. 4.7 shows a propagating Ca^{2+} wave that results when the cytosol, modeled using (4.21), is excited with the same initial conditions as in Fig. 4.4. In this case, $[Ca^{2+}]_i$ returns to basal as the Ca^{2+} transient propagates past a given point, and the cellular response is referred to as a *traveling pulse.* Traveling pulses produced by excitable dynamics are often called "trigger waves," because a perturbation at one end of the excitable medium triggers a signal that may propagate with undiminished amplitude.

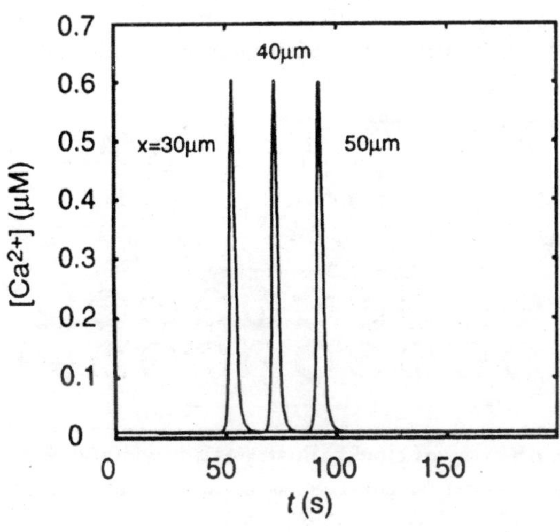

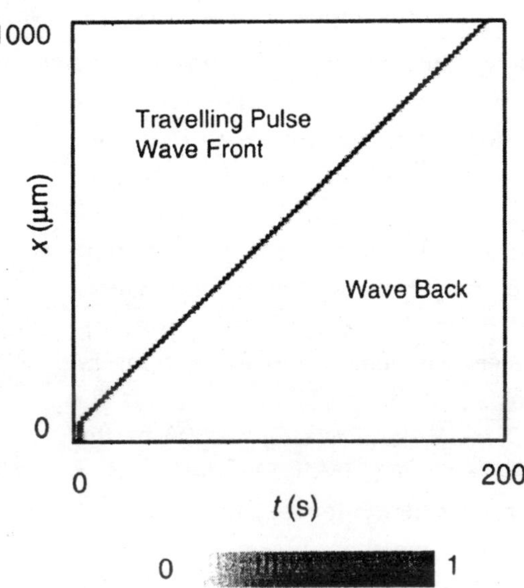

Fig. 4.7. **$[Ca^{2+}]_i$ time courses and array plot similar to Fig. 4.4 except that parameters are changed so that the $(w, [Ca^{2+}]_i)$ phase plane is excitable as in (B) of Fig. 4.6. The simulation is implemented using (4.21) and results in a traveling pulse similar to Ca^{2+} responses in the immature *Xenopus* oocyte. The slope of the front gives a velocity of 5 μm/s.**

SIMULATION OF A KINEMATIC WAVE

In previous sections we have discussed how the cytoplasm's physiological state influences Ca^{2+} wave propagation. When the cytoplasm is bistable, a likely wave phenomenon is a traveling front; conversely, an excitable cytoplasm will likely support trigger waves, *i.e.*, traveling pulses. These observations suggest that we consider the case of an oscillatory cytoplasm. As discussed already, IP_3-mediated whole-cell Ca^{2+} responses are oscillatory in many cell types. Fig. 4.8 demonstrates that it is an easy matter to adjust IP_3R parameters (Ca^{2+}-dependence of activation and inactivation) in our model so that the $(w, [Ca^{2+}]_i)$ phase plane is oscillatory and a large-amplitude (0.01 μM < $[Ca^{2}+]_i$ < 0.65 μM) stable limit cycle exists.

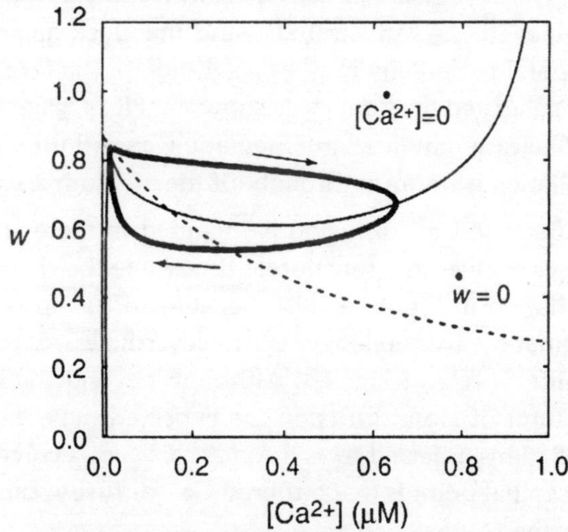

Fig. 8.8. **$(w, [Ca^{2+}]_i)$ phase plane when the physiological state of the cytoplasm is oscillatory. Parameters as in Fig. 4.6 except d_2 = 0.5 μM and d_5 = 0.15 μM. In the Li-Rinzel reduction of the DeYoung-Keizer IP_3R model, d_5 and d_2 affect IP_3R activation and inactivation, respectively.**

Even in the absence of diffusion an oscillatory medium will support *kinematic waves.* Fig. 4.9 illustrates this possibility using an array of clocks. Depending on initial conditions (*i.e.,* the time to which the clocks are initially set), the absence of physical coupling between the clocks does not preclude the appearance of a traveling wave.

Regardless of initial condition, all the clocks will eventually return to their initial state. Thus, kinematic waves are distinct from traveling fronts and pulses in that they can occur with arbitrary shape. Although a kinematic wave can be of arbitrary.shape, deceptively shaped waves may look like traveling waves when there is a certain relationship between wavelength and the oscillatory frequency of the medium. To see this, let $c(x, 0) = c_0(x)$ and $w(x, 0) = w_0(x)$ be the initial Ca^{2+} profile for our model tuned to exhibit limit cycle oscillations with period τ. If we choose $(c_0(x), w_0(x))$ "on" the limit cycle (for all x), then in the absence of Ca^{2+} diffusion the $[Ca^{2+}]_i$ will be temporally periodic at every spatial location, *i.e.,* $c(x, t + \tau) = c(x, t)$. If the initial condition also happens to be spatially periodic, so that $c_0(x + \lambda) = c_0(x)$ (and similarly for w_0), then $[Ca^{2+}]_i$ will remain spatially periodic, *i.e.,* $c(x + \lambda, t) - c(x, t)$, where λ is the spatial period (the wavelength). As illustrated using the clock analogy in Fig. 4.9, this simultaneous spatial and temporal periodicity implies a solution of the form $c(z) = c(x + vt)$ if $v = \lambda/\tau$. Using the parameters that generated the limit cycle shown in Fig. 4.8, Fig. 4.10 demonstrates that oscillatory IP_3-mediated Ca^{2+} release may lead to kinematic waves. Initial conditions are chosen so that the phase of the oscillation is uniform throughout the medium, except for a central region that is phase advanced.

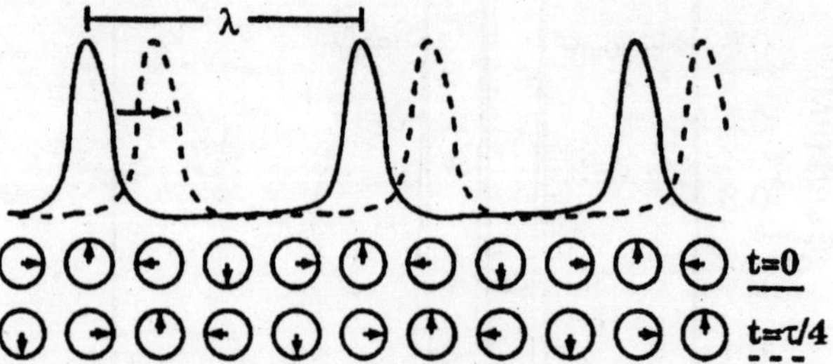

Fig. 4.9. An array of clocks illustrates a kinematic wave. The peak of the wave coincides with the hand of the clock pointing to 12 o'clock. The temporal period (τ) is the time it takes for the hand at a given location to make one complete clockwise revolution. The minimum distance between clocks with the same phase is the spatial period, or wavelength (τ). A wave takes *T* seconds to travel one wavelength, so that the wave speed is $v = \lambda/\tau$.

Because Ca^{2+} diffusion is included in these simulations following (4.21), the distribution of phases evolves as a function of time (note the changing spatial profile of the wave peak with each oscillation in Figure 4.10B). Because diffusion is now coupling cytoplasmic oscillations, pure kinematic waves no longer exist. Nevertheless, at each spatial location the solution remains "near" the limit cycle of Fig. 4.4. Although beyond the scope of this chapter, it can be shown that in the limit of slow diffusion (as expected here due to Ca^{2+} buffers), the relative phases of the oscillations, referred to as the *phase gradient,* evolve according to Burgers equation. The important conceptual point is that buffered Ca^{2+} diffusion causes the phase of the oscillations to synchronize over time.

Figure 4.10A illustrates this with time courses $[Ca^{2+}]_i\ (x_i, t)$ from two spatial positions. Notice that the oscillations at these two locations synchronize; the (apparent) velocity of the repetitive wave between these two points increases. Indeed, the velocity of a kinematic wave can be arbitrarily large because it is not a propagating signal. In the array plot of Fig. 4.10B, the diminution of the phase gradient with each cycle is striking.

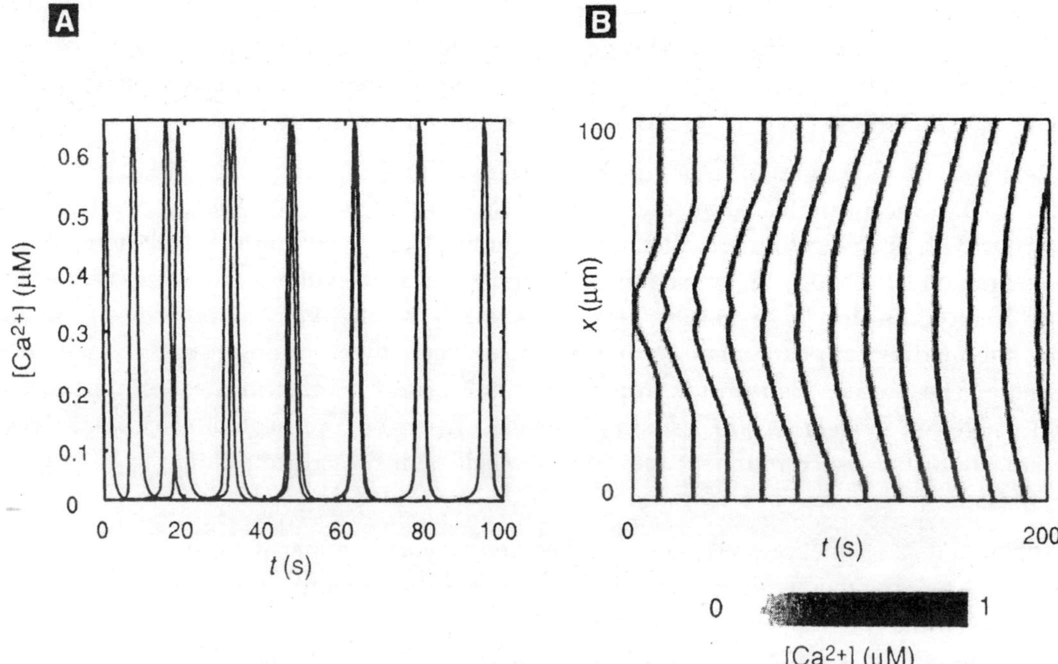

Fig. 4.10. $[Ca^{2+}]_j$ time courses and array plot similar to Fig. 4.7 except that the (w, $[Ca^{2+}]_j$) phase plane is oscillatory as in Fig. 4.4. The simulation is implemented using (4.21) and results in kinematic wave (also called a phase wave). (A) Time courses $c(x_i, t)$ at $x_i = 50$ and 63 μm show that diffusion causes oscillations at different regions to synchronize. (B) An array plot illustrates the smoothing of the relative phases of the oscillations.

The presence of endogenous and exogenous Ca^{2+} buffers slows Ca^{2+} diffusion. If the physiological state of a cell's cytoplasm is oscillatory, observed wave phenomena may very well be kinematic in nature, and the influence of Ca^{2+} diffusion on propagation of the wave may be primarily through the evolution of the phase gradient. Conversely, an excitable cytbsol absolutely requires diffusion to support a propagating fronts (*i.e.*, trigger waves). While kinematic waves can exist even in the absence of diffusion, the wave speed of a propagating front scales with the effective diffusion coefficient in a predictable manner (at least in the limit of low affinity Ca^{2+} buffers). These distinctions are important and one may easily be deceived. For example, the observation of repetitive waves (*i.e.*, a *wave train)* may suggest an oscillatory medium. On the other hand, such a wave train may be due to periodic forcing (or recurrent entry) of an excitable medium, as in target patterns and spiral waves imaged in the immature *Xenopus* oocyte.

SPARK-MEDIATED CALCIUM WAVES

Throughout this discussion we have, without exception, made the so-called *continuum approximation.* That is, we have made no attempt to explicitly model the stochastic dynamics of individual Ca^{2+} ions undergoing Brownian motion. Indeed, an averaging process is implicit in any continuum description of matter. Consider that small molecules (such as free Ca^{2+} ions) in micromolar solution are essentially point particles with mean intermolecular spacing of approximately 100 nm; that is, in a volume of $(100 \text{ nm})^3$ a single solute molecule will be found on average. On a distance scale of 10 nm (a factor of ten smaller) the medium is highly heterogeneous. If thermal motion of both solvent and solute could be stopped, only one out of a thousand volume elements with dimension $(10 \text{ nm})^3$ would be found to contain a solute molecule.

Clearly, at distance scales of 10 nm (or less) a continuum description is invalid. On the other hand, if a fluorescent molecule at micromolar concentrations is optically detected using a microscope with micron resolution, the relevant volume ($1\ \mu m^3 = 10^9\ nm^3$) is likely to contain a thousand molecules. Though the actual number of molecules in the volume will vary as a function of time due to diffusion, these density fluctuations will be small. Thus, on a distance scale of a micron the medium appears spatially homogeneous, and the continuum approximation is legitimate. The phrase "continuum approximation" can also refer to the fact that in our spatial whole-cell models we are assuming that the ER can be represented as a compartment continuously distributed throughout the cytosol, albeit with prescribed volume fraction given by f_{ER} in (4.8).

Although it can sometimes be rigorously justified, anyone familiar with the mathematics of diffusion in heterogeneous media will recognize that this assumption is made largely for convenience. Intracellular heterogeneities abound, and only sometimes can these heterogeneities be "averaged out" in a way that justifies a bidomain description (*i.e.,* a mathematical technique known as homogenization). One important cellular inhomogeneity known to have a profound impact on propagating IP_3-mediated Ca^{2+} waves is the distribution of IP_3Rs. In the immature *Xenopus* oocyte, for example, IP_3Rs occur in clusters of 10-100 with intercluster spacing on the order of a few microns.

Under some conditions this organization of Ca^{2+} release sites reveals itself in cauliflower-like wave fronts. Localized Ca^{2+} elevations due to the activation of a single Ca^{2+} release site (a Ca^{2+} puff) and even a single channel (a Ca^{2+} blip) are now regularly observed (Yao *et al.* 1995; Parker *et al.* 1996). Localized Ca^{2+} elevations due to intracellular Ca^{2+} release have been observed in other cell types as well. When a Ca^{2+} release event is mediated by ryanodine receptors (RyRs), as in cardiac myocytes, the phenomenon is referred to as a Ca^{2+} spark. Spark-mediated waves in cardiac myocytes are amenable to modeling because Ca^{2+} waves often propagate along the longitudinal axis of a myocyte, *i.e.,* perpendicular to the orientation of the sarcomeric Z-lines. The Z-lines, transverse tubules, and Ca^{2+} release sites are oriented in the transverse direction and are regularly spaced with a separation of 2 μm.

The ultrastructure of cardiac myocytes thus lends itself to one-dimensional modeling in which the cellular heterogeneity along the longitudinal axis is represented. Fig. 4.11 shows some examples of propagating spark-mediated waves and isolated sparks in cardiac myocytes. One such modeling approach (Keizer *et al.* 1998) is to augment the continuum bidomain description (4.8) with a spatially periodic maximum conductance for Ca^{2+} release, say, $v_{RyR}(x)$ in analogy to v_{IP_3R} in the IP_3-mediated wave model. Numerical studies show that spark-mediated Ca^{2+} waves propagate in a saltatory manner, similar to experiment. In the following section we will consider a related and even more idealized model of spark-mediated Ca^{2+} waves, the fire-diffuse-fire model.

THE FIRE-DIFFUSE-FIRE MODEL

The essence of minimal models of spark-mediated Ca^{2+} waves is to model Ca^{2+} release sites as idealized point sources, that is, an array of Dirac delta functions, denoted by $\delta(x - x_i)$, where x_i is the spatial position of the ith release site. While the Dirac delta function is actually not a function at all (rather, it is a distribution, the limit of a sequence of functions), it is often thought of as a sharply peaked function that is zero everywhere except $x = x_i$. In addition, the delta function is normalized so that

$$\int_{-\infty}^{\infty} \delta(x - x_i)dx = 1, \qquad \text{...(4.22)}$$

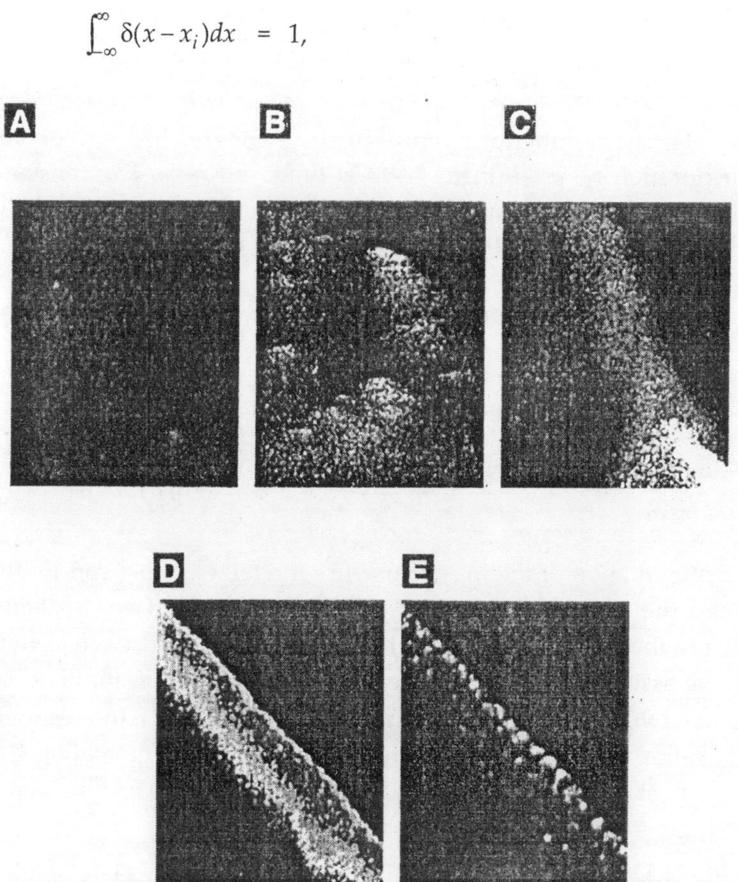

Fig. 4.11. Confocal line-scan images of isolated Ca^{2+} sparks, wave initiation, and a spark-mediated propagating Ca^{2+} wave in cardiac myocytes. Time is vertical (bar is 100 ms except for second panel which is 200 ms) and space is horizontal (bar is 5 μm). Image subtraction shows spatially localized Ca^{2+} release originating from sites separated by ≈ 2 μm.

and it has the so-called sifting property

$$\int_{-\infty}^{\infty} f(x)\delta(x - x_i)dx = f(x_i). \qquad \text{...(4.23)}$$

Armed with delta functions, we can model spark-mediated Ca^{2+} waves by modifying the propagating front model (4.15) so that Ca^{2+} release occurs only at regularly spaced release sites, that is,

$$\frac{\partial c}{\partial t} = D\frac{\partial^2 c}{\partial x^2} + f(c)\sum_i \delta(x - x_i), \qquad ...(4.24)$$

where x_i is the location of the *i*th release site with kinetics given by *f(c)* (Keener 2000).

We can further idealize this model for spark-mediated waves by removing the Ca^{2+} dependence of the release rate *f(c)*. Instead, assume that the source strength $f_i(t)$ of each release site is a square pulse function of time,

$$f_i(t) = \frac{\sigma}{\tau_R} H(t - t_i)H(t_i + \tau_R - t).$$

In this expression *H(t)* is the Heaviside step function, so that $H(t) = 0$ for $t < 0$, $H(t) = 1$ for $t \geq 0$. The product of these Heaviside functions represents the source turning "on" at time $t = t_i$ and remaining on for duration τ_R, *i.e.*, it turns "off" at time $t = t_i + \tau_R$. The constant σ represents the source amplitude and has units of μM. μm because from (4.22) we see that the delta function has units of 1/μm. The normalization factor $(1/\tau_R)$ is chosen so that

$$\int_{-\infty}^{\infty} f_i(t)\, dt = \sigma \qquad ...(4.25)$$

Substituting this form for the Ca^{2+} release rate into (4.24) gives the "fire-diffuse-fire" model of spark-mediated Ca^{2+} wave propagation,

$$\frac{\partial c}{\partial t} = D\frac{\partial^2 c}{\partial x^2} + \frac{\sigma}{\tau_R}\sum_i \delta(x - x_i)H(t - t_i)H(t_i + \tau_R - t). \qquad ...(4.26)$$

To finish our presentation of the fire-diffuse-fire model, we must specify the location of each release site (x_i) and the times (t_i) at which Ca^{2+} release begins. If we further assume a regular array of release sites, then $x_i = id$ where d is the site spacing. Because we have Ca^{2+}-induced Ca^{2+} release in mind, we assume that Ca^{2+} release at site i begins when the local Ca^{2+} concentration $c(x_i, t)$ achieves a fixed threshold Ca^{2+} concentration c_θ (here 0.1 μM). Interestingly, Fig. 4.12 shows that the fire-diffuse-fire model supports continuous as well as propagating Ca^{2+} signals. In Fig. 4.12A, the time constant for Ca^{2+} release, is τ_R, is 1 s and the propagating signal is similar to the traveling front solutions presented earlier in this chapter.

Conversely, Fig. 4.12B presents a simulation using $\tau_R = 10$ ms. Here spark-like Ca^{2+} releases lead to a propagating signal that is distinctly saltatory. Note that the continuous wave is traveling at 11.3 μm/s. while the saltatory wave is traveling at 67 μm/s. The long duration of Ca^{2+} release in the continuous case appears to slow the velocity of the propagating signal. The fire-diffuse-fire model can be analyzed to give insight into the continuous and saltatory limits of Ca^{2+} wave propagation.

As the reader may expect, it is not τ_R alone but rather a dimensionless parameter that determines the existence and form of propagating signals. Indeed, it can be shown that the relevant dimensionless quantity is $D\tau_R/d^2$. The continuous limit corresponds to $D\tau_R/d^2 \gg 1$ and the saltatory limit to $D\tau_R/d^2 \ll 1$. Below we study these two limits separately to determine how diffusion influences wave velocity in both limits.

ANALYSIS OF THE CONTINUOUS LIMIT

The continuous limit pertains when $D\tau_R/d^2 \gg 1$, that is, when diffusion is fast and the release time long compared to the intersite separation. This limit could be achieved in a simulation by increasing the density of release sites $(d \to 0)$ while simultaneously decreasing the release rate (σ) so that the release per unit length (σ/d) is constant.

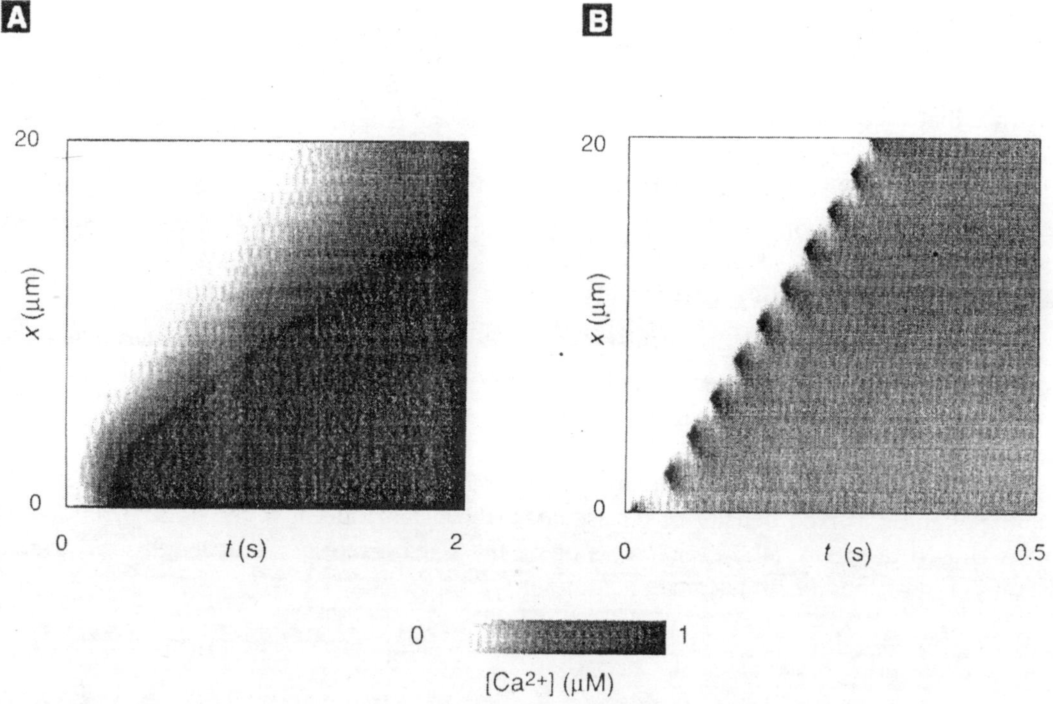

Fig. 4.12. Simulation of Fire-Diffuse-Fire model in both continuous (A) and saltatory (B) regimes using (4.26). The separation between release sites is d = 2 µm. The continuous wave (τ_R = 1 s) is traveling 11.3 µm/s, while the saltatory wave (τ_R = 10 ms) is traveling 67 µm/s.

Because the thresholding that determines Ca^{2+} release times (t_i) is the only nonlinear part of our model (4.26), we can convolve the release rates $f_i(t)$ with the diffusion kernel to write an explicit expression for $c(x, t)$:

$$c(x, t) = \frac{\sigma}{\tau_R} \sum_{i=-(N-1)}^{N-1} \int_{t_i}^{\min(t, t_i+\tau_R)} \frac{dt'}{\sqrt{4\pi D(t-t')}} \exp\left[-\frac{(x-x_i)^2}{4D(t-t')}\right], \text{ for } t > t_{N-1}, \qquad ...(4.27)$$

where we have assumed a symmetric profile $c(x, t)$ and the firing of $N - 1$ sites on either side of the origin. Equation (4.27) is not yet a solution because we have the unknown parameters t_i to determine. To find the time at which site N fires we evaluate (4.27) at $(t = t_N, x = x_N)$ and set $c(x_N, x_N) = c_\theta$. This yields

$$c_\theta = \frac{\sigma}{\tau_R} \sum_{i=-(N-1)}^{N-1} \int_{t_i}^{\min(t_N, t_i+\tau_R)} \frac{dt'}{\sqrt{4\pi D(t_N-t')}} \exp\left[-\frac{(x_N-x_i)^2}{4D(t_N-t')}\right],$$

an expression that can be solved for t_N. At this point we seek traveling-wave like solutions with regular firing times. This implies that $t_i = i\Delta$, where Δ is the time interval between adjacent site firings and a velocity of propagation given by $v = d/\Delta$. By substituting $\Delta = t_N/N$ and solving for Δ it can be shown that the velocity of such a propagating wave has the following dependence on model parameters:

$$v \approx \sqrt{\left(\frac{\sigma}{dc_\theta}\right)\left(\frac{D}{\tau_R}\right)}. \qquad \text{...(4.28)}$$

Notice that the factor of $\sqrt{D/\tau_R}$ is similar to what we have seen before for traveling fronts and pulses. The existence of traveling wave solutions depends on $\sqrt{\sigma/dc_\theta}$.

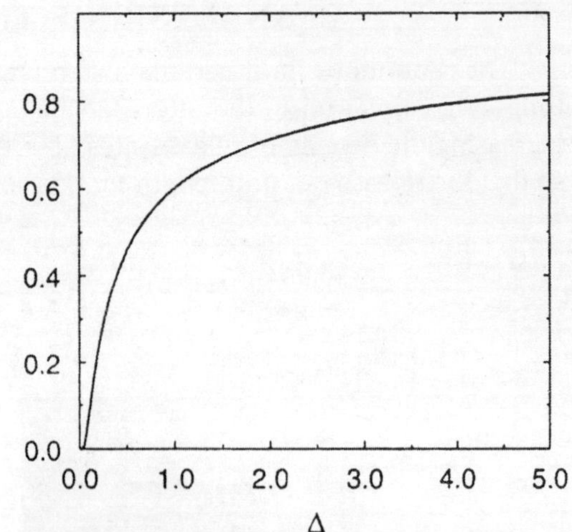

Fig. 4.13. Relationship between $\alpha = c_\theta d/\sigma$ and the dimensionless firing interval ($\Delta = 1/v$) for the fire-diffuse-fire model in the saltatory limit. Since $g(\Delta) < 1$, waves do not propagate for $\alpha > 1$. From Keizer *et al.* (1998).

Analysis of the Saltatory Limit

The saltatory limit of the fire-diffuse-fire model corresponds to $D\tau_R/d^2 \ll 1$ in (4.26), that is, when diffusion is slow and release time short compared to the intersite separation. This limit could be achieved in a simulation by decreasing the time constant for Ca^{2+} release (τ_R) while maintaining a fixed density of release sites (d constant). Because the normalization factor ($1/\tau_R$) is chosen to satisfy (4.25) regardless of τ_R, the release rate per unit length (σ/d) remains constant as $\tau_R \to 0$, and (4.26) becomes

$$\frac{\partial c}{\partial t} = D\frac{\partial^2 c}{\partial x^2} + \sigma \sum_i \delta(x - x_i)\,\delta(t - t_i). \qquad \text{...(4.29)}$$

We now nondimensionalize space ($x^* = x/d$, $x_i^* = x_i/d = i$), time ($t^* = tD/d^2$, $t_i^* = t_iD/d^2$), and concentration ($c^* = c/c_\theta$) and drop asterisks to write

$$\frac{\partial c}{\partial t} = \frac{\partial^2 c}{\partial x^2} + \frac{1}{\alpha}\sum_i \delta(x - x_i)\,\delta(t - t_i), \qquad \text{...(4.30)}$$

where $\alpha = c_\theta d/\sigma$. Using the diffusion kernel we obtain an implicit expression for $c(x, t)$,

$$c(x, t) = \sum_i H(t - t_i)\sqrt{\frac{1}{4\pi\alpha^2(t - t_i)}}\exp\left[-\frac{(x - i)^2}{4(t - t_i)}\right], \qquad \text{...(4.31)}$$

where the presence of the Heaviside function indicates no contribution from sites that have not yet fired.

Assuming that the sites for which $-N < i < N$ have fired at times t_i through t_N (left and right pairs

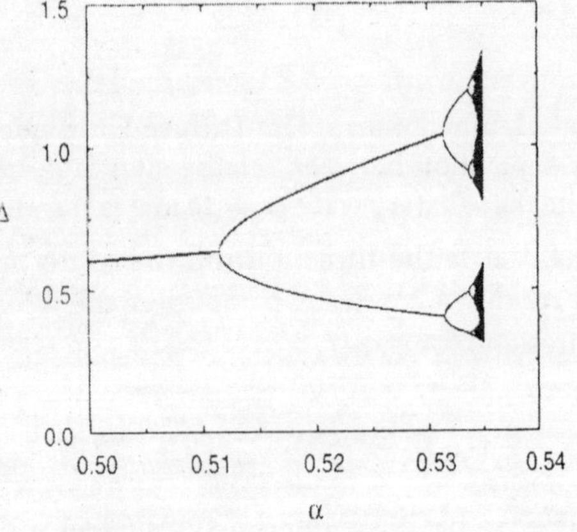

Fig. 4.14. For a range of values for α, the saltatory limit of the fire-diffuse-fire model is simulated by successively calculating the dimensionless firing interval (Δ) for $n = 1, 2, \ldots$ using (4.32) and the criterion $c(n, t_i) = 1$. The wave was initiated by simultaneously firing all sites for $-15 < n < 14$. A period doubling cascade begins at $\alpha \approx 0.512$ and terminates near $\alpha \approx 0.535$, beyond which waves do not propagate.

simultaneously) we want to determine t_{N+1}. The dimensionless threshold for firing is now $c = 1$, so from (4.31) we have

$$c(x_{N+1}, t_{N+1}) = 1 = \sum_{i=-N}^{N} \sqrt{\frac{1}{4\pi\alpha^2(t_{N+1}-t_i)}} \exp\left[\frac{(N+1-i)^2}{4(t_{N+1}-t_i)}\right]. \quad ...(4.32)$$

Because we are interested only in long time solutions, we consider the large N limit and neglect all the terms in the above sum with $i \le 0$, that is, we are following a rightward traveling wave that eventually is not influenced by the sites to the left of the origin. Under this approximation (4.32) simplifies to

$$\alpha = \sum_{i=1}^{N} \sqrt{\frac{1}{4\pi(t_{N+1}-t_i)}} \exp\left[\frac{(N+1-i)^2}{4(t_{N+1}-t_i)}\right]. \quad ...(4.33)$$

This expression is an implicit map for t_{N+1} as a function of all the previous firing times, $(t_N, t_{N-1}, ..., t_1) \to t_{N+1}$. traveling-wave-like solutions correspond to fixed points of this map with regular firings, that is, $t_{N+1} = t_N + \Delta$ with constant Δ (giving a dimensionless velocity of $v = 1/\Delta$). Substituting $\Delta = t_{N+1} - t_N$ in (4.33) we obtain

$$\alpha = \sum_{i=1}^{N} \sqrt{\frac{1}{4\pi\Delta\,(N+1-i)}} \exp\left[\frac{(N+1-i)}{4\Delta}\right]. \quad ...(4.34)$$

Defining $n = N + 1 - i$ and taking the limit $N \to \infty$ we obtain

$$\alpha = \sum_{n=1}^{\infty} \sqrt{\frac{1}{4\pi n\Delta}} \exp(-n/4\Delta) \equiv g(\Delta). \quad ...(4.35)$$

It can be shown that $0 \le g(\Delta) \le 1$. The first equality holds in the high-velocity limit ($\Delta \to 0$) and the second for low velocity ($\Delta \to \infty$). Since $g(\Delta)$ is monotonic, we can numerically calculate a unique solution $\Delta = g^{-1}(\alpha)$ when $0 < \alpha < 1$, *i.e.*, the range of $g(\Delta)$. At this point it is worth remembering that $\alpha = c_\theta d/\sigma$. Thus, a necessary condition for the existence of a velocity is that $\alpha < 1$. If the sites are too far apart or too weak or the threshold too high, there can be no propagating waves. Interestingly, it can be shown that propagation failure occurs through a sequence of instabilities as a is increased well before the condition $\alpha = 1$ is reached (see further reading). When α is small ($\sigma/dc_\theta \gg 1$), propagating wave solutions exist and are stable. It can be shown that the dimensional velocity v in this saltatory propagation limit is given by

$$v \approx \frac{4D}{d} \ln\left(\frac{\sigma}{dc_\theta}\right); \quad ...(4.36)$$

that is, it scales linearly with D, quite unlike propagating waves in the continuous limit, which have velocity that scales with $\sqrt{D}$ as in (4.28). If fact, whenever τ_R is sufficiently small (even if α is not particularly small), the saltatory wave velocity predicted by the fire-diffuse-fire model scales linearly with D.

MODELING LOCALIZED CALCIUM ELEVATIONS

Our discussion of spark-mediated Ca^{2+} wave propagation began with a discussion of cellular heterogeneity. During this discussion of Ca^{2+} wave phenomena it is important to remember that many cellular processes (including synaptic transmission, activity-dependent synaptic plasticity, and regulation of neuronal excitability) can be initiated by changes in intracellular Ca^{2+} concentration in the absence of a global response such as a Ca^{2+} oscillation or wave. For this reason, Ca^{2+} sparks, puffs, and localized Ca^{2+} elevations near voltage gated plasma membrane Ca^{2+} channels (sometimes called Ca^{2+} microdomains) are cellular signals of great interest. Localized Ca^{2+} elevations are not only the "building blocks" of global Ca^{2+} release events, but also highly specific regulators of cellular function.

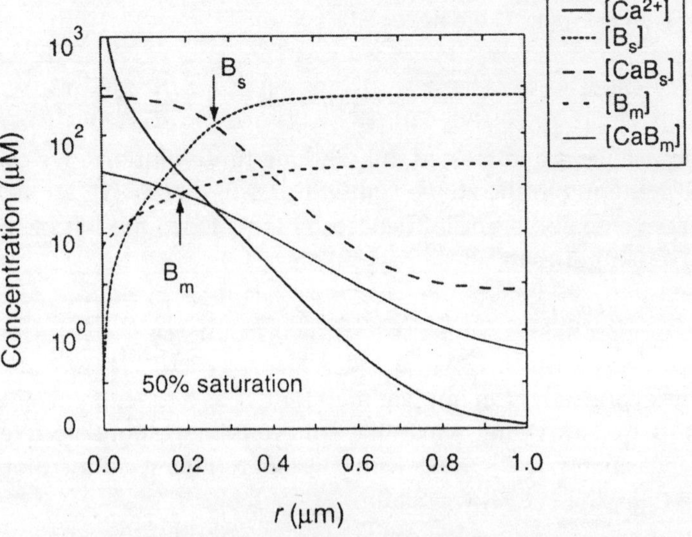

Fig. 4.15. Representative calculation of Ca^{2+} profile near a point source for free Ca^{2+}. Source amplitude (σ) and elapsed time are 5 pA and 10 ms, respectively. From Smith *et al.* (1998).

As discussed above in the context of experimental observations of Ca^{2+} waves, the interpretation of microfluorometric measurements of Ca^{2+} puffs and sparks is complicated by the diffusion of Ca^{2+}, endogenous buffers, and indicator, all of which contribute to the dynamics of a fluorescence signal during and after Ca^{2+} release. While in the case of global Ca^{2+} responses and Ca^{2+} waves the equilibrium relation (4.1) will likely hold between Ca^{2+} and indicator, this is not so easily demonstrated in the case of localized Ca^{2+} elevations (though it remains true in some cases). A further complicating factor in interpreting *confocal* microfluorometric measurements is the optical blurring that occurs due to limited spatial resolution.

The reaction-diffusion equations for the buffered diffusion of intracellular Ca^{2+}, (4.10)-(4.12), are the starting point of a theoretical understanding the dynamics of localized Ca^{2+} elevations. In the simplest scenario, a Ca^{2+} puff or spark is due to Ca^{2+} release through one channel or a tight cluster of channels. If Ca^{2+} is released from intracellular Ca^{2+} stores deep within a large cell (so that the plasma membrane is far away and doesn't influence the time-course of the event), and the intracellular milieu is homogeneous and isotropic, then we have spherical symmetry. In this case, the evolving profiles of Ca^{2+} and buffer (though a function of time and distance from the source) will not be functions of the polar (ϕ) or azimuthal (θ) angle. In the case of such spherical or radial symmetry the Laplacian reduces to

$$\nabla^2 = \frac{1}{r^2}\frac{\partial}{\partial r}\left[r^2\frac{\partial}{\partial r}\right] = \frac{\partial^2}{\partial r} + \frac{2}{r}\frac{\partial}{\partial r}. \qquad ...(4.37)$$

Fig. 4.15 shows a spherically symmetric calculation of a localized Ca^{2+} elevation using the full equations for the buffered diffusion of Ca^{2+}, (4.10)-(4.12), with parameters consistent with measurements of the effective diffusion coefficient in *Xenopus* oocyte cytoplasm. Fig. 4.15 is a numerically calculated snapshot of the concentration profiles for each species after an elapsed time of 10 ms. The concentration $[Ca^{2+}]_i$ is elevated near the source *(thick solid line)*. Because released free Ca^{2+} reacts with buffer, the concentration of bound buffer *(thin solid and thick dashed lines)* is elevated near the source. Conversely, the concentration of free buffer *(thin dotted and dashed lines)* decreases near the source. In this simulation, 250 μM stationary buffer was included in addition to 50 μM mobile buffer (both with K of 10 μM).

A source amplitude of 5 pA was used, corresponding to a tight cluster of IP_3Rs. Interestingly, Fig. 4.15 shows that the mobile buffer is less easily saturated than stationary buffer, in spite of the fact that the stationary buffer is at fivefold higher concentration (note arrows). Simulations such as these have played a role in understanding the dynamics of puffs and sparks. Fig. 4.16 shows a Ca^{2+} spark simulated using parameters consistent with experimental observation in cardiac myocytes.

Such simulations confirm that the timecourse of observed fluorescence can be explained by a 2 pA, 15 ms Ca^{2+} -release event from a tight cluster of RyRs located on the sarcoplasmic reticulum membrane. Parameter studies using this model indicate that Ca^{2+} spark properties (such as brightness, full width at half maximum, and decay time constant) are very dependent on indicator dye parameters (such as association rate constant, concentration, and diffusion coefficient). These relationships are not always intuitive. For example, increasing indicator dye concentration decreases the brightness of the simulated Ca^{2+} spark in Fig. 4.16. This is partly due to the fact that spark brightness is a normalized measure (peak/basal fluorescence), and partly due to the fact that high concentrations of indicator perturb the underlying free Ca^{2+} signal.

STEADY-STATE LOCALIZED CALCIUM ELEVATIONS

Numerical simulations like those in Fig. 4.15 confirm that localized Ca^{2+} elevations achieve steady-state values very rapidly (within microseconds) near point sources. Steady-state solutions to the full equations are thus of interest because they allow estimates of "domain" Ca^{2+} concentration near open Ca^{2+} channels. These steady-state solutions lend themselves to analysis, giving insight into the limiting (long time) "shape" of localized Ca^{2+} elevations. In the case of one mobile buffer, steady-state solutions to (4.10)-(4.12) will satisfy the following boundary value problem :

$$0 = D_C \nabla^2 [Ca^{2+}]_i - k^+ [B][Ca^{2+}]_i + k^- ([B]_T - [B]), \qquad ...(4.38)$$

$$0 = D_B \nabla^2 [B] - k^+ [B][Ca^{2+}]_i + k^- ([B]_T - [B]), \qquad ...(4.39)$$

with associated boundary conditions

$$\lim_{r\to 0} \left\{ -4\pi r^2 D_C \frac{d[Ca^{2+}]_i}{dr} \right\} = \sigma, \quad \lim_{r\to\infty} [Ca^{2+}]_i = [Ca^{2+}]_\infty,$$

$$\lim_{r\to 0} \left\{ -4\pi r^2 D_B \frac{d[B]}{dr} \right\} = 0, \quad \lim_{r\to\infty} [B] = [B]_\infty = \frac{K[B]_T}{K + [Ca^{2+}]_\infty}. \qquad ...(4.40)$$

Here we have eliminated the equation for [CaB] (rather than [B] as before) and written D_C and D_B for the diffusion coefficients of free Ca^{2+} and free buffer, respectively.

Fixed buffers, while important for the time-dependent evolution of localized Ca^{2+} elevations, have no influence on steady states. This can be seen by inspecting (4.10)-(4.12), where $D_{CB} = 0$ implies $R = -k_+[B][Ca^{2+}]_i + k_-[CaB] = 0$ at steady state.

There are many advantages to nondimensionalizing equations before preceding to analyze them. A convenient nondimensionalization of (4.38)-(4.40) begins by scaling $[Ca^{2+}]_i$ and [B] by representative concentrations, the dissociation constant (K) and total concentration of buffer ($[B]_T$). This gives two dimensionless dependent variables, $\hat{c}$ and $\hat{b}$, given by $\hat{c} = [Ca^{2+}]_i/K$ and $\hat{b} = [B]/[B]_T$.

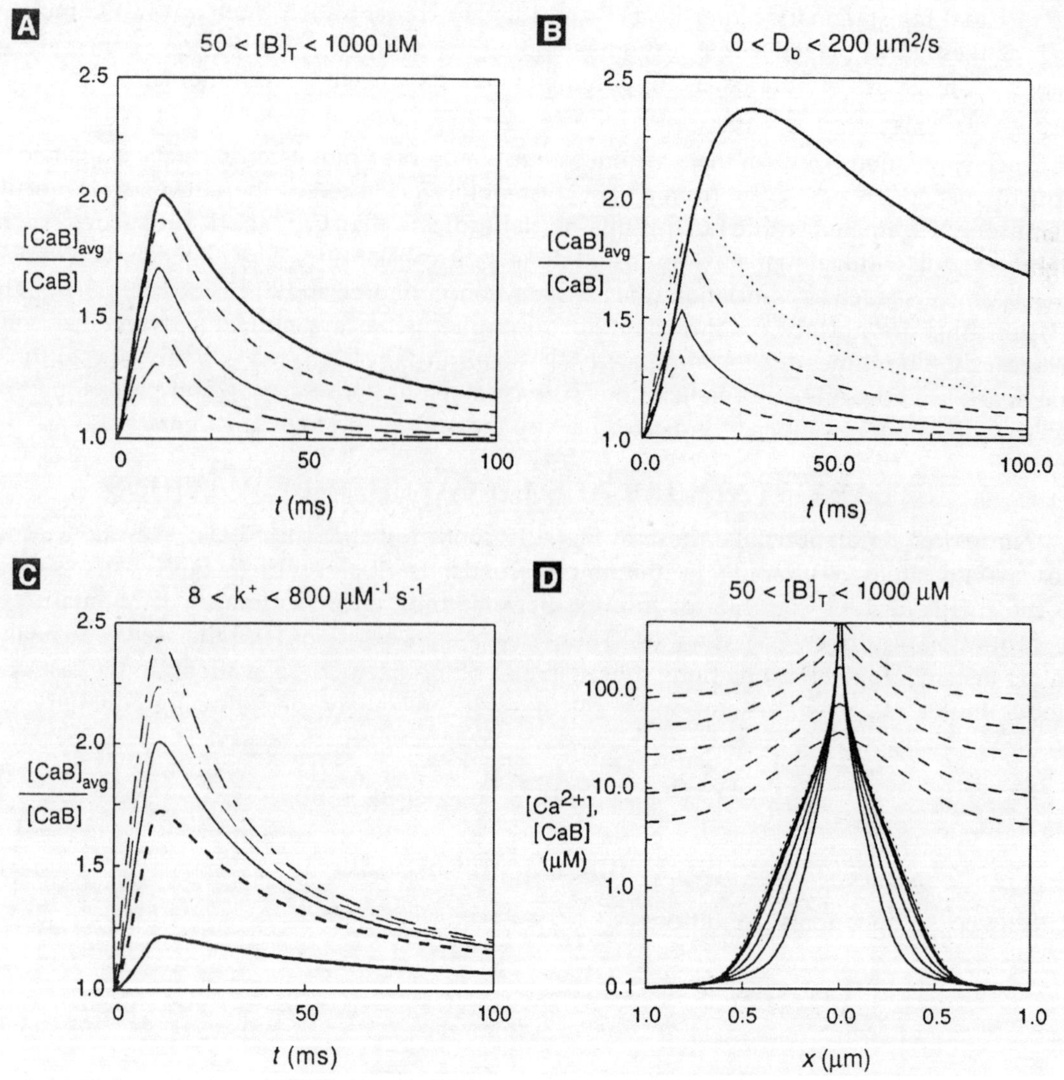

Fig. 4.16. Effects of indicator dye parameters on Ca^{2+} spark properties. Source amplitude is 2 pA for 15 ms and simulated fluo-3 has K of 1.13 μM. (A, B, C) Time course of normalized blurred fluorescence signal estimated according to (4.2) with $\eta_B = 0$. (D) Snapshot of $[Ca^{2+}]_i$ and [CaB] profiles before termination of Ca^{2+} release. Dotted line shows the Ca^{2+} profile with no fluo-3. For details of the simulated confocal point spread function and buffer parameters.

The reader can confirm that nondimensionalizing the independent variable ($\hat{\rho} = r/L$ with $L = \sigma/4\pi D_C K$) simplifies (4.38) and (4.39) to

$$\hat{\varepsilon}_c \nabla^2_{\hat{\rho}} \hat{c} - (\hat{c}\hat{b} + \hat{b} - 1) = 0, \quad \text{...(4.41)}$$

$$\hat{\varepsilon}_b \nabla^2_{\hat{\rho}} \hat{b} - (\hat{c}\hat{b} + \hat{b} - 1) = 0, \quad \text{...(4.42)}$$

where the subscript on the Laplacian indicates that differentiation is with respect to $\hat{\rho}$.

Two dimensionless diffusion coefficients ($\hat{\varepsilon}_c$ and $\hat{\varepsilon}_b$) appear in these equations. In terms of the original dimensional parameters of the problem, they are given by $\hat{\varepsilon}_c = \varepsilon\alpha$ and $\hat{\varepsilon}_b = \varepsilon D$, where $\alpha = K/[B]_T$ is a buffering factor (small when buffer is at high concentration compared to the dissociation constant), $D = D_B/D_C$ is a relative diffusion coefficient between buffer and Ca^{2+}, and the common factor ε is given by

$$\varepsilon = (4\pi)^2 D^3_C K/\sigma^2 k^+, \quad \text{...(4.43)}$$

a quantity that is small for strong sources and/or fast buffers.

THE STEADY-STATE EXCESS BUFFER APPROXIMATION (EBA)

The dimensionless steady-state equations for the buffered diffusion of Ca^{2+} near a point source, (4.41) and (4.42), are nonlinear, and no general analytical solution is known for these equations. However, we can begin to understand the behavior of solutions (and the effect of Ca^{2+} buffers on Ca^{2+} domains) by considering (4.41) and (4.42) in limiting parameter regimes. The first such limit we will consider is called the "excess buffer approximation." If buffer is in excess, then the parameter $\alpha = K/[B]_T$ will be very small, and $\hat{\varepsilon}_c \approx \varepsilon\alpha$ will also be small when $\varepsilon = O(1)$. Therefore, we consider in detail (4.41) and (4.42) when $\hat{\varepsilon} \approx 0$, which in physical terms implies that the diffusion coefficient of c is small compared to the size of the reaction terms in (4.41).

The mathematically inclined reader will notice that this is a singular perturbation problem. Because this technique goes beyond the scope of this chapter, we present only a heuristic analysis here. The interested reader is invited to consult Smith *et al.* (2001) for a more rigorous treatment.

With this caveat, we formally set $\hat{\varepsilon}_c = 0$ in (4.41), giving

$$\hat{c}\hat{b} + \hat{b} - 1 = 0, \quad \text{...(4.44)}$$

which implies that (4.42) simplifies to

$$\nabla^2 \hat{b} = 0$$

When combined with the boundary conditions for b (see Exercise 13), this equation implies

$$\hat{b} = \hat{b}_\infty, \quad \text{...(4.45)}$$

where $\hat{b}_\infty = 1/(1 + \hat{c}_\infty)$ is the fraction of free buffer far from the source. Thus, our assumption that buffer is in excess ($\hat{\varepsilon}_c = 0$) implies that the buffer is not perturbed from its equilibrium value $\hat{b}_\infty$. Substituting (4.45) into (4.41) gives

$$\hat{\varepsilon}_c \nabla^2 \hat{c} - \hat{b}_\infty (\hat{c} - \hat{c}_\infty) = 0.$$

This is a linear equation satisfied by

$$\hat{c} = \frac{1}{\hat{\rho}} e^{-\hat{\rho}/\Lambda} + \hat{c}_\infty, \qquad \text{...(4.46)}$$

where the dimensionless space constant $\Lambda = \sqrt{\hat{\varepsilon}_c/\hat{b}_\infty}$ and $\hat{c}_\infty = [\mathrm{Ca}^{2+}]_i/K$ are chosen to satisfy the boundary conditions on $\hat{c}$ that can be derived from (4.40).

When this result is expressed in dimensional form, we have

$$[\mathrm{Ca}^{2+}]_i = \frac{\sigma}{4\pi D_C r} e^{-r/\lambda} + [\mathrm{Ca}^{2+}]_\infty, \qquad \text{...(4.47)}$$

where λ is the characteristic length constant for the mobile Ca^{2+} buffer given by $\lambda = \sqrt{D_C/k^+[\mathrm{B}]_\infty}$. This *excess buffer approximation,* first derived by Neher (Neher 1986), is valid when mobile buffer is in high concentration and/or when the source amplitude is small, that is, $\lim_{r\to 0}[\mathrm{B}] \approx [\mathrm{B}]_\infty$. Note that λ decreases with increasing association rate constant (k^+) and free buffer concentration far from the source ($[\mathrm{B}]_\infty$). When a buffer is in excess, we thus expect further increases in concentration to restrict localized Ca^{2+} elevations. In addition, buffers with fast-reaction kinetics are expected to restrict localized Ca^{2+} elevations more than slow buffers.

THE STEADY-STATE RAPID BUFFER APPROXIMATION (RBA)

The *steady-state rapid buffer approximation* near a point source for Ca^{2+} can be derived by noticing in (4.43) that rapid buffer (large k^+) leads to small values of ε. This results in small values for both $\hat{\varepsilon}_c$ and $\hat{\varepsilon}_b$, which in physical terms implies that the diffusion coefficient of both $\hat{c}$ and $\hat{b}$ are small compared to the size of the reaction terms in (4.41) and (4.42).

If we formally set $\hat{\varepsilon}_c = \hat{\varepsilon}_b = 0$ in these equations, we find that, as before, (4.44) holds. Solving for $\hat{b}$, we find that at every spatial location $\hat{b}$ is given by

$$\hat{b} = \frac{1}{1+\hat{c}}, \qquad \text{...(4.48)}$$

or in dimensional terms,

$$[\mathrm{B}] = \frac{K[\mathrm{B}]_\mathrm{T}}{K + [\mathrm{Ca}^{2+}]_i}. \qquad \text{...(4.49)}$$

These equations are statements of *local equilibrium,* the fundamental assumption used in deriving the rapid buffer approximation in the context of traveling waves.

We proceed with this derivation of the steady-state RBA by subtracting (4.42) from (4.41) to give

$$\nabla^2_{\hat{\rho}}\left(\hat{\varepsilon}_c\hat{c} - \hat{\varepsilon}_b\hat{b}\right) = 0, \qquad \text{...4.50)}$$

In physical terms this expression is equivalent to the statement that at steady state the flux of total Ca^{2+}, diffusing in both free and bound forms across any spherical surface centered on the source is equal to the flux entering through the source. Integrating twice with respect to p and using the boundary conditions to determine the integration constants gives

$$\hat{\varepsilon}_c\hat{c} - \hat{\varepsilon}_b\hat{b} = \frac{\hat{\varepsilon}_c}{\hat{\rho}} + \hat{\varepsilon}_c\hat{c}_\infty - \hat{\varepsilon}_b\hat{b}_\infty. \qquad ...(4.51)$$

Substituting (4.48) into this equation gives

$$\hat{\varepsilon}_c\hat{c} - \hat{\varepsilon}_b\left(\frac{1}{1+\hat{c}}\right) = \frac{\hat{\varepsilon}_c}{\hat{\rho}} + \hat{\varepsilon}_c\hat{c}_\infty - \hat{\varepsilon}_b\hat{b}_\infty, \qquad ...(4.52)$$

which upon solving for $\hat{c}$ and converting back into dimensional form gives the steady-state RBA

$$[\mathrm{Ca}^{2+}]_i = \frac{1}{2D_C}\left(-D_C K + \frac{\sigma}{4\pi r} + D_C[\mathrm{Ca}^{2+}]_\infty - D_B[\mathrm{B}]_\infty\right.$$

$$\left. + \sqrt{\left(D_C K + \frac{\sigma}{4\pi r} + D_C[\mathrm{Ca}^{2+}]_\infty - D_B[\mathrm{B}]_\infty\right)^2 + 4D_C D_B[\mathrm{B}]_\mathrm{T} K}\right). \qquad ...(4.53)$$

The steady-state RBA tends to be valid when ε is small, *e.g.*, when buffers have large association and dissociation rate constants (k^+ and k^-). Interestingly, a sufficiently large source amplitude (σ) can compensate for modest binding rates, also causing ε to be small and the steady-state RBA to be valid.

COMPLEMENTARITY OF THE STEADY-STATE EBA AND RBA

The fundamental assumptions used in deriving the steady-state excess and rapid buffer approximations are significantly different. In the case of the RBA (4.53) we assumed that buffer and Ca^{2+} were in local equilibrium, (4.49). Because $[Ca^{2+}]_i \to \infty$ as $r \to 0$ in (4.53), we see that according to the steady-state RBA

$$\lim_{r\to 0}[\mathrm{B}] \approx 0 \text{ (RBA)}. \qquad ...(4.54)$$

Thus, the steady-state RBA cannot be valid unless the source is strong enough to saturate the buffer. On the other hand, in our derivation of the EBA we assumed that the buffer is not perturbed from its equilibrium value, (4.45). If this is true even near the source, then

$$\lim_{r\to 0}[\mathrm{B}] \approx [\mathrm{B}]_\infty \text{ (EBA)}. \qquad ...(4.55)$$

Thus, we expect the EBA and RBA approximations to be complementary, in the sense that the steady-state solution to the full equations for the buffered diffusion of Ca^{2+} near a point source (the correct answer) cannot simultaneously be EBA-like, as in (4.55), and RBA-like, as in (4.54). In the process of extending both the EBA and RBA to higher order, this expectation has been confirmed.

MEMBRANOUS TRANSPORT

5

Ionic channels are not the only mechanism that cells use to transport impermeant species across membranes. Cells have developed a great variety of transport proteins for moving both ions and molecules from one cellular compartment to another. For example, to maintain the concentration imbalance of Na^+, K^+, and Ca^{2+} across the plasma membrane it is necessary to pump ions against significant concentration gradients. In the case of Ca^{2+} ions the ratio of concentrations outside to inside is greater than four orders of magnitude (ca. 2 mM outside and 0.1 μ, M inside). In addition to pumps, there are numerous specific cotransporters and exchangers that allow ions and small molecules to be transported selectively into internal compartments or out of the cell.

Unlike ionic channels, for which the driving force is a passive combination of electrical potential and ionic concentration differences, most transporters and pumps expend considerable energy. In many animal cells, for example, it has been estimated that nearly 25% of the ATP that is utilized is devoted to maintaining low cytoplasmic Na^+ and high cytoplasmic K^+ concentrations via Na^+/K^+ pumps. In this chapter we provide an overview of some of the mechanisms, other than ionic channels, that cells use to pump and transport small molecules and ions. We first introduce the ideas behind passive transport, using a passive glucose transporter (GLUT) as an example. We then introduce analytic, diagrammatic, and numerical methods for calculating rates of transport and apply them to a simplified model of the GLUT transporter.

Using the cotransport of glucose and Na^+ as an example, we then discuss how to create models of transporters and how this transporter functions physiologically in intestinal epithelial cells. A great deal is known about the kinetic steps involved in the pumping of Ca^{2+} by Ca^{2+} pumps in internal stores, and we use this mechanism to illustrate how phosphorylation by ATP drives the pumping mechanism. In the final section of the chapter we focus on the cyclic nature of these mechanisms and describe additional transporters that operate via comparable kinetic cycles.

PASSIVE TRANSPORT

We must be a little careful about what we call passive transport, because there is no free lunch. The following example is not without its energy costs, but it uses background thermal energy rather than the explicit energy in a particular molecule such as ATP. Glucose is a six-carbon sugar that is a major fuel for intermediary metabolism in animals. It is derived from carbohydrates in the gut and is transported through epithelial cells in the intestines into the blood stream and thence to the brain, pancreas, liver, muscle, and other organs. There glucose is taken up and metabolized via glycolytic enzymes.

Although uptake from the gut involves active cotransport of glucose with sodium ions, peripheral tissues such as fat, muscle, and liver transport glucose via a class of passive membrane transporters referred to as GLUT transporters. Because glucose is a major energy source for cells, understanding the rate of glucose transport into cells via GLUT is important physiologically. At least six isotypes of GLUT transporters have been isolated, GLUT1 through GLUT6, each of which is prevalent in one or more types of tissue (Bell *et al.* 1993). GLUT2, for example, is found in glucose-sensing pancreatic beta cells that secrete insulin from pancreatic tissue, as well as in liver cells. Extensive kinetic experiments have lead to a cartoon description of the steps involved in the tranport processes.

The transitions of the transporter itself, facing the inside of the cell to facing the outside of the cell, is driven by heat or thermal fluctuation. Thus the heat of the system and the concentration gradient of glucose are the only driving forces in the system. Because no other energy is needed, it is called passive transport. It is similar in this sense to diffusion, which we discuss.

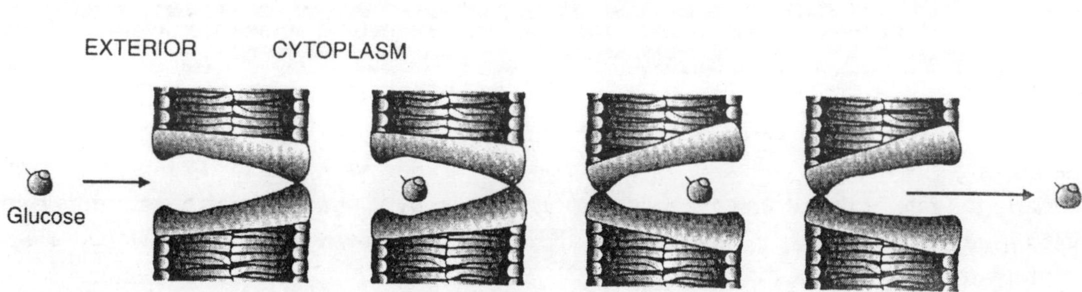

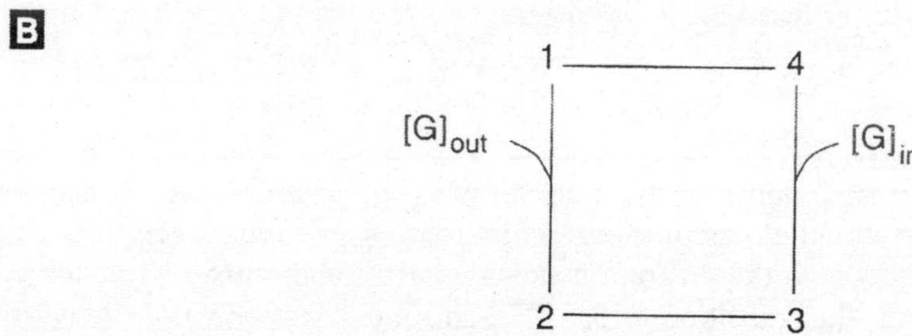

Fig. 5.1. (A) Cartoon of four states of a GLUT transporter, showing the empty pore facing the exterior of the cell, glucose bound facing the exterior, glucose bound facing the interior, and the open pore facing the interior of the cell. (B) Four-state kinetic diagram of a GLUT transporter based on the cartoon in (A).

Fig. 5.1A shows four different states of the transporter. State S_1 has an empty binding site for glucose exposed to the exterior of the cell. When glucose binds to this state, the transporter makes a transition to state S_2, with glucose bound and facing the exterior. In S_2, a glucose molecule is bound to the transporter, which is still facing the exterior. State S_3 is the state with the transporter then facing the interior. When glucose dissociates from GLUT and ends up inside the cell, the transporter is left in state S_4. Finally, the cycle can repeat if S_4 makes the conformational transition to S_1. All of these processes are reversible.

This kinetic "cartoon" is easily translated into a conventional kinetic model of the sort often employed in biochemistry. Here the model takes the form of the diagram in Fig. 5.IB. The labeled corners in Fig. 5.IB correspond to the states S_1-S4 of a GLUT transporter described in the previous paragraph, and the lines represent elementary molecular processes. The transition from S_1 to S_2 is a *bimolecular* process, because it requires the interaction of a glucose molecule (indicated as $[G]_{out}$ in the diagram) and the GLUT molecule in S_1. The transition from S_2 to S_1, on the other hand, involves only the GLUT molecule and is therefore unimolecular: Thus only S_2 appears at the end of the line connecting S_1 to S_2. This illustrates an important aspect of transitions between molecular states: They are reversible, corresponding to the property of microscopic reversibility of molecular processes.

The rates of the elementary processes depicted in the kinetic diagram are determined again by the law of mass action as we discussed in Chapter 1. Thus the rate of the transition from S_i to S_2 is given by $J_{12} = k_{12}\,[G]_{out}\,x_1$, where the square brackets denote concentration, and x_1 represents the fraction of the GLUT molecules in S_1, $x_1 = N_1/N$, where N is the total number of transporters. The factor k_{12} is the rate constant, in this case bimolecular, with practical units of's^{-1} mM"^{1}. Similarly the rate of the reverse reaction, $2 \rightarrow 1$, is given by $J_{21} = k_{21}, x_2$ with k_{21} a unimolecular rate constant (with units s^{-1}). Table 5.1 lists the forward and reverse rate expressions for all of the processes in the kinetic diagram

Table 5.1 Rate Expressions for Glut Transporter

Forward Process	*Rate*	*Reverse Process*	*Rate*
$S_1 \rightarrow S_2$	$k_{12}\,[G]_{out}\,x_1$	$S_2 \rightarrow S_1$	$k_2 x_2$
$S_2 \rightarrow S_3$	$k_{23}x_2$	$S_3 \rightarrow S_2$	$k_{32}x_3$
$S_3 \rightarrow S_4$	$k_{34}x_3$	$S_4 \rightarrow S_3$	$k_{43}\,[G]_{in}\,x_4$
$S_4 \rightarrow S_1$	$k_{41}X_4$	$S_1 \rightarrow S_4$	$k_{14}\,x_1$

Having established the correspondence of the diagram with rate expressions, we can write down the differential equations that the diagram represents. To do so we must keep track of the change that each elementary process in the diagram makes for each state. Thus the fraction x_1 of transporters in S_1 decreases with the transition to S_2 or S_4, and increases with transitions from S_1 or S_4. Using this idea, in conjunction with the kinetic diagram and Table 5.1, the ordinary differential equations follow for the rate of change in the number of states:

$$dx_1/dt = -\,k_{12}[G]_{out}\,x_1 + k_{21}x_2 + k_{41}x_4 - k_{14}x_1,$$

$$dx_2/dt = k_{12}\,[G]_{out}\,x_1 - k_{21}x_2 - k_{23}x_2 + k_{32}x_3,$$

$$dx_3/dt = k_{23}x_2 - k_{32}x_3 - k_{34}x_3 - k_{34}x_3 + k_{43}\,[G]_{in}\,x_4,$$

$$dx_4/dt = k_{34}x_3 - k_{43}\,[G]_{in}\,x_4 - k_{41}x_4 + k_{14}x_1. \qquad ...(5.1)$$

Because the kinetic model involves only interconversion of GLUT states, the total number of transporters should be preserved. Again we use a conservation law, $N_1 + N_2 + N_3 + N_4 = N$ or $x_1 + x_2 + x_3 + x_4 = 1$, to ensure that transporters are neither created nor destroyed. This condition can be checked by adding together the expressions on the right-hand side of (5.1). It is easily verified that all of the terms cancel, leading to the result $d(x_1 + x_2 + x_3 + x_4)/dt = 0$, which shows that the sum of the fractions of transporters in different states does not change. That is, $x_1 + x_2 + x_3 + x_4$ has a constant value, which in this case is 1.

We see again that one of the dependent variables can be eliminated using the conservation law by writing $x_4 = 1 - x_1 - x_2 - x_5$. The differential equation for x_4 becomes redundant and the number of differential equations to be solved is reduced to only three along with the algebraic equation for x_4:

$$\begin{aligned} dx_1/dt &= m_{11}x_1 + m_{12}x_2 + m_{13}x_3 + k_{41}, \\ dx_2/dt &= m_{21}x_1 + m_{22}x_2 + m_{23}x_3, \\ dx_3/dt &= m_{31}x_1 + m_{32}x_2 + m_{33}x_3 + k_{43}[\mathrm{G}]_{in}, \\ x_4 &= 1 - x_1 - x_2 - x_5. \end{aligned} \qquad ...(5.2)$$

where the 3 × 3 array m_{ij} is a matrix whose elements can be found by substituting $x_4 = 1 - x_1 - x_2 - x_3$ into the first three equations of (5.1). We leave it to the reader to verify, for example, that $m_{11} = -(k_{12}[\mathrm{G}]_{out} + k_{14})$ and $m_{33} = -(k_{32} + k_{34} + k_{43}[\mathrm{G}]_{in})$.

With the formulation of the model equations, the first steps in the modeling process are completed. What remains is the analysis of the equations. We will leave the analysis of this model for an exercise, and we will use a reduced model to demonstrate several ways to analyze the transport rates for transporters.

TRANSPORT RATES

In the previous section we described a four-state model of a GLUT-type glucose transporter. Like all models of transporters, the "states" are distinct molecular arrangements of the transporter protein and the small molecules *(ligands)* that interact with it. These states are the basic kinetic ingredients of the kinetic mechanism, and we identify the states by numbering them $S_1, S_2, ..., S_M$, where M is the total number. For the GLUT transporter, we summarize the mechanism by a kinetic diagram with lines between the states representing possible transitions. Each line stands for a forward and reverse step that can be either unimolecular or bimolecular. Which states are connected together and the nature of the transitions connecting the states must be determined experimentally, and the resulting diagrams summarize succinctly a great deal of kinetic information.

As we have seen, these state diagrams can easily be translated into differential equations that describe how the number of transporters in each state changes with time.

What is important physiologically is not the rate at which states of a transporter change with time but rather the rate at which the ions or molecules that are transported get across the membrane. This can be determined from the kinetic mechanism but requires additional analysis, because the transport rate is a property of the entire mechanism rather than an individual step. To make this distinction clear, consider the simplified three-state version of the mechanism for a GLUT transporter in Fig. 5.2. In this simplification, S_1 and S_4 in the four-state diagram in Fig. 5.1 have been treated as a single state. The reasons why we can make this type of simplification of a mechanism are described more fully.

The three transitions represented are the binding of glucose to the transporter from the exterior ($S_1 - S_2$) and interior ($S_1 - S_3$) of the cell and the conformation change in which the glucose moves from the exterior to the interior $(S_2 - S_3)$. Using x_i to represent the fraction of the total number of transporters in state $S_i = S_1, S_2, S_3$, the kinetic equation for this diagram are

$$dx_1/dt = -J_{12} + J_{31},$$
$$dx_2/dt = J_{12} - J_{23},$$
$$dx_3/dt = J_{23} - J_{31}, \qquad (5.3)$$

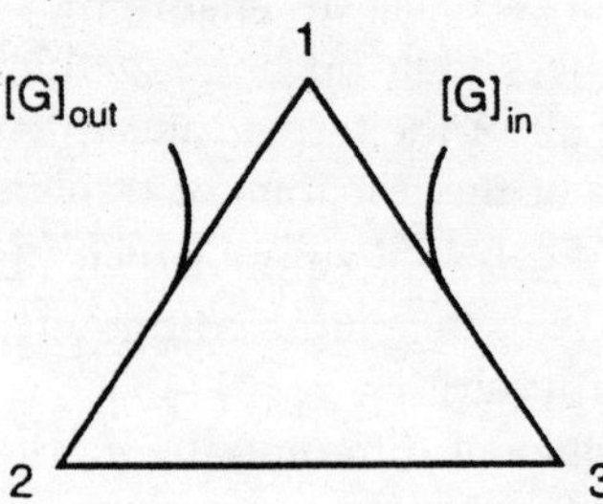

Fig. 5.2. A simplified three-state diagram for the GLUT transporter. S_1 and S_4 in the four-state diagram in Fig. 5.1 have been treated as a single state.

Here the J_{ij} are the *net rates* for the indicated transition, with the $i \rightarrow j$ direction taken as positive. Thus,

$$J_{12} = k^*_{12}[G]_{out}x_1 - k_{21}x_2, \qquad (5.4)$$
$$J_{23} = k_{23}x_2 - k_{32}x_3, \qquad (5.5)$$
$$J_{31} = k_{31}x_3 - k^*_{13}[G]_{in}x_1 \qquad (5.6)$$

In these equations we use a superscript * to indicate a bimolecular rate constant. Net rates often appear in analyzing transport rates, and we will refer to the J_{ij} as *fluxes.* We can see by adding equations (5.3) together that the total number of tranporters is conserved: $x_1 + x_2 + x_3 = 1$ is already satisfied.

Experimentally, the rate at which glucose is transported into the cell is determined by the rate at which the concentration of glucose accumulates inside the cell in the absence of metabolism. Because $[G]_{in}$ is measured in units of millimolar, its rate of change of can be determined from the diagram and the rate equations (5.3) to be

$$\frac{d[G]_{in}}{dt} = \frac{\text{millimoles of transporter}}{\text{cellular volume}} . J_{31} = \left(\frac{10^3 N}{V_{in} A}\right) . J_{31} = R_{in}, \qquad (5.7)$$

where A is Avogadro's number, V_{in} is the cellular volume in liters, and the factor 10^3 N/A converts the total number of transporters N to millimoles. A related measure of the transport rate is the rate of change of $[G]_{out}$, which give a measure of the transport rate R_{out} based on the loss of glucose from outside the cell. In analogy to (5.7), R_{out} is easily seen to be given by

$$-\frac{d[G]_{out}}{dt} = \left(\frac{10^3 N}{V_{out} A}\right) . J_{12} = R_{out} \qquad (5.8)$$

In order to simplify the interpretation of experiments, V_{out} is usually chosen to be much greater than V_{in}, and so to a good approximation $[G]_{out}$ can be taken as a constant.

These two measures of the rate of transport of glucose are generally not proportional to one another, because $J_{12} \neq J_{31}$. Inspection of (5.3) shows, however, that the two fluxes are equal at steady-state, in which case

$$J_{12}^{SS} = J_{23}^{SS} = J_{31}^{SS} = J^{SS} \qquad (5.9)$$

This state condition can be inferred directly from the diagram in Fig. 5.2 by noting that the total flux into and out of each state must vanish for the number of transporters in each state to be steady. Thus at steady-state J^{SS} provides a unique measure of the transport rate, which can be written

$$R^{SS} = \left(\frac{10^3 N}{V_{in} A}\right).J^{SS} \tag{5.10}$$

This rate is achieved, however, only after a transient period during which the states of the transporter come to steady-state.

To calculate the transport rate using (5.10) it is necessary to calculate the value of J^{ss}, and this, in turn, requires the steady-state values of the x_i. There are three ways that this can be done: numerically, by solving the differential equations; with linear algebra, which gives an analytical expression for the x_i, or using diagrammatic methods. The first method is explored in the exercises.

Algebraic Method

We can obtain the steady-state value of the transport rate by solving the linear equations for the x_i. Substituting the expressions for the fluxes given in (5.4) – (5.6) into (5.3) and eliminating, x_3 using the conservation condition $x_3 = 1 - x_1 - x_2$ gives the 2 × 2 linear equations

$$dx/dt = \hat{A}\boldsymbol{x} + \boldsymbol{y} \tag{5.11}$$

with
$$\hat{A} = \begin{pmatrix} -(k_{12}+k_{13}+k_{31} & k_{21}-k_{31} \\ k_{12}-k_{32} & -(k_{21}+k_{23}+k_{32}) \end{pmatrix} \text{ and } y = \begin{pmatrix} k_{31} \\ k_{32} \end{pmatrix} \tag{5.12}$$

To simplify notation we have introduced the *pseudo-unimolecular* rate constants

$$k_{12} = k^{*}{}_{12}[\mathrm{G}]_{\mathrm{out}} \text{ and } k_{13} = k^{*}{}_{13}[\mathrm{G}]_{\mathrm{in}} \tag{5.13}$$

The steady state of (5.11) is determined by the algebraic equation

$$\hat{A}\boldsymbol{x}^{SS} = -\boldsymbol{y} \tag{5.14}$$

This equation has the solution

$$\boldsymbol{x}^{ss} = -\hat{A}^{-1}\boldsymbol{y} \tag{5.15}$$

with $\hat{A}^{-1}$ the inverse matrix of $\hat{A}$. Using the explicit expression for the $\hat{A}^{-1}$ given in Appendix A, (5.15) gives

$$\begin{aligned} x_1^{ss} &= (a_{22}y_1 - a_{12}y_2)\det\hat{A}, \\ x_2^{ss} &= (-a_{21}y_1 + a_{11}y_2)\det\hat{A}, \\ x_3^{ss} &= 1 - x_1^{ss} - x_2^{ss} \end{aligned} \tag{5.16}$$

To evaluate the transport rate using the algebraic method we need to substitute the expressions in (5.16) into one of the expressions in (5.9) for J^{ss}. For example, using the first expression gives

$$J_{12}^{ss} = \frac{1}{\det\hat{A}}(k_{12}(a_{22}y_1 - a_{12}y_2) - k_{21}(-a_{21}y_1 + a_{11}y_2)) \tag{5.17}$$

To get an explicit expression in terms of the ra constants k_{ij} using (5.17) we need to substitute tl expressions for the matrix elements a_{ij} given in (5.1 and then calculate the determinant of $\hat{a}$. The resultir expression is messy and offers numerous opportuniti for making algebraic mistakes. An alternative is to u the diagrammatic method, which circumvents all these algebraic difficulties.

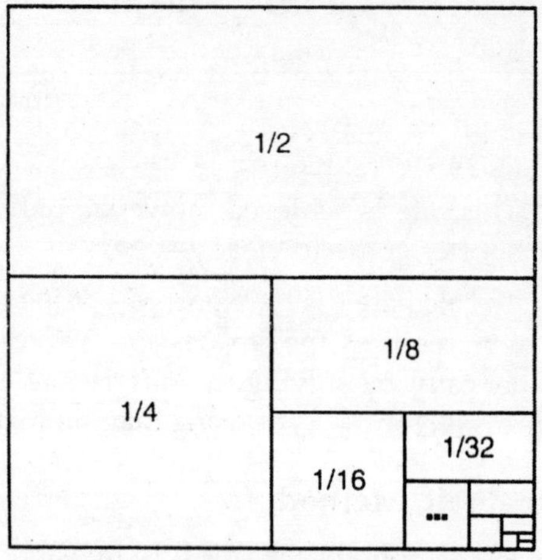

Fig. 5.3. A diagram representing the geometric series in (5.18).

Diagrammatic Method

Diagrams that represent mathematical expressioı are in common use in both quantum electrodynami ("Feynman diagrams") and statistical mechanics but a significantly less familiar in biology. Although tl diagrammatic method for obtaining J^{ss} involves a few new ideas, it leads to vastly simpler, more transparent expressions for the fluxes than those provided by the algebraic method. To help motivate the use of diagrams, consider the following infinite sum:

$$\frac{1}{2}+\frac{1}{4}+\ldots+\frac{1}{2^n}+\ldots=\sum_{n=1}^{\infty}\frac{1}{2^n} \qquad (5.18)$$

Even if the reader has previously encountered this geometric series, few probably remember that the sum converges exactly to the value one. On the other hand, a simple glance at the diagram in Fig. 5.3 makes the answer clear immediately.

Diagrams of the sort that are used in solving for the fluxes for the three-state GLUT transporter are shown in Fig. 5.4. In general, a diagram is a set of vertices (representing the states) and lines representing unimolecular (or pseudo-unimolecular) transitions between states. A *complete diagram* for the GLUT transporter, which includes all of the lines and all of the vertices in the model, is shown in Fig. 5.4A.

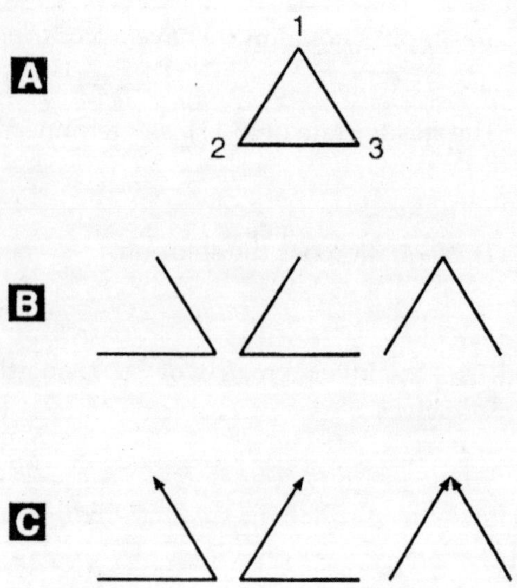

Figure 5.4 (A) The complete diagram, (B) partial diagrams, and (C) directed diagrams for state Si of the 3-state GLUT transporter model.

Note that it differs from the kinetic diagram in Fig. 5.1, because the glucose concentrations have been absorbed into the pseudounimolecular rate constants as in (5.13). The three *partial diagrams* for the model (with the vertices unlabeled) are shown in Fig. 5.4B. Partial diagrams are obtained from the complete diagram by removing lines, and they have the maximum number of lines possible without forming a cycle. A third type of diagram, called a, *directional diagram* can be constructed from the partial diagrams. Directional diagrams have arrowheads attached to the lines such that all of the lines "flow" into a single vertex. The directional diagrams for state Si are given in Fig. 5.4C. Note that for ease in writing, the arrowhead on a line will be dropped whenever its direction is obvious, as in the first and second directional diagrams in Fig. 5.4C. Three comparable directional diagrams can be drawn for states S_2 and S_5.

These diagrams represent algebraic expressions just as the areas in the diagram for the geometric series in Fig. 5.3 represent fractions. For these diagrams each line with an arrowhead (a *directed line*) represents the unimolecular or pseudounimolecular rate constant for the indicated transition. For example, the two lines in the third diagram of Fig. 5.4C represent k_{21} and k_{31}. Diagrams with several directed lines represent the product of all the indicated rate constants. Thus the directional diagrams stand for products of two rate constants. For example, the first directional diagram in Fig. 5.4C is shorthand notation for the product $k_{23}k_{31}$, whereas the second and third diagrams represent $k_{21}k_{32}$ and $k_{21}k_{31}$, respectively.

There is a general theorem that connects the directional diagrams with the fractional occupancy of states in the kinetic diagram. In particular, the fractional occupancy of state S_i, is given by the expression

$$x_i^{ss} = \frac{\text{sum of all state } S_i \text{ directional diagrams}}{\text{sum of all directional diagrams}} \tag{5.19}$$

For x_1^{ss} this yields the expressions given in Fig. 5.5, where we have adopted the symbol Σ to represent the sum of all directional diagrams. The division by Σ ensures that $\Sigma_i x_i^{ss} = 1$.

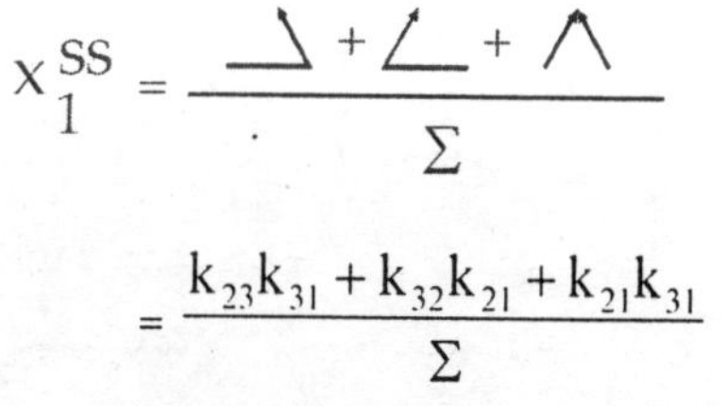

$$X_1^{SS} = \frac{\text{◺} + \text{∠} + \wedge}{\Sigma} = \frac{k_{23}k_{31} + k_{32}k_{21} + k_{21}k_{31}}{\Sigma}$$

Fig. 5.5. A diagrammatic expression for the fractional occupancy of GLUT transporters in state S_1 at steady state.

Although the expression for x_i^{ss} in (5.19) can be proven for any mechanism that can be represented by a kinetic diagram, we give a proof only for the three-state GLUT model. Because the steady-state solution for this model is unique, we need only show that the expression in (5.19) leads to the equality of all the fluxes J_{ij} at steady-state. Rather than write out the algebra, we use the diagrams themselves to complete the proof. This is illustrated in Fig. 5.6, where J_{12}^{ss} is calculated. The second equality uses (5.19), and in the third we have used the directed lines corresponding to the rate constants k_{21} and k_{21} to add extra directed lines to the diagram. Two pairs of terms cancel to give the final equality, in which only the difference of the two *cyclic diagrams* appears. A cyclic diagram is derived from a partial diagram with one additional flux added to produce a cycle. Similar manipulations show that the third equality also holds for J_{23}^{ss} and J_{31}^{ss}. Thus $J_{12}^{ss} = J_{23}^{ss} = J_{31}^{ss}$ which is the condition for steady state. The final equality in Fig. 5.6 is a corollary that can be generalized for any kinetic diagram, i.e.,

$$J_{ij}^{ss} = \frac{\text{sum of differences of cyclic diagrams with } i, j \text{ in the cycle}}{\Sigma} \tag{5.20}$$

Thus the steady-state flux for the three-state model is given by the difference of the two cyclic fluxes (counterclockwise − clockwise) divided by the sum over all partial diagrams for the complete diagram.

Rate of the GLUT Transporter

We need one more key fact about cyclic diagrams in order to simplify the expression for the transport rate: The product of the bimolecular and unimolecular rate constants in the counterclockwise direction of a cycle equals that of those in the clockwise direction. This is called the *thermodynamic restriction* on the rate constants because it is a consequence of the laws of chemical thermodynamics. To see why the thermodynamic restriction is true, consider the situation in which no transport occurs, i.e., $[G]_{out} = [G]_{in}$. If we revert to the notation for chemical reactions, then the three steps in the cycle for the three-state GLUT transporter can be written

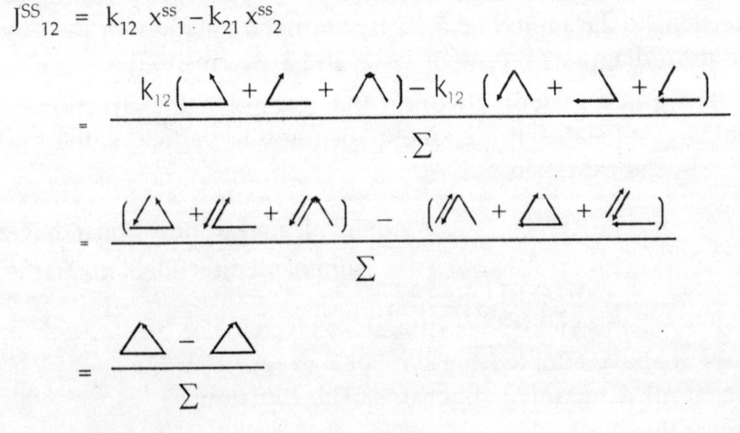

$$J^{SS}_{12} = k_{12}\, x^{ss}_{1} - k_{21}\, x^{ss}_{2}$$

$$= \frac{k_{12}(\; + \; + \;) - k_{12}(\; + \; + \;)}{\Sigma}$$

$$= \frac{(\; + \; + \;) - (\; + \; + \;)}{\Sigma}$$

$$= \frac{\; - \;}{\Sigma}$$

Fig. 5.6. Calculation of J^{ss}_{12} using diagrams.

$$G_{out} + S_1 = S_2 \text{ with equilibrium constant } K_{12} = k_{21}/k^{*}_{12},$$
$$S_2 = S_3 \text{ with equilibrium constant } K_{23} = k_{32}/k_{23},$$
$$S_3 = S_1 + G_{in} \text{ with equilibrium constant } K_{31} = k^{*}_{13}/k_{31}. \quad (5.21)$$

where S_1, S_2, and S_3 represent the three states of the transporter. It is easy to show that the equilibrium constants K_{ij} for the "reactions" are the ratios of the rate constants, as indicated next to each reaction in (5.21). If we add these three chemical reactions together we get the net reaction

$$G_{out} = G_{in} \quad (5.22)$$

A basic property of equilibrium constants is that when reactions are added, the equilibrium constants are multiplied. Therefore, the equilibrium constant for the net reaction (5.22) is .

$$K_{net} = K_{12}K_{23}K_{31} = \frac{k_{21}k_{32}k^{*}_{13}}{k^{*}_{12}k_{23}k_{31}} \quad (5.23)$$

But at chemical equilibrium the concentrations of product (G_{in}) and reactant (G_{out}) in (5.22) are equal, so that $K_{net} = [G]^{eq}_{in}/[G]^{eq}_{out} = 1$. Using this fact in (5.23) and rearranging gives the following thermodynamic restriction on the rate constants:

$$k^{*}_{13}k_{32}k_{21} = k^{*}_{12}k_{23}k_{31} \quad (5.24)$$

When constructing models, it is essential that the thermodynamic restriction on rate constants be satisfied for all cycles. Otherwise, the model will violate the second law of thermodynamics.

Using the results in the previous sections we can write an explicit expression for the rate of the three-state GLUT transporter. Combining (5.10) with the final equation in Fig. 5.6, we obtain

$$R^{ss} = \frac{10^3 N}{V_{in} A} \cdot \frac{k^*_{12}[\mathrm{G}]_{\mathrm{out}} k_{23} k_{31} - k^*_{13}[\mathrm{G}]_{\mathrm{in}} k_{32} k_{21}}{\Sigma} = \frac{10^3 N k^*_{12} k_{23} k_{31} ([\mathrm{G}]_{\mathrm{out}} - [\mathrm{G}]_{in})}{V_{in} A \,\Sigma} \tag{5.25}$$

where in the second equality we have used (5.24), the thermodynamic restriction on the rate coefficients. Again, V_{in} is the volume. According to (5.25), the steady-state transport rate is positive when the concentration of glucose outside of the cell exceeds that inside, and vanishes when the two concentrations are the same. This is a consequence of the thermodynamic restriction on the rate constants and is just what is expected for a passive transport mechanism. In the next section we consider the Na^+/glucose cotransporter, which utilizes a gradient of Na^+ to transport glucose from a low concentration to a higher concentration.

For a *symmetric* transporter there is no difference between the kinetic steps occurring inside and outside of the cell. This means that the rate constants for the transitions $2 \rightarrow 3$ and $2 \leftarrow 3$ are the same and that the association and dissociation rate constants are the same inside and outside as well. In this case there are only three different rate constants:

$$k^*_{12} = k^*_{13} = k^+ \text{ (glucose association)},$$
$$k_{21} = k_{31} = k^- \text{ (glucose dissociation)},$$
$$k_{23} = k_{32} = k \text{ (transport)}. \tag{5.26}$$

It is not difficult to evaluate the sum Σ of the directed diagrams explicitly in this case, which is

$$\Sigma = k^+(2k + k^-)\,(K + [\mathrm{G}]_{\mathrm{out}} + [\mathrm{G})_{\mathrm{in}}), \tag{5.27}$$

where we have written the dissociation constant $K = (k^-/k^+)$. Thus for the symmetric GLUT transporter model the transport rate can be written

$$R^{ss} = \frac{R_{\max}([\mathrm{G}]_{\mathrm{out}} - [\mathrm{G}]_{\mathrm{in}})}{K + ([\mathrm{G}]_{\mathrm{out}} + [\mathrm{G}]_{\mathrm{in}})}, \tag{5.28}$$

where the maximal rate is

$$R_{\max} = \frac{10^3 N k k^-}{(2k + k^-) V_{\mathrm{in}} A}. \tag{5.29}$$

Equations (5.28) and (5.29) provide explicit expressions for the transport rate for the symmetric transporter in terms of the rate constants for the model.

Glucose uptake can be measured experimentally using 3-O-methyl glucose, a non-metabolizable analogue of glucose. This further simplifies the expressions, because the concentration of the analogue is initially zero, $[\mathrm{G}]_{\mathrm{in}} = 0$, inside the cell. As a practical matter experiments involve large numbers of cells rather than a single cell. However, both N and V_{in} increase in proportion to the number of cells, so that the value of $R_{\max}$ is still characteristic of a single cell. So for this type of experiment the rate expression in (5.28) can be written

E —L→ E-L E-L ——— E*-L E-ATP —ADP→ E~P E —$h\nu$→ E* E-L —M→ M-E-L

Fig. 5.7. Elementary kinetic processes for transporters representing ligand (*L*) binding, ligand transport, phosphorylation, light excitation, and multiple ligand binding.

$$R^{ss} = \frac{R_{\max}[\mathrm{G}]_{\mathrm{out}}}{K + [\mathrm{G}]_{\mathrm{out}}} \tag{5.30}$$

One way to analyze the experimental rate of glucose uptake is using an Eadie-Hofstee plot. The Eadie-Hofstee plot is a graph of the experimental rate of glucose uptake R versus $R/[G]_{out}$ for a range of values of $[G]_{out}$. According to (5.30) this plot should give a straight line with y-intercept equal to R_{max} and slope equal to K. This can be seen by first rearranging (5.30) to get

$$R/[G]_{out} = \frac{R_{max}}{K + [G]_{out}}. \tag{5.31}$$

Then we multiply both sides by $K + [G]_{out}$ and divide by $R/[G]_{out}$ rearranging to obtain:

$$[G]_{out} = \frac{R_{max}}{R/[G]_{out}} - K. \tag{5.32}$$

If this expression for $[G]_{out}$ is substituted in the second factor in the identity

$$R = \frac{R}{[G]_{out}}.[G]_{out}, \tag{5.33}$$

we obtain

$$R = R_{max} - \frac{R}{[G]_{out}}.K, \tag{5.34}$$

which is the Eadie-Hofstee expression for the rate. Exercise 9 illustrates how transport rates can be simulated for the four-state model of a GLUT transport and how to analyze the results using an Eadie-Hofstee plot.

THE NA$^+$ /GLUCOSE COTRANSPORTER

A great variety of specialized proteins have evolved to transport specific substances across membranes in cells. Whereas the mechanisms of these transporters differ in detail, they also share a number of common features. For example, all of the known transporters bind the ligand or ligands that they transport, and of course, they must dissociate them as well. These steps must occur on both sides of the membrane for transport to occur, so there must be a process or processes in which the ligands are transported across the membrane. Fig. 5.7 illustrates some of the elementary kinetic processes that are found for transporters, including chemical modification of the transporter by phosphorylation, light-induced conformational changes, and multiple ligand binding.

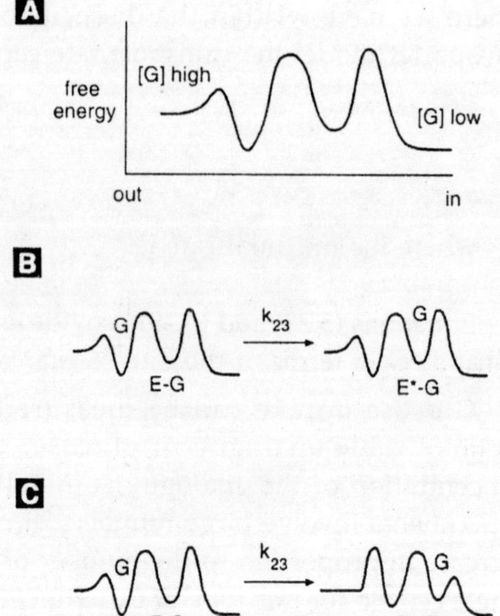

Fig. 5.8. (A) Schematic representation of the free energy profile across a membrane for a transporter protein. Two possibilities for state changes for the transport step 2 to 3 for the GLUT transporter are indicated: (B) simple barrier crossing; and (C) barrier crossing via a conformational transition.

From the point of view of chemical physics, the transport step can be viewed as energetic rearrangements that involve both the transporter protein and the ligand being transported. Fig. 5.8A is a schematic representation of what the energy profile across a GLUT transporter might look like when the glucose concentration outside is high and inside is low. The energy profile is the Gibbs free energy, rather than the potential energy.

For the average kinetic events that we are considering the influence of entropy effects must be taken into account. The reason that the free energy of glucose is higher outside is simply that the concentration of glucose is higher outside the cell than inside. The peaks of the free energy represent barriers to the movement of glucose across the transporter. Two possibilities for the transition from state 2 to state 3 are shown in Fig. 5.8B and Fig. 5.8C. The first represents a barrier crossing in which the transition 2 $\rightarrow$ 3 does not influence the shape of the energy profile. In the second, on the other hand, the energy profile is different after the transition, as might be the case if the transition involved a conformational change. Although understanding the transport step is an important feature of building a model of a transporter, it describes neither how a transporter works nor the rate of transport, which was seen in Section 5.2 to be a property of the complete model, not a single step.

To illustrate how a complete model of a transporter is created, we consider the Na^+/glucose cotransporter from intestinal epithelial cells. This transporter utilizes a concentration gradient of Na^+ to transport glucose from the intestine into the epithelial cells that line the gut. This is "uphill" transport, because the concentration of glucose in the epithelial cells exceeds that in the intestine. As shown schematically in Fig. 5.9, the cotransporter works in concert with a Na^+/K^+ ATPase and passive transport of glucose by GLUT transporters, both at the basolateral side of the epithelium, to move glucose from the intestine to the blood stream. The Na^+/K^+ ATPase helps eliminate the Na^+ that accompanies glucose uptake during cotransport, thereby maintaining a low concentration of Na^+ inside of the cell.

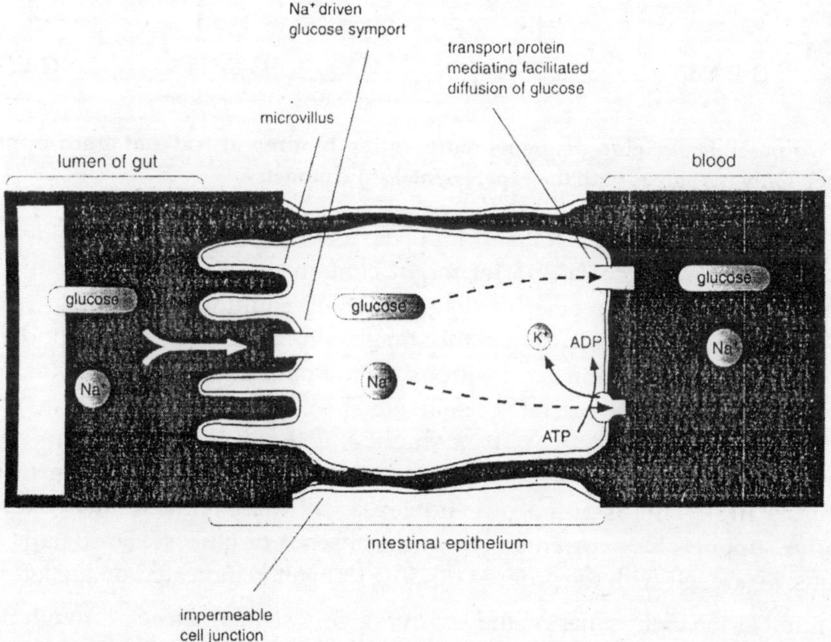

Fig. 5.9. A cartoon representing the cotransport of glucose and Na^+ into intestinal epithelium, followed by the passive transport of glucose into the blood. Energy stored in the gradient of Na^+ (higher in the lumen of the gut) is utilized to transport glucose from a high concentration to a low concentration. Na^+ that accumulates in the epithelial cells is removed by active transport into the blood by the Na^+/K^+ ATPase.

A model for any transporter must incorporate a number of basic experimental facts. One of these is *stoichiometry*, which for the cotransporter is the number of Na^+ ions transported per glucose molecule. Experimental measurements on the Na^+/glucose cotransporter from intestine yield a stoichiometry of 2 Na^+ to 1 glucose. Another important fact about the cotransporter is the absolute requirement for Na^+. If Na^+ is absent from the external medium, glucose is not transported. In addition, the cotransporter is *electrogenic*, because transport generates an electrical current due to the transport of Na^+.

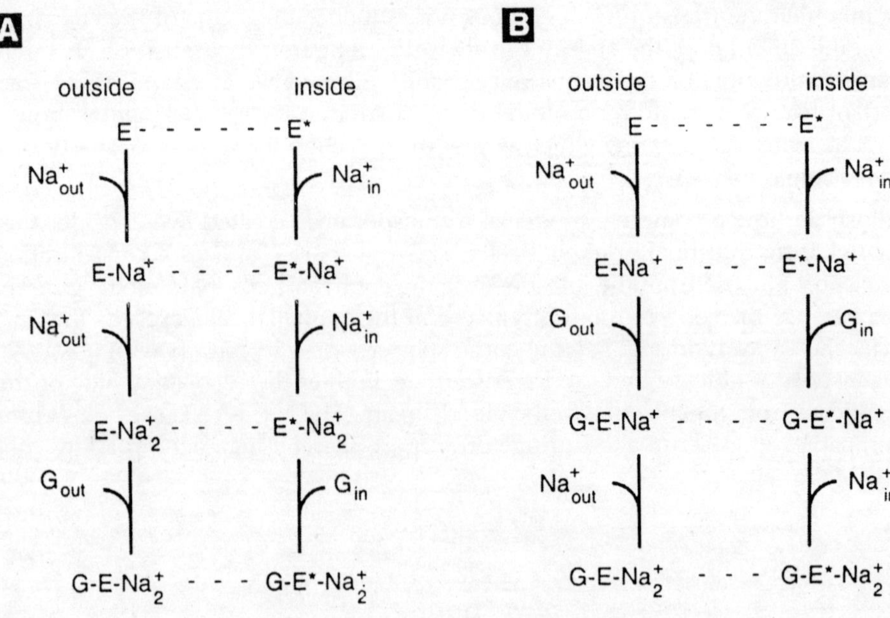

Fig. 5.10. Two possible skeleton diagrams representing binding of Na^+ and glucose for the Na^+/glucose cotransporter. Only (A) agrees with the experimental stoichiometry.

These observations require 2 Na^+ and 1 glucose association steps on each side of the membrane. A partial skeleton for the cotransporter might contain the kinetic steps connected together as shown in Fig. 5.10A. It is also possible that the second sodium binds *after* the glucose as in Fig. 5.10B. However, this can be ruled out if the states with Na^+ and glucose bound from the outside are connected by conformational transitions to comparable states inside (as indicated by the dashed lines). In that case the six state cycle in Fig. 5.10B (E to E-Na^+ to G-E-Na^+ to G-E*-Na^+ to E*-Na^+ to E* to E) would transport only a single Na^+ for every glucose molecule, implying a stoichiometry at steady state less than 2:1. The third possibility, not shown in Fig. 5.10, is that glucose binds first. This is ruled out, however, by the experimental observation that the cotransporter supports Na+ currents even in the absence of glucose. For details see (Parent *et al.* 1992a). Thus we are left with the ordered binding of ligands indicated on the left in Fig. 5.10.

If we number the eight states on the left in Fig. 5.10 sequentially S_1 through S_8, starting at "E" and moving counterclockwise, there are a number of possibilities for conformational changes connecting the left and right sides of the diagram. Fig. 5.11 illustrates six alternatives. Alternative (B) is easily eliminated, because it does not transport glucose. Although diagram (C) does transport glucose (S_4 to S_5), it does not includes steps that transport only Na^+, and therefore conflicts with the fact that the transporter produces a Na^+ current in the absense of glucose. Diagram (D) can be ruled out because it has the wrong stoichiometry (1 Na^+: 1 glucose). This leaves as possible mechanisms diagram (A), which is fully connected, and diagrams (E) and (F), each of which is missing Na^+ transport steps.

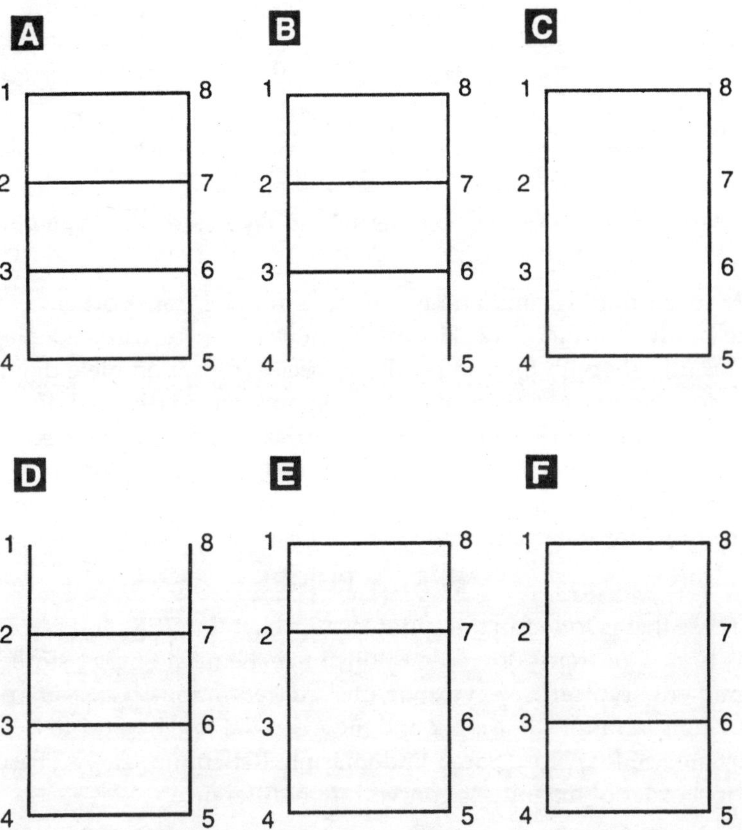

Fig. 5.11. Six possible diagrams for the Na^+/glucose transporter with transport steps included. Only (A), (E), and (F) are compatible with experiment.

All three of these diagrams are compatible with the experimental evidence, and all three can be "reduced" to a diagram with the 6-state skeleton givenin Fig. 5.12A. This method of reducing diagrams uses the *rapid equilibrium approximation* that applies to steps for which the forward and reverse rates are rapid with respect to other steps in the diagram. The details of how this method works are explained, although the basic idea can be seen by comparing Fig. 5.12A and Fig. 5.12B.

The experimental values of rate constants for the six-state model have been assigned by Parent and colleagues. Step S_4 to S_5 in the six-state model is the dissociation of glucose inside the cell, and this step is extremely fast. This permits the two states to be approximated as a single combined state (state $S_{4/5}$ in Fig. 5.12B and reduces the diagram to five states as shown. We must be careful in doing so to readjust the rates to account for the reduction. The details of the process for doing this is given, but it is not difficult. In short, only a portion of the combined state $S_{4/5}$ reacts to the other states. The portion to be used in each reaction is determined as a result of the reduction using simple algebra.

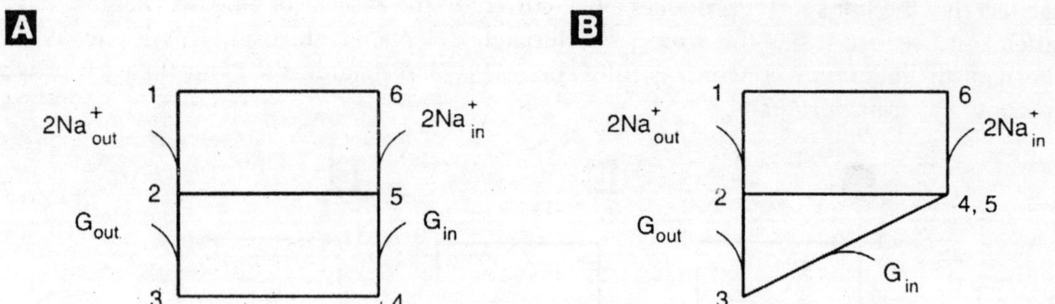

Fig. 5.12. (A) Six-state and (B) five-state simplification of the eight-state diagram for the Na⁺/glucose cotransporter.

It is possible to write diagrammatic expressions for the transport rate for either the five-state or six-state models in Fig. 5.12. However, the number of directed diagrams and cyclic diagrams increases quickly with the complexity of cycles in the complete diagram. For example, for the five-state model there are 6 pairs of cyclic diagrams and 55 directed diagrams. Nonetheless, the general expressions in (5.19) and (5.20) remain valid and can be used to obtain explicit expressions for the steady-state fluxes. The diagrammatic method does not, however, provide information about the transient time-dependence of the fluxes. This is most conveniently obtained by numerical integration of the equations.

SERCA PUMPS

The Ca^{2+} -ATPase that is found in the endoplasmic reticulum (ER) and sarcoplasmic reticulum (SR) of muscle is typical of transporters that utilize the chemical energy stored in ATP to pump ions against a gradient. Typical free cytosolic Ca^{2+} concentrations $[Ca^{2+}]_i$ are of the order of 0.1 μm, whereas Ca^{2+} concentrations in the ER and SR are in the range of 0.1 to 1 mM. These pumps, which are abbreviated SERCA for "Sarco-Endoplasmic Reticulum Ca^{2+} ATPase," therefore have to surmount a 3 to 4 order of magnitude concentration difference.

Topologically, the SR and ER are equivalent to the "outside" of the cell, and both compartments function to store Ca^{2+} for a variety of cellular processes. In muscle, Ca^{2+} release from the SR is involved in triggering muscle contraction, whereas Ca^{2+} release from the ER is involved in stimulating hormone secretion and other intracellular signaling cascades. Pumping of Ca^{2+} by SERCA is the primary mechanism by which SR and ER Ca^{2+} stores are maintained. A different type of Ca^{2+} pump (PMCA), which is found in the plasma membrane, functions to pump Ca^{2+} out of the cell. Although there are several isotypes of SERCA found in different tissues, the rate at which they pump Ca^{2+} has a simple dependence on $[Ca^{2+}]_i$. The pumping rate can be measured using vesicles prepared from either SR or ER membranes. The rate of vesicle accumulation of $^{45}Ca^{2+}$, a radioactive isotope of Ca^{2+}, can then be used to determine the pumping rate.

Experimentally, the rate has a sigmoidal dependence on $[Ca^{2+}]_i$ with a Hill coefficient close to two, i.e.,

$$R = \frac{R_{max}[Ca^{2+}]_i^2}{K^2 + [Ca^{2+}]_i^2} \tag{5.35}$$

The Hill coefficient is related to the stoichiometry of the SERCA pump, which is known to be $2Ca^{2+}$:1ATP. X-ray diffraction of SERCA pumps in bilayers has produced a low-resolution structure with three segments: a stalk region just outside the bilayer close to the binding sites for Ca^{2+}, a head region that contains the ATP binding site, and a large transmembrane region through which the Ca^{2+} is transported. Binding experiments have revealed two binding sites for Ca^{2+}.

$2Ca^{2+}$ I ⟶ A Ca^{2+} I ⟶ A Ca^{2+} I* ⟶ A

Fig. 5.13. Two simplified models explaining the Ca^{2+} dependence of the experimental SERCA pump rate. (A) A single binding event versus (B) two sequential binding events.

A simple model that is consistent with the Ca^{2+} dependence of the pump rate can be constructed using only two states: an inactive state *I* and an active state a connected by the mechanism shown in Fig. 5.13 A. This model leads to the rate expression in (5.35) if we assume that the two states rapidly equilibrate and that only the active state transports Ca^{2+}. Rapid equilibration implies the balance of the forward and reverse rates in Fig. 5.13 A. This leads to the equilibrium condition

$$k^- / k^+ = K_{eq} = \frac{[Ca^{2+}]_i^2[I]}{[A]}, \tag{5.36}$$

where [I] and [A] are the per unit area concentrations of SERCA pumps in the two states, and the equilibrium constant K_{eq} is the ratio of the rate constants. Solving (5.36) for [I], substituting that expression into the conservation condition [I] + [A] = *N*, and then solving for [A] gives the concentration of active SERCAs:

$$[A] = \frac{N[Ca^{2+}]_i^2}{K^2 + [Ca^{2+}]_i^2} \tag{5.37}$$

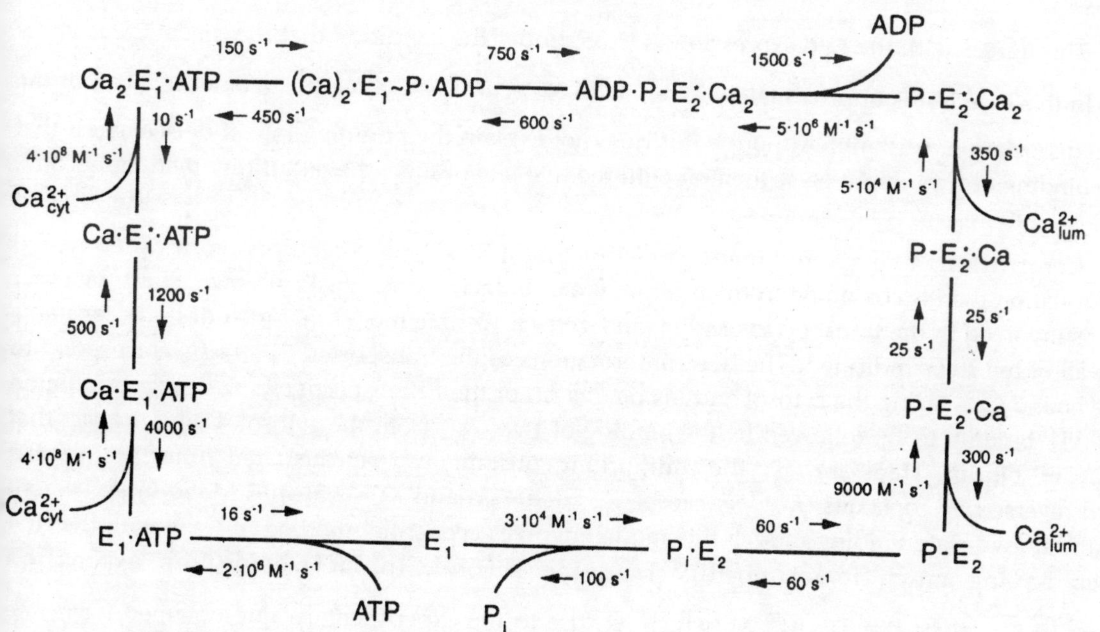

g. 5.14 A twelve-state model of the SERCA pump. Note the two sequential Ca^{2+} binding steps on the left-nd side. Although the cycle is driven by the hydrolysis of ATP, all of the steps in the diagram contribute the steady-state rate.

Here we have defined $K = \sqrt{K_{eq}}$, which has the units of concentration. As can be seen from (5.37), the numerical value of K equals the concentration of Ca^{2+} at which half of the SERCAs are in the active state. A small value of K is said to correspond to a *high affinity* binding site, and a large value to a *low affinity* site. If the rate constant for the active state to transport Ca^{2+} is k, then (5.37) gives the transport rate

$$R = k[\mathrm{A}] = \frac{R_{\max}[\mathrm{Ca}^{2+}]_i^2}{K^2 + [\mathrm{Ca}^{2+}]_i^2} \tag{5.38}$$

with $R_{\max} = kN$.

Although this mechanism agrees with the measured transport rate and provides an expression for R_{max}, there are several things wrong with it. First, it assumes the simultaneous binding of two Ca^{2+}, which is highly improbable. Second, it doesn't provide an explanation for the transport rate constant k. Third, it doesn't explain how ATP might be involved. We can eliminate the first criticism by expanding the model to include sequential binding of two Ca^{2+}, as indicated in Fig. 5.13B. For this mechanism there are two simultaneous binding equilibria:

$$K_1 = \frac{[\mathrm{Ca}^{2+}]_i[\mathrm{I}]}{[\mathrm{I}^*]} \text{ and } K_2 = \frac{[\mathrm{Ca}^{2+}]_i[\mathrm{I}^*]}{[\mathrm{A}]} \tag{5.39}$$

In analogy to what was done for the previous model, these equations can be combined with the conservation condition [I] + [I*] + [A] = N to obtain

$$[\mathrm{A}] = \frac{N}{1 + (K_1K_2/[\mathrm{Ca}^{2+}]_i^2) + K_2/[\mathrm{Ca}^{2+}]_i} \tag{5.40}$$

This agrees with the rate expression in (5.35) under the condition that $K_2 << [Ca^{2+}]_i$ and $K_2 << K_1$. In this case [A] is approximately given by (5.37) with $K = \sqrt{K_1K_2}$, the geometric mean of the two dissociation constants. Although this does not explain the pumping rate, it does suggest that the binding of Ca^{2+} might be sequential with the first site of much lower affinity than the second, i.e., $K_1 >> K_2$.

Constructing a complete kinetic model of the SERCA pump requires more experimental information than is contained in the pumping rate. In fact, a great deal is known about the other steps involved in the transport cycle. Fig. 5.14 gives a 12-state model that includes rate constants for all of the steps indicated. The two conformations of the transporter, E_1 and E_2, correspond to the bound Ca^{2+} facing the cytosol and inside *(lumen)* of the ER, respectively. The cycle is initiated by ATP binding to E_1, followed by the binding of two Ca^{2+} from the cytosol. Using the fact that the Ca^{2+} binding steps are fast, the equilibrium constants can be calculated from the forward and reverse rate constants $(K = k^-/k^+)$ to be $K_1 = 1.10^{-5}$ M and $K_2 = 2.5 \cdot 10^{-8}$ M. So in agreement with the two-state binding model, this model involves sequential binding of Ca^{2+}, with the first step having much lower affinity than the second. In fact, using the expression $K = \sqrt{K_1K_2}$, gives $K = 5 \cdot 10^{-7}$ M, which is close to the experimental value obtained from rate measurements in vesicles. The cycle in Fig. 5.14 is driven by the phospho-rylation of SERCA, which facilitates the conformational transition that exposes bound Ca^{2+} to the lumen.

TRANSPORT CYCLES

Like enzymes, transporters are unchanged by the transport process. Indeed, transporters can be thought of as enzymes whose primary purpose is to alter the location of a molecule rather than its chemical state. If we take the more general point of view suggested by nonequilibrium thermodynamics, an enzyme and a transporter are simply different classes of the same generic type of protein that catalyze a change in free energy. As we noted in the previous section, a transporter accomplishes this by altering the concentration that the molecule experiences.

An enzyme, on the other hand, alters the chemical bonds in the molecule. The catalytic nature of a transporter is apparent in the cyclic structure of the transport mechanism. The GLUT transporter, the Na^+/glucose cotransporter, and the SERCA pump described in all function in cycles that leave the transporter unchanged. Three additional examples of transport cycles that have been used to explain experimental transport rates are given in Fig. 5.15 and Fig. 5.16: a P-type proton pump, the adenine nucleotide transporter from mitochondria, and bacteriorhodopsin, a light-driven proton pump. In a transport cycle a ligand is moved from one cellular compartment to another by some type of driving force.

In passive transport, like that for the GLUT transporter, the driving force is simply the concentration difference of ligand. For transporters that involve more than one ligand, such as the Na^+/glucose cotransporter or the adenine nucleotide transporter in Fig. 5.14B, the driving force is a combination of ligand concentration differences. For ATP-dependent pumps, on the other hand, the driving force is the chemical energy stored in the terminal phosphate bond of ATP, which is transferred in a phosphorylation step (cf. Fig. 5.14 and Fig. 5.15A) to the transporter. Phosphorylation maintains the high-energy state, and the high-energy phosphate bond facilitates conformational transitions that lead to the transport and dissociation of the transported ion. A third form of driving force is light, as indicated in the transport cycle for bacteriorhodopsin in Fig. 5.16. In this case energy from a photon excites a state of the transporter causing a trans to cis conformational change in the structure of retinal that is otherwise inaccessible to thermal motion, thereby releasing a proton at the exterior face of the membrane.

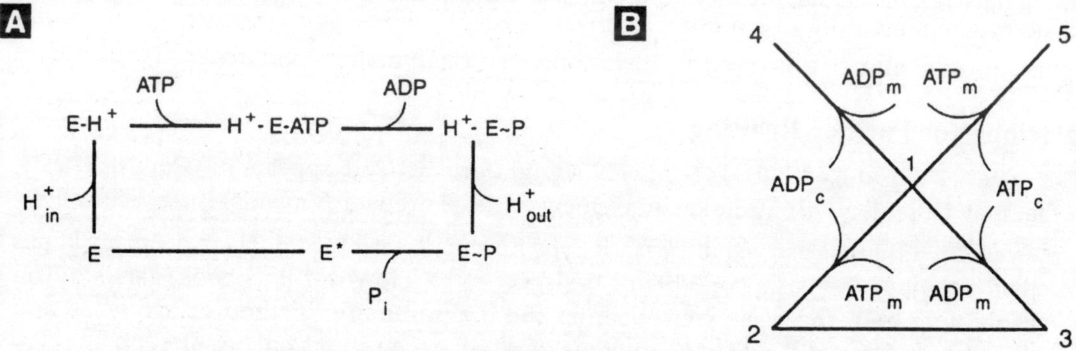

Fig. 5.15. (A) A six-state cycle for a P-type proton ATPase. The phosphorylation step precedes and facilitates dissociation of the proton outside of the cell. (B) Five-state cycle for the adenine nucleotide translocator from mitochondria. The cycle 1-2-3-1 exchanges an ADP^{3-} from the cytoplasm for an ATP^{4-} from the mitochondria. Because the two ions have different charges, the translocator is electrogenic. Steps 1-4 and 1-5 slow the translocation rate by tying up the translocator in states not involved in the cycle.

As long as we are interested in the average properties of a transporter, it is correct to picture transport cycles as occurring in a fixed direction governed by the driving forces. Dynamic changes in an individual transporter molecule, on the other hand, are stochastic. This is a result of the microscopic reversibility of the kinetic steps in a cycle. Although it is more probable that an individual GLUT transporter will move glucose from a high concentration to a low concentration, the reverse will occur with nonvanishing probability. In fact, any step in a transport cycle can and will occur in the opposite direction to the average transport rate. This reversibility has been demonstrated experimentally for a number of transporters, one of the most convincing being the reversal of SERCA pumps to produce ATP by reversing the Ca^{2+} gradient. Thus the dynamic changes in an individual transporter molecule consist of a series of random positive and negative steps around the cycle that over time lead to an average transport rate in the direction dictated by the driving forces.

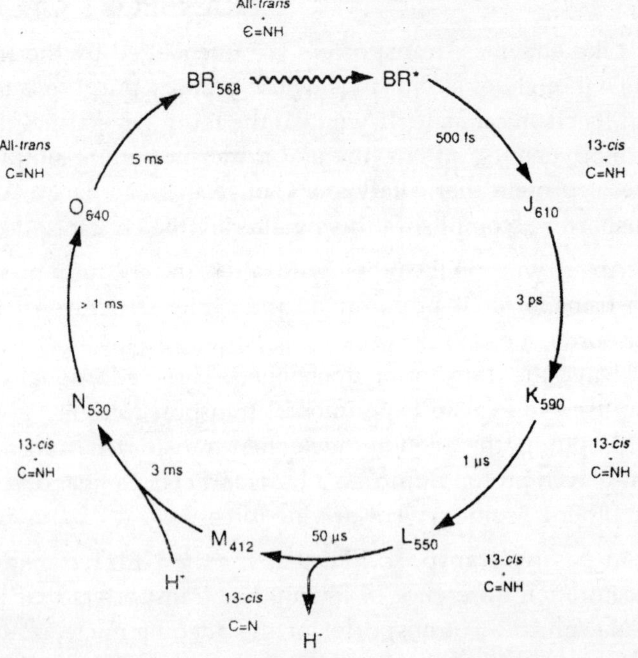

Fig. 5.16. An irreversible diagram for the light-driven proton pump, bacteri-orhodopsin, a 27 kD protein from the salttolerent bacterium, *Halobacteriumhalo-bium.* Light absorbed in the step BR_{56}8-BR* drives the trans to cis conformational change in retinal, leading to the release of a proton.

Cycles are a common feature of other cellular processes with some of the most complex cycles governing muscle contraction, the rotary motion of flagella, and other so-called molecular motors. In metabolism the operation of the F_0F_1 ATPase, which is responsible for converting proton gradients and ADP into ATP, is governed by a combined catalytic-transport cycle.

Suggestions for Further Reading

The material introduced here is relatively straightforward and appears in many fine texts. Some that may be particularly useful at an elementary level are worth mentioning specifically:

- *Free Energy Transduction and Biochemical Cycle Kinetics,* Terrell L. Hill. There are many books that introduce the basic concepts of transport rates. However, this book stands out in explaining both the basic concepts and the diagrammatic method in a concise and complete fashion. Hill also has numerous more advanced books on the subject.
- *The Fluctuating Enzyme,* edited by G. Rickey Welch. For a somewhat higher level discussion of transporter kinetics, see the chapter "Not Just Catalysts: Molecular Machines in Bioenergetics" by G. Rickey Welch and Douglas B Kell.
- *Electrogenic Ion Pumps,* Peter Lauger. This chapter gives a development of the basic physical chemical basis behind many of the electrogenic ion pumps covered here.

MULTIPLE ALIGNMENT

6

The goal of protein sequence comparison is to discover ''biological'' (*i.e.*, structural or functional) similarities among proteins. Biologically similar proteins may not exhibit a strong sequence similarity, and one would like to recognize the structural/functional resemblance even when the sequences are very different. If sequence similarity is weak, pairwise alignment can fail to identify biologically related sequences because weak pairwise similarities may fail the statistical test for significance. Simultaneous comparison of many sequences often allows one to find similarities that are invisible in pairwise sequence comparison.

To quote Hubbard et aL, 1996 "pairwise alignment whispers... multiple alignment shouts out loud." Straightforward dynamic programming solves the multiple alignment problem for k sequences of length n. Since the running time of this approach is $O((2n)^k)$, a number of different variations and some speedups of the basic algorithm have been devised. However, the exact multiple alignment algorithms for large k are not feasible, and many heuristics for suboptimal multiple alignment have been proposed. A natural heuristic is to compute $\binom{k}{2}$ optimal pairwise alignments of the k strings and combine them together in such a way that induced pairwise alignments are close to the optimal ones.

Unfortunately, it is not always possible to combine pairwise alignments into multiple alignments, since some pairwise alignments may be incompatible. As a result, many multiple alignment algorithms attempt to combine some compatible subset of optimal pairwise alignments into a multiple alignment. This can be done for some small subsets of all $\binom{k}{2}$ pairwise alignments. The problem is deciding which subset of pairwise alignments to choose for this procedure. The simplest approach uses pairwise alignment to iteratively add one string to a growing multiple alignment.

Feng and Doolittle, 1987 use the pair of strings with greatest similarity and "merge" them together into a new string following the principle "once a gap, always a gap." As a result, the multiple alignment of k sequences is reduced to the multiple alignment of $k-1$ sequences. Many other iterative multiple alignment algorithms use similar strategies.

Although the Feng and Doolittle, 1987 algorithm works well for close sequences, there is no "performance guarantee" for this method.

The first "performance guarantee" approximation algorithm for multiple alignment, with approximation ratio $2-\frac{2}{k}$, was proposed by Gusfield, 1993. The idea of the algorithm is based on the notion of compatible alignments and uses the principle "once a gap, always a gap."

Feng and Doolittle, 1987 and Gusfield, 1993 use optimal pairwise (2-way) alignments as building blocks for multiple k-way alignments. A natural extension of this approach is to use optimal 3-way (or l-way) alignments as building blocks for k-way alignments. However, this approach faces some combinatorial problems since it is not clear how to define compatible l-way alignments and how to combine them. Bafna *et al.*, 1997 devised an algorithm for this problem with approximation ratio $2 - \frac{l}{k}$. Biologists frequently depict similarities between two sequences in the form of *dot-matrices.* A dot-matrix is simply a matrix with each entry either 0 or 1, where a 1 at position (i, j) indicates some similarity between the i-th position of the first sequence and the j-th position of the second one.

The similarity criteria vary from being purely combinatorial (e.g., a match of length m with at most k mismatches starting at position i of the first sequence and j of the second one) to using correlation coefficients between physical parameters of amino acids. However, no criterion is perfect in its ability to distinguish "real" (biologically relevant) similarities from chance similarities (noise).

In biological applications, noise disguises the real similarities, and the problem is determining how to filter noise from dot-matrices. The availability of several sequences sharing biologically relevant similarities helps in filtering noise from dot-matrices. When k sequences are given, one can calculate $\binom{k}{2}$ pairwise dot-matrices. If all k sequences share a region of similarity, then this region should be visible in all $\binom{k}{2}$ dot-matrices. At the same time, noise is unlikely to occur consistently among all the dot-matrices. The practical problem is to reverse this observation: given the $\binom{k}{2}$ dot-matrices, find similarities shared by all or almost all of the k sequences and filter out the noise. Vingron and Argos, 1991 devised an algorithm for assembling a k-dimensional dot-matrix from 2-dimensional dot-matrices.

SCORING A MULTIPLE ALIGNMENT

Let A be a finite *alphabet* and let $a_1,..., a_k$ be k sequences (strings) over A. For convenience, we assume that each of these strings contains n characters. Let A' denote AU{–}, where '–' denotes space. An *alignment* of strings $a_1,...,a_k$ is specified by a $k \times m$ matrix A, where $m \geq n$. Each element of the matrix is a member of A', and each row i contains the characters of a_i in order, interspersed with $m - n$ spaces. We also assume that every column of the multiple alignment matrix contains at least one symbol from A.

The score of multiple alignment is defined as the sum of scores of the columns and the optimal alignment is defined as the alignment that minimizes the score. The score of a column can be defined in many different ways. The intuitive way is to assign higher scores to the columns with large variability of letters. For example, in the *multiple shortest common supersequence* problem, the score of a column is defined as the number of different characters from A in this column. In the *multiple longest common subsequence* problem, the score of a column is defined as -1 if all the characters in the column are the same, and 0 otherwise.

In the more biologically adequate *minimum entropy* approach, the score of multiple alignment is defined as the sum of entropies of the columns. The entropy of a column is defined as

$$-\sum_{x \in A'} p_x \log p_x$$

where p_x is the frequency of letter $x \in A'$ in a column i. The more variable the column, the higher the entropy. A completely conserved column (as in the multiple LCS problem) would have minimum entropy 0.

The minimal entropy score captures the biological notion of good alignment, but it is hard to efficiently analyze in practice. Below we describe *Distance from Consensus* and *Sum-of-Pairs (SP)* scores, which are easier to analyze.

- *Distance from Consensus.* The consensus of an alignment is a string of the most common characters in each column of the multiple alignment. The *Distance from Consensus* score is denned as the total number of characters in the alignment that differ from the consensus character of their columns.
- *Sum-of-Pairs (SP-score).* For a multiple alignment A = (a_{ih}), the induced score of pairwise alignment A_{ij} for sequences a_i and a_j is

$$s(A_{ij}) = \sum_{h=1}^{m} d(a_{ih}, a_{jh}),$$

where d specifies the *distance* between elements of A'. The *Sum-of-Pairs score* (SP-score) for alignment A is given by $\Sigma_{ij}\, s(A_{ij})$. In this definition, the score of alignment A is the sum of the scores *of projections* of A onto all pairs of sequences a_i and a_j. We assume the metric properties for distance d, so that $d(x, x) = 0$ and $d(x, z) \le d(x, y) + d(y, z)$ for all x, y, and z in A'.

ASSEMBLING PAIRWISE ALIGNMENTS

Feng and Doolittle, 1987 use the pair of strings with greatest similarity and "merge" them together into a new string following the principle "once a gap, always a gap." As a result, the multiple alignment of k sequences is reduced to the multiple alignment of $k - 1$ sequences (one of them corresponds to the merged strings). The motivation for the choice of the closest strings at the early steps of the algorithm is that close strings provide the most reliable information about alignment.

Given an alignment A of sequences $a_1,...a_k$ and an alignment A' of some subset of of the sequences, we say that A is *compatible* with A' if A aligns the characters of the sequences aligned by A' in the same way that A' aligns them. Feng and Doolittle, 1987 observed that given any tree in which each vertex is labeled with a distinct sequence a_i, and given pairwise alignments specified for each tree edge, there exists a multiple alignment of the k sequences that is compatible with each of the pairwise alignments. In particular, this result holds for a star on k vertices, i.e., a tree with *center* vertex and $k - 1$ leaves.

Lemma 7-1 *For any star and any specified pairwise alignments $A_1, ..., A_{k-1}$ on its edges, there is an alignment A for the k sequences that is compatible with each of the alignments $A_1,..., A_{k-1}$.*

Given a star G, define a *star-alignment* A_G as an alignment compatible with optimal pairwise alignments on the edges of this star. The alignment A_G optimizes $k - 1$ among $\binom{k}{2}$ pairwise alignments in SP-score. The question of how good the star alignment is remained open until Gusfield, 1993 proved that if the star G is chosen properly, the star alignment approximates the optimal alignment with ratio $2 - \frac{2}{k}$.

Let G(V, E) be an (undirected) graph, and let $\gamma(i, j)$ be a (fixed) shortest path between vertices $i \ne j \in V$ of length $d(i, j)$. For an edge $e \in E$, define the communication cost $c(e)$ as the number of shortest paths $\gamma(i > j)$ in G that use edge *e*. For example, the communication cost of every edge in a star with k vertices is $k - 1$. Define the communication cost of the graph G as $c(G) = \Sigma_{e \in E} c(e) = \Sigma_{i \neq j \in V} d(i, j)$. The *complete graph* on k vertices H_k has a minimum communication cost of $c(G) = \frac{k(k-1)}{2}$ among all k-vertex graphs. We call $b(G) = \frac{c(G)}{c(H_k)} = 2\frac{c(G)}{k(k-1)}$ *the normalized communication cost* of G. For a star with k vertices, $b(G) = 2 - \frac{2}{k}$. Feng and Doolittle, 1987 [100] use a tree to combine pairwise alignments into a multiple alignment, and it turns out that the normalized communication cost of this tree is related to the approximation ratio of the resulting heuristic algorithm.

Define $C(G) = (c_{ij})$ as a $k \times k$ matrix with $c_{ij} = c(e)$ if (i, j) is an edge e in G and $c_{ij} = 0$ otherwise. The *weighted sum-of-pairs score* for alignment A is

$$\sum_{ij} c_{ij}.s(A_{ij})$$

For notational convenience, we use the *matrix dot product* to denote scores of alignments. Thus, letting $S(A) = (s(A_{ij}))$ be the matrix of scores of pairs of sequences, the weighted sum-of-pairs score is $C(G) \cdot S(A)$. Letting E be the unit matrix consisting of all Is except the main diagonal consisting of all 0s, the (unweighted) sum-of-pairs score of alignment A is $E \cdot S(A)$.

The pairwise scores of an alignment inherit the triangle inequality property from the distance matrix. That is, for any alignment A, $s(A_{ij}) \leq s(A_{ik}) \leq s(A_{kj})$, for all i, j, and k. This observation implies the following:

Lemma 7.2. *For any alignment A of k sequences and a star G,* $E \cdot S(A) \leq C(G) \cdot S(A)$.

APPROXIMATION ALGORITHM FOR MULTIPLE ALIGNMENTS

Let $\mathcal{G}$ be a collection of stars in a k-vertex graph. We say that the collection G is *balanced* if $S_{G \subset \mathcal{G}} C(G) = pE$ for some scalar $p > 1$. For example, a collection of k stars with k different center vertices is a balanced collection with $p = 2(k - 1)$. Since $C(G)$ is non-zero only at the edges of the star G, and since star alignment A_G induces optimal alignments on edges of G,

$$C(G) \cdot S(A_{\mathcal{G}}) \leq C(G) \cdot S(A)$$

for any alignment A.

Lemma 7.3. *If* $\mathcal{G}$ *is a balanced set of stars, then*

$$\min_{G \in \mathcal{G}} C(G) \cdot S(A_G) \leq \frac{p}{|\mathcal{G}|} \min_{A} E \cdot S(A)$$

Proof. We use an averaging argument.

$$\min_{G \in \mathcal{G}} C(G) \cdot S(A_G) \leq \frac{1}{|\mathcal{G}|} \Sigma_{G \in \mathcal{G}} C(G) \cdot S(A_G)$$

$$\leq \frac{1}{|\mathcal{G}|} . \Sigma_{G \in \mathcal{G}} C(G) \cdot S(A) = \frac{p}{|\mathcal{G}|} . E \cdot S(A)$$

Here the inequality holds for an arbitrary alignment A, and in particular, it holds for an optimal alignment.

Lemmas 7.2 and 7.3 motivate the **Align** algorithm:

1. Construct a balanced set of stars, $\mathcal{G}$.
2. For each star G in $\mathcal{G}$, assemble a star alignment A_G.
3. Choose a star G such that $C(G) \cdot S(A_G)$ is the minimum over all stars in $\mathcal{G}$.
4. Return A_G.

Theorem 7.1. *(Gusfield, 1993 Given a balanced collection of stars $\mathcal{G}$,* Align *returns an alignment with a performance guarantee of $2 - 2/k$ in $O(k \cdot n^2 \cdot |\mathcal{G}|)$ time.*

Proof. Note that $\frac{p}{|\mathcal{G}|} = \frac{C(G) \cdot E}{E \cdot E} = 2 - \frac{2}{k}$. **Align** returns the alignment A_G that is optimal for a star $G \in \mathcal{G}$, and for which the smallest score, $\min_{G \in \mathcal{G}} C(G) \cdot S(A_G)$, is achieved. Lemmas 7.2 and 7.3 imply that $E \cdot S(A_G) \leq C(G) \cdot S(A_G) \leq \left(2 - \frac{2}{k}\right) \cdot \min_A E \cdot S(A)$.

ASSEMBLING *L*-WAY ALIGNMENTS

An *l-star* $G = (V, E)$ on k vertices is defined by $r = \frac{k-1}{l-1}$ cliques of size l whose vertex sets intersect in only one *center* vertex. For a 3-star with $k = 2t + 1$ vertices, $c(G) = (2t - 1)\,2t + t$ and $b(G) = 2 - \frac{3}{k}$. For an l-star with $k = (l - 1)t + 1$ vertices, $c(G) = ((l - 1)t + 1 - l + 1)(l - 1)\,t + t\left(\frac{l(l-1)}{2} - l + 1\right)$ and $b(G) = 2 - \frac{1}{k}$. The communication cost of an edge e in an l-star G with center c is

$$c(e) = \begin{cases} k - l + 1, & \text{if } e \text{ is incident to } c \\ 1, & \text{otherwise} \end{cases}$$

Note that for the communication cost matrix of an l-star G,

$$C(G) \cdot E = (k - l + 1) \cdot (k - 1) + \frac{k-1}{l-1}\binom{l-1}{2} = \binom{k}{2}\left(2 - \frac{l}{k}\right)$$

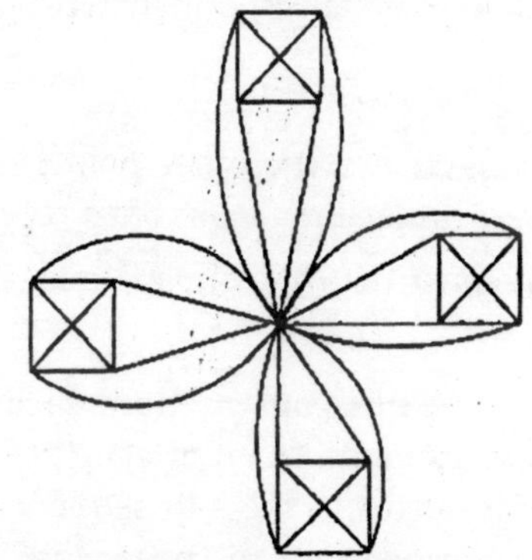

Fig. 7.1. 5-star with four cliques.

Let $A_1, \ldots, A_r$ be alignments for the r cliques in the l-star, with each A_i aligning l sequences. A construction similar to Feng and Doolittle, 1987 implies an analog of lemma 7.1:

Lemma 7.4. *For any l-star and any specified alignments $A_1, \ldots, A_r$ for its cliques, there is an alignment A for the k sequences that is compatible with each of the alignments $A_1, \ldots, A_r$.*

One can generalize the Align algorithm for l-stars and prove analogs of lemmas 7.2 and 7.3. As a result, theorem 7.1 can be generalized for l-stars, thus leading to an algorithm with the performance guarantee equal to the normalized communication cost of l-stars, which is $2 - \frac{l}{k}$. The running time of this algorithm is $O(k(2n)^l |\mathcal{G}|)$. Therefore, the problem is finding a small balanced set of l-stars $\mathcal{G}$.

We have reduced the multiple alignment problem to that of finding an optimal alignment for each clique in each l-star in a balanced set $\mathcal{G}$. How hard is it to find a balanced set $\mathcal{G}$? A trivial candidate is simply the set of all l-stars, which is clearly balanced by symmetry. For $l = 2$, Gusfield, 1993 [144] exploited the fact that there are only k 2-stars to construct a multiple alignment algorithm with an approximation ratio of $2 - \frac{2}{k}$. This is really a special case, as for $l > 2$, the number of l-stars grows exponentially with k, making the algorithm based on generation of all l-stars computationally infeasible. Constructing a *small* balanced set of l-stars *is* not trivial. Pevzner, 1992 solved the case of $l = 3$ by mapping the problem to maximum matching in graphs. Bafna *et al.*, 1997 further designed a $2 - \frac{l}{k}$ approximation algorithm for arbitrary l.

DOT-MATRICES AND IMAGE RECONSTRUCTION

Given k sequences, it is easy to generate $\binom{k}{2}$ pairwise dot-matrices. It is much harder to assemble these pairwise dot-matrices into a k-dimensional dot-matrix and to find regions of similarity shared by all or almost all k sequences. To address this problem, Vihinen, 1988 and Roytberg, 1992 proposed "superimposing" pairwise dot-matrices by choosing one reference sequence and relating all others to it. Below we describe the Vingron and Argos, 1991 algorithm for assembling a k-dimensional dot-matrix from 2-dimensional dot-matrices.

We represent the problem of assembling pairwise similarities in a simple geometric framework. Consider M integer points in k-dimensional space,

$$\left(i_1^1,\dots,i_k^1\right),\dots,\left(i_1^M,\dots,i_k^M\right),$$

for which we do not know the coordinates. Suppose we observe the projections of these points onto each pair of dimensions s and t, $1 \le s < t \le k$:

$$\left(i_s^1,\dots,i_t^1\right),\dots,\left(i_s^M,\dots,i_t^M\right)$$

as well as some other points *(noise)*. Suppose also that we cannot distinguish points representing *real* projections from ones representing noise. The *k-dimensional image reconstruction* problem is to reconstruct M k-dimensional points given $\binom{k}{2}$ projections (with noise) onto coordinates 3 and t for $1 \le s < t < k$.

In this construction, each similarity (consensus element) shared by k biological sequences corresponds to an integer point $(i_1,\dots i_k)$ in k-dimensional space, where i_s is the coordinate of the consensus in the s-th sequence, $1 \le s \le k$. In practice, it is hard to find the integer points $(i_1,\dots i_k)$ corresponding to consensuses. On the other hand, it is easy to find (though with considerable noise) the projections (i_s, i_t) of all consensuses $(i_1,\dots i_k)$ onto every pair of coordinates s and t. This observation establishes the link between the multiple alignment problem and the k-dimensional image reconstruction problem.

From the given dots in the side-planes, we propose to keep only those that fulfill the following criterion of consistency: the point (i, j) in projection s, t is called *consistent* if for every other dimension u there exists an integer m such that (i, m) belongs to the projection s, u and (j, m) belongs to the projection t, u. Obviously each "real" point, i.e., each one that was generated as a projection of a k-dimensional point, is consistent. In contrast, random points representing noise are expected to be inconsistent. This observation allows one to filter out most (though possibly not all) of the noise and leads to the Vingron and Argos, 1991 [344] algorithm that multiplies and compares dot-matrices.

MULTIPLE ALIGNMENT VIA DOT-MATRIX MULTIPLICATION

We model the collection of $\binom{k}{2}$ dot-matrices as a *k-partite graph* $G(V_1 \cup V_2 \cup ... \cup V_k, E)_t$ where V_i is the set of positions in the i-th sequence. We join the vertex $i \in V_s$ with $j \in V_t$ by an (undirected) edge e if there exists a dot at position (i, j) of the dot-matrix comparing sequences s and t. An edge $e \in E$ will be written as $e = (s, i | t, j)$ to indicate that it joins vertices $i \in V_s$ and $j \in V_t$. We denote a *trigle* formed by three edges $(s,i | t, j),(t, j | u, m)$, and $(s, i | u, m)$ as $(s, i | t, j | u, m)$. We now define an edge $(s, i | t, j)$ to be *consistent* if for every $u \neq s. t, 1 \leq u \leq k$ there exists a triangle $(s, i | t, j | u, m)$ for some m. A subset $E' \subseteq E$ is called consistent if for all edges $(s, i(i, j) \in E'$ there exist triangles $(s, i | t, j | u, m), \forall u \neq s, t, 1 \leq u \leq k$, with all edges of these triangles in E'. The k-partite graph G is defined as *consistent* if its edge-set is consistent. Clearly, if $G'(V, E')$ and $G''(V, E'')$ are consistent graphs, then their *union* $G(V, E' \cup E'')$ is consistent. Therefore, we can associate with any k-partite graph a unique maximal consistent subgraph. Our interest is in the following:

Graph Consistency Problem. Find the maximal consistent subgraph of an n-partite graph.

The dot-matrix for sequences s and t is defined as *adjacency* matrix A_{st}:

$$(A_{st})_{ij} = \begin{cases} 1, \text{ if } (s, i | t, j) \in E \\ 0, \text{ otherwise} \end{cases}$$

Each such matrix corresponds to a subset of E and we will apply the operations $\cup$, $\cap$, and $\subset$ to the matrices A_{st}. We reformulate the above definition of consistency in terms of boolean multiplication (denoted by "o") of the adjacency matrices. A k-partite graph is consistent if and only if

$$A_{st} \subseteq A_{su} \circ A_{ut} \quad \forall s\, t, u : 1 \leq s, t, u \leq k, s \neq t \neq u \tag{7.1}$$

Characterization suggests the following simple procedure to solve the consistency problem: keep only those Is in the adjacency matrix that are present both in the matrix itself and in all products A_{su} o A_{ut} Doing this once for all adjacency matrices will also change the matrices used for the products. This leads to the iterative matrix multiplication algorithm (superscripts distinguish different iterations) that starts with the adjacency matrices $A_{st}^{(0)} := A_{st}$ of the given k-partite graph and defines:

$$A_{st}^{(l+1)} = A_{st}^{(l)} \cap \Big(\bigcap_{u \neq s,t} A_{su}^{(l)} \circ A_{ut}^{(l)} \Big)$$

Once this is done for all indices s and t, the process is repeated until at some iteration m $A_{st}^{(l+1)} = A_{st}^{(l)}$ for all $1 \leq s, t \leq k$.

The dot-matrix multiplication algorithm (Vingron and Argos, 1991 [344]) converges to the maximal consistent subgraph and requires $O(L^3k^3)$ time per iteration (L is the length of the sequences). Since the number of iterations may be very large, Vingron and Pevzner, 1995 [345] devised an algorithm running in time $O(L^3k^3)$ overall, which is equivalent to the run-time of only one iteration of the matrix-multiplication algorithm. In practical applications the input data for the algorithm are sparse matrices, which makes the algorithm even faster. Expressed in the overall number M of dots in all $\binom{k}{2}$ dot-matrices, the running time is $O(kLM)$.

SOME OTHER PROBLEMS AND APPROACHES

Multiple Alignment via Evolutionary Trees

It often happens that in addition to sequences $a_1,..., a_k$, biologists know (or assume that they know) the evolutionary history (represented by an *evolutionary tree)* of these sequences. In this case, $a_1,..., a_k$ are assigned to the leaves of the tree, and the problem is to reconstruct the ancestral sequences (corresponding to internal vertices of the tree) that minimize the overall number of mutations on the edges of the tree. The score of an edge in the tree is the edit distance between sequences assigned to its endpoints, and the score of the evolutionary tree is the sum of edge scores over all edges of the tree. The optimal multiple alignment for a given evolutionary tree is the assignment of sequences to internal vertices of the tree that produces the minimum score. Wang et al, 1996 and Wang and Gusfield, 1996 developed performance guarantee approximation algorithms for evolutionary tree alignment.

Cutting Corners in Edit Graphs

Carrillo and Lipman, 1988 and Lipman et al, 1989 suggested a branch-and-bound technique for multiple alignment. The idea of this approach *is* based on the observation that if one of the pairwise alignments imposed by a multiple alignment is bad, then the overall multiple alignment won't have a good score. This observation implies that "good" multiple alignment imposes "good" pairwise alignments, thus limiting a search to the vicinity of a main diagonal in a *k*-dimensional alignment matrix.

COMPUTATIONAL APPLICATIONS OF MODELING

7

Molecular modeling can be defined as an application of computers to generate, manipulate, calculate and predict realistic molecular structures and associated properties. The computational approaches, in particular molecular mechanics, to obtain energetic and structural information for biomolecules are presented. Methodologies for energy calculation/geometry optimization, dynamic simulation and conformational search are discussed. The applications of molecular modeling packages, Chem3D and HyperChem, are described.

INTRODUCTION TO MOLECULAR MODELING

Molecular modeling can be considered as a range of computerized techniques based on theoretical chemistry methods and experimental data that can be used either to analyze molecules and molecular systems or to predict molecular, chemical, and biochemical properties. It serves as a bridge between theory and experiment to:

1. Extract results for a particular model.
2. Compare experimental results of the system.
3. Compare theoretical predictions for the model.
4. Help understanding and interpreting experimental observations.
5. Correlate between microscopic details at atomic and molecular level and macroscopic properties.
6. Provide information not available from real experiments.

Thus molecular modeling can be defined as the generation, manipulation, calculation, and prediction of realistic molecular structures and associated physicochemical as well as biochemical properties by the use of a computer. It is primarily a mean of communication between scientist and computer, the imperative interface between human-comprehensive symbolism, and the mathematical description of the molecule. The endeavor is made to perceive and recognize a molecular structure from its symbolic representations with a computer. Thus functions of the molecular modeling include:

- **Structure retrieval or generation:** Crystal structures of organic compounds can be found in the Cambridge Crystallographic Datafiles. Those that does not exist may be generated by 3D rendering software. The 3D structural coordinates of biomacromolecules can be retrieved from Protein Data Bank.
- **Structural visualization:** Computer graphics is the most effective means for visualization and interactive manipulation of molecules and molecular systems. Numerous software

programs (e.g., Cn3D, RasMol and KineMage, are available for visualization, management, and manipulation of molecular structures.

- **Energy calculation and minimization:** One of the fundamental properties of molecules is their energy content and energy level. Three major theoretical computational methods of their calculation include empirical (molecular mechanics), semiempirical, and *ab initio* (quantum mechanics) approaches. Energy minimization results in geometry optimization of the molecular structure.
- **Dynamics simulation and conformation search:** Solving motion of nuclei in the average field of the electrons is called *quantum dynamics*. Solution to the Newton's equation of motion for the nuclei is known as *molecular dynamics*. Integration of Newton's equation of motion for all atoms in the system generates molecular trajectories. *Conformation search* is carried out by repeating the process by rotating reference bonds (dihedral angles) of the molecule under investigation for finding lowest energy conformations of molecular systems.
- **Calculation of molecular properties:** Methods of estimating or computing properties (i.e., interpolating properties, extrapolating properties, and computing properties). Some computing properties are boiling point, molar volume, solubility, heat capacity, density, thermodynamic quantities, molar refractivity, magnetic susceptibility, dipole moment, partial atomic charge, ionization potential, electrostatic potential, van der Waals surface area, and solvent accessible surface area.
- **Structure superposition and alignment:** Computing activities and properties of molecules often involves comparisons across a homologous series. Such techniques require superposition or alignment of structures.
- **Molecular interactions, docking:** The intermolecular interaction in a ligand-receptor complex is important and requires difficult modeling exercises. Usually the receptor (e.g., protein) is kept rigid or partially rigid while the conformation of ligand molecule is allowed to change.

The advent of high-speed computers, availability of sophisticated algorithms, and state-of-the-art computer graphics have made plausible the use of computationally intensive methods such as quantum mechanics, molecular mechanics, and molecular dynamics simulations to determine those physical and structural properties most commonly involved in molecular processes. The power of molecular modeling rests solidly on a variety of well-established scientific disciplines including computer science, theoretical chemistry, biochemistry, and biophysics. Molecular modeling has become an indispensable complementary tool for most experimental scientific research.

Computational biochemistry and computer-assisted molecular modeling have rapidly become a vital component of biochemical research. Mechanisms of ligand-receptor and enzyme-substrate interactions, protein folding, protein-protein and protein-nucleic acid recognition, and *de novo* protein engineering are but a few examples of problems that may be addressed and facilitated by this technology.

ENERGY MINIMIZATION, DYNAMICS SIMULATION, AND CONFORMATIONAL SEARCH

Energy Calculation

Computational approaches to potential energy may be divided into two broad categories: quantum mechanics and molecular mechanics. The basis for this division depends on the incorporation of the Schrodinger equation or its matrix equivalent. It is now widely recognized that both methods reinforce one another in an attempt to understand chemical and biological behavior at the molecular level. From a purely practical standpoint, the complexity of the problem, time constraints, computer size, and other limiting factors typically determine which method is feasible. Quantum mechanics (QM) can be further divided into *ab initio* and semiempirical methods. The *ab initio* approach uses the Schrodinger equation as the starting point with post-perturbation calculation to solve electron correlation.

Various approximations are made that the wave function can be described by some functional form. The functions used most often are a linear combination of Slater-type orbitals (STO), exp($-ax$), or Gaussian-type orbitals (GTO), exp($-ax^2$). In general, *ab initio* calculations are iterative procedures based on self-consistent field (SCF) methods. Self-consistency is achieved by a procedure in which a set of orbitals is assumed and the electron-electron repulsion is calculated. This energy is then used to calculate a new set of orbitals, and these in turn are used to calculate a new repulsion energy.

The process is continued until convergence occurs and self-consistency is achieved. On the other hand, the term semiempirical is usually reserved for those calculations where families of difficult-to-solve integrals are replaced by equations and parameters that are fitted to experimental data. Semiempirical methods describe molecules in terms of explicit interactions between electrons and nuclei and are based on the principles:

- Nuclei and electrons are distinguished from each other.
- Electron-electron and electron-nuclear interaction are explicit.
- Interactions are governed by nuclear and electron charges—that is, potential energy and electron motions.
- Interactions determine the space distribution of nuclei and electrons and their energies.

For the best result, the molecule being computed should be similar to molecules in the database used to parameterize the method. However, if the molecule being computed is significantly different from anything in the parameterization set, erratic results may be obtained. Semiempirical calculations have been successful in dealing with organic compounds. For large biomolecules, semiempirical calculations cannot be applied effectively; in these cases the methods referred to as molecular mechanics (MM) can be used to model their structures and behaviors.

Molecular mechanics utilizes simple algebraic expressions for the total energy of a compound without computing a wave function or total electron density. The fundamental assumption of MM or its tool, empirical force field (EFF or simply force field, FF), is that data determined experimentally for small molecules can be extrapolated to larger molecules. It is aimed at quickly providing energetically favorable conformations for large systems. Molecular mechanical method is based on the following principles:

1. Nuclei and electrons are lumped together and treated as unified atom-like particles.
2. Atom-like particles are treated as spherical balls.
3. Bonds between particles are viewed as springs.
4. Interactions between these particles are treated using potential functions derived from classical mechanics.

5. Individual potential functions are used to describe different types of interactions.
6. Potential energy functions rely on empirically derived parameters that describe the interactions between sets of atoms.
7. The potential functions and the parameters used for evaluating interactions are termed a force field.
8. The sum of interactions determines the conformation of atom-like particles.

Therefore, MM energies have no meaning as absolute quantities. They are used for comparing relative strain energy between two or more conformations.

In molecular mechanical calculations, the force fields generally take the form

$$E_{\text{total}} = E_r + E_\theta + E_\phi + E_{nb} + [\text{special terms}]$$

in which the successive terms, expressing the total energy (E_{total}), are energies associated with bond stretching (E_r), bond angle bending (E_θ), bond torsion (E_ϕ), nonbond interactions (E_{nb}), plus specific terms such as hydrogen bonding (E_{hb}) in biochemical systems. Most MM equations are similar in the types of terms they contain. There are some differences in the forms of the equations that can affect the choice of FF and parameters for the systems of interest. Examples are (*a*) MM2/3 for organic compounds and (*b*) AMBER or CHARMm for biological molecules.

For most FF, the internal energy terms are similar, namely,

$$E_r = \Sigma K_r\ (r - r_0)^2$$

$$E_\theta = \Sigma K_\theta\ (\theta - \theta_0)^2$$

and

$$E_\phi = \Sigma K_\theta\ [1 + \cos\ (n\phi - \phi_0)]$$

where K_r, K_ϕ, and K_θ are force constants for bond, angle, and dihedral angle, respectively. r_0, θ_0, and ϕ_0 define the equilibrium distance, equilibrium angle, and phase angle for the given type. n is the periodicity of the Fourier term. These parameter values are derived from model molecules and vary among different FF. However, different potential functions may be used by different FF for E_{nb} and special terms such as E_{hb}. Lennard-Jones 6-12 potential is the most commonly used for van der Waals interactions (E_{vdw}) in E_{nb}, such that

$$E_{vdw} = \Sigma\Sigma(A_{ij} / r_{ij}^{12} - B_{ij} / r_{ij}^{6})$$

Because biochemical molecules are often charged, an electrostatic energy (E_{elec}) term is added to E_{nb}:

$$E_{elec} = \Sigma\Sigma(q_i q_j)/(Dr_{ij})$$

or separately to take into account the interactions between nonbound but interacting atoms i and j with the distance of r_{ij}. A_{ij} and B_{ij} are van der Waal parameters. D is a molecular dielectric constant (vary from 1 *in vaccuo* to 80 in water) that accounts for the environmental attenuation of electrostatic interaction between the two atoms with the point charge q_i and q_j. The hydrogen bonding E_{hb} term differs. Of the two most commonly used force fields in biochemistry, AMBER introduces the 10-12 potential, that is,

$$E_{hb} = \Sigma\Sigma(C_{ij} / r_{ij}^{12} - D_{ij} / r_{ij}^{10})$$

while CHARMm consider both the distance and angle of the hydrogen bond interactions among three atoms (A for acceptor, H for hydrogen, and D for donor). You can choose to calculate all nonbonded interactions or to truncate (cut off) the nonbonded interaction calculations using a switched or shifted function. Useful guidelines for nonbonded interactions are as follows:

- Calculate all nonbonded interactions for small and medium-sized molecules.
- Use either switched function or shifted function to decrease computing time for macromolecules such as proteins and nucleic acids.
- Switched function is a smooth function, applied from the inner radius (K_{on}) to the outer radius (K_{off}), that gradually reduces nonbonded interaction to zero. The suggested outer radius is approximately 14 Å, and the inner radius is approximately 4 Å less than the outer radius.
- Shifted function is smooth function, applied over a whole nonbonded distance, from zero to outer radius, that gradually reduces nonbonded interaction to zero.

Thus different FF are designed for different systems and purposes. The databases used also differ. The users of MM software should be aware of these differences because a particular FF may work extremely well within one molecular structure class but may fail when applied to other types of structures.

Molecular mechanics is an extremely widely used method for generating molecular models for a multitude of purposes in chemistry and biochemistry. The first major reason for the popularity of MM is its speed, which makes it computationally feasible for routine usage. The alternative methods for generating molecular geometries, such as *ab initio* or semiempirical molecular orbital calculations, consume much larger amounts of computer time, making them much more expensive to use.

The economy of MM makes studies of relatively large molecules such as biomacromolecules feasible on a routine basis. Thus, MM has become a primary tool of computational biochemists. MM is relatively simple to understand. The total strain energy is broken down into chemically meaningful components that correspond to an easily visualized picture of molecular structure. Molecular mechanics also has some limitations. The potential pitfall to be aware of is that MM routines will generate a conformation for which the strain energy is minimized.

However, the minimum found during a calculation may not be the global minimum. It is relatively easy for the procedure to become trapped in a local energy minimum. There are schemes for minimizing the risk of such entrapment in local minima. For example, the calculation can be done a number of times starting from different initial geometries to see if the final geometry found remains the same. Another obvious drawback of the MM approach is that it cannot be used to study any molecular system where electronic effects are dominant. Here, quantum mechanical approaches that explicitly account for the electrons in molecules must be used. To study electronic behavior in biomolecules, QM and MM are combined into one calculation (QM/MM) that models a large molecule (e.g., enzyme) using MM and one crucial section of the molecule (e.g., active site) with QM. This is designed to give results that have good speed where only the region needs to be modeled quantum mechanically. Normally, a single-point energy calculation is for stationary point on a potential energy surface.

The calculation provides an energy and the gradient of that energy. The gradient is the root-mean-square (RMS gradient) of the derivative of the energy with respect to Cartesian coordinates, that is,

$$\text{RMS Gradient} = (3N)^{-1} [\Sigma(\delta E/\delta X)^2 + (\delta E/\delta Y)^2 + (\delta E/\delta Z)^2]^{1/2}$$

At a minimum, the gradient is zero. Thus the size of the gradient can provide qualitative information to assess if a structure is close to a minimum.

Energy Minimization and Geometry Optimization

The basic task in the computational portion of MM is to minimize the strain energy of the molecule by altering the atomic positions to optimal geometry. This means minimizing the total nonlinear strain energy represented by the FF equation with respect to the independent variables, which are the Cartesian coordinates of the atoms. The following issues are related to the energy minimization of a molecular structure:

- The most stable configuration of a molecule can be found by minimizing its free energy, G.
- Typically, the energy E is minimized by assuming the entropy effect can be neglected.
- At a minimum of the potential energy surface, the net force on each atom vanishes, therefor the stable configuration.
- Because the energy *zero* is arbitrary, the calculated energy is relative. It is meaningful only to compare energies calculated for different configurations of chemically identical systems.
- It is difficult to determine if a particular minimum is the *global minimum,* which is the lowest energy point where force is zero and second derivative matrix is positive definite. *Local minimum* results from the net zero forces and positive definite second derivative matrix, and *saddle point* results from the net zero forces and at least one negative eigenvalue of the second derivative matrix.

The most widely used methods fall into two general categories: (1) steepest descent and related methods such as conjugate gradient, which use first derivatives, and (2) Newton-Raphson procedures, which additionally use second derivatives. The steepest descent method depends on (1) either calculating or estimating the first derivative of the strain energy with respect to each coordinate of each atom and (2) moving the atoms. The derivative is estimated for each coordinate of each atom by incrementally moving the atom and storing the resultant strain energy change. The atom is then returned to its original position, and the same calculation is repeated for the next atom.

After all the atoms have been tested, their positions are all changed by a distance proportional to the derivative calculated in step 1. The entire cycle is then repeated. The calculation is terminated when the energy is reduced to an acceptable level. The main problem with the steepest descent method is that of determining the appropriate step size for atom movement during the derivative estimation steps and the atom movement steps.

The sizes of these increments determine the efficiency of minimization and the quality of the result. An advantage of the first-derivative methods is the relative ease with which the force field can be changed. The conjugate gradient method is a first-order minimization technique. It uses both the current gradient and the previous search direction to drive the minimization. Because the conjugated gradient method uses the minimization history to calculate the search direction and contains a scaling factor for determining step size, the method converges faster and makes the step sizes optimal as compared to the steepest descent technique. However, the number of computing cycles required for a conjugated gradient calculation is approximately proportional to the number of atoms (N), and the time per cycle is proportional to N^2.

The Fletcher-Reeves approach chooses a descent direction to lower energy by considering the current gradient, its conjugate, and the gradient for the previous step. The Polak-Ribiere algorithm improves on the Fletcher-Reeves approach by additional consideration of the previous conjugate and tends to converge more quickly. The Newton-Raphson methods of energy minimization utilize the curvature of the strain energy surface to locate minima.

The computations are considerably more complex than the first-derivative methods, but they utilize the available information more fully and therefore converge more quickly. These methods involve setting up a system of simultaneous equations of size $(3N - 6)$ $(3N - 6)$ and solving for the atomic positions that are the solution of the system. Large matrices must be inverted as part of this approach. The general strategy is to use steepest descents for the first 10-100 steps (500-1000 steps for proteins or nucleic acids) and then use conjugate gradients or Newton-Raphson to complete minimization for convergence (using RMS gradient or/and energy difference as an indicator).

For most calculations, RMS gradient is set to 0.10 (you can use values greater than 0.10 for quick, approximate calculations). The calculated minimum represents the potential energy closest to the starting structure of a molecule. The energy minimization is often used to generate a structure at a stationary point for a subsequent single-point calculation or to remove excessive strain in a molecule, preparing it for a molecular dynamic simulation.

Dynamics Simulation

Molecules are dynamic, undergoing vibrations and rotations continually. The static picture of molecular structure provided by MM therefore is not realistic. Flexibility and motion are clearly important to the biological functioning of proteins and nucleic acids. These molecules are not static structures, but exhibit a variety of complex motions both in solution and in the crystalline state. The most commonly employed simulation method used to study the motion of protein and nucleic acid on the atomic level is the *molecular dynamics* (MolD) method. It is a simulation procedure consisting of the computation of the motion of atoms in a molecule according to Newton's laws of motion.

The forces acting on the atoms, required to simulate their motions, are generally calculated using molecular mechanics force fields. Rather than being confined to a single low-energy conformation, MolD allows the sampling of a thermally distributed range of intramolecular conformation. Molecular dynamics calculations provide information about possible conformations, thermodynamic properties, and dynamic behavior of molecules according to Newtonian mechanics. A simulation first determines the force on each atom (F_t) as the function of time, equal to the negative gradient of the potential energy (V) with respect to the position (x_i) of atom i:

$$F_i = \delta V / \delta x_i$$

The acceleration a_i of each atom is determined by

$$a_i = F_i / m_i$$

The change in velocity v_i, is equal to the integral of acceleration over time. In MolD, one numerically and iteratively integrates the classical equations of motion for every explicit atom N in the system by marching forward in time by tiny time increments, Δt. A number of algorithms exist for this purpose and the simplest formulation is shown below:

$$x_i(t + \Delta t) = x_i(t) + v_i(t)\Delta t$$

$$\mathrm{v}_i(t + \Delta t) = v_i(t) + a_i(t)\Delta t = v_i(t) + \{F(x_1 \ldots x_N, t)/m\}\Delta t$$

The kinetic energy *(K)* is defined as

$$K = 1/2\Sigma m_i v_i$$

The total energy of the system, called the Hamiltonian (*H*), is the sum of the kinetic (*K*) and potential *(V)* energies:

$$\mathrm{H}(r, p) = K(p) + V(r)$$

where *p* is the momenta of the atoms and *r* is the set of Cartesian coordinates.

The time increment must be sufficiently small that errors in integrating 6*N* equations (3*N* velocities and 3*N* positions) are kept manageably small, as manifested by conservation of the energy. As a result, Δt must kept on the order of femtoseconds (10^{-15} *S*). Furthermore, because the forces *F* must be recalculated for every time step, MolD is a computation intensive task. Thus, the overall time scale accessible to MolD calculations is on the order of picoseconds (1 ps = 1 × 10^{-12} s). The molecular simulation approximates the condition in which the total energy of the system does not change during the equilibrium simulation.

One way to test for success of a dynamics simulation and the length of the time step is to determine the change in kinetic and potential energies between time steps. In the microcanonical ensemble (constant number, volume, and energy), the change in kinetic energy should be of the opposite sign and exact magnitude as the potential energy. The MolD normally consists of three phases: heating, equilibration, and cooling.

To perform MolD, the structure is submitted to a minimization procedure to relieve any strain inherent in the starting positions of the atoms. The next step is to assign velocities to all the atoms. These velocities are drawn from a low-temperature Maxwellian distribution. The system is then equilibrated by integrating the equations of motion while slowly raising the temperature and adjusting the density. The temperature is raised by increasing the velocities of all of atoms. There is a simple analytical function expressing the relationship between kinetic energies of the atoms and the temperature of the system:

$$T(t) = 1/\{k_B(3N - n)\}\sum_{i=1}^{N} m_i \mid v_i \mid$$

where T(*t*) = temperature of the system at time *t*

(3*N* — *n*) = number of degrees of freedom in the system

v_i = velocity of atom *i* at time *t*

k_B = Boltzmann constant

m_i = mass of atom *i*

N = number of atoms in the system

This process of raising the temperature of the system will cover a time interval of 10-50 ps. The period of heating to the temperature of interest is followed by a period of equilibration with no temperature changes. The stabilization period will cover another time interval of 10-50 ps. The mean kinetic energy of the system is monitored; and when it remains constant, the system is ready for study. The structure is in an equilibrium state at the desired temperature. The MolD experiment consists of allowing the molecular system to run free for a period of time, saving all the information about the atomic positions, velocities, and other variables as a function of time. This (voluminous) set of data is called *trajectory*.

The length of time that can be saved during a trajectory sampling is limited by the computer time available and the speed of the computer. Once a trajectory has been calculated, all the equilibrium and dynamic properties of the system can be calculated from it. Equilibrium properties are obtained by averaging over the property during the time of the trajectory. Plots of the atomic positions as a function of time schematically depict the degree to which molecules are moving during the trajectory. The RMS fluctuations of all of the atoms in a molecule can be plotted against time to summarize the aggregate degree of fluctuation for the entire structure.

The methods of MolD are becoming an important component for the study of protein structures in an effort to rationalize structural basis for protein activity and function. Molecular dynamics simulations are efficient for searching the conformational space of medium-sized molecules, especially ligands in free and complexed states. Quenched dynamics is a combination of high-temperature molecular dynamics and energy minimization. For a conformation in a relatively deep local minimum, a room temperature molecular dynamics simulation may not overcome the barrier.

To overcome barriers, conformational searches use elevated temperature (> 600 K) at constant energy. To search conformational space adequately, simulations are run for 0.5-1.0 ps each at high temperature. For a better estimate of conformations the quenched dynamics should be combined with simulated annealing, which is a cooling simulation. Cooling a molecular system after heating or equilibration can (1) reduce stress on molecules caused by a simulation at elevated temperatures and take high-energy conformational states toward stable conformations and (2) overcome potential energy barriers and force a molecule into a lower energy conformation. Quenched dynamics can trap structures in local minimum.

The molecular system is heated to elevated temperatures to overcome potential energy barriers and then cooled slowly to room temperature. If each structure occurs many times during the search, one is assured that the potential energy surface of that region has been adequately sought. The molecular dynamics is useful for calculating the time-dependent properties of an isolated molecule. However, molecules in solution undergo collisions with other molecules and experience frictional forces as they move through the solvent.

Langevin dynamics simulates the effect of molecular collisions and the resulting dissipation of energy that occur in real solvents without explicitly including solvent molecules by adding a frictional force (to model dissipative losses) and a random force (to model the effect of collisions) according to the Langevin equation of motion:

$$a_i = F_i/m_i - \gamma v_t + R_i/m_i$$

where γ is the friction coefficient of the solvent and K, is the random force imparted to the solute atom by the solvent. The friction coefficient determines the strength of the viscous drag felt by atoms as they move through the medium, and y is the friction coefficient related to the diffusion constant (D) of the solvent by $\gamma = k_B T/mD$. At low values of the friction coefficient, the dynamic aspects dominate and Newtonian mechanics is recovered as $\gamma \rightarrow 0$. At high values of γ, the random collisions dominate and the motion is diffusion-like. Monte Carlo simulations are commonly used to compute the average ther-modynamic properties of a molecule or a molecular system especially the structure and equilibrium properties of liquids and solutions. They have also been used to conduct conformational searches under nonequilibrium conditions.

Unlike MolD or Langevin dynamics, which calculate ensemble averages by calculating averages over time, Monte Carlo calculations evaluate ensemble averages directly by sampling configurations from the statistical ensemble. To generate trajectories that sample commonly occurring configurations, the Metropolis method is generally employed. Thermodynamically, the probability of finding a system in a state with ΔE above the ground state is proportional to $\exp(-\Delta E//kT)$. Thus, if the energy change associated with the random movement of atoms is negative, the move is accepted. If the energy change is positive, the move is accepted with probability $\exp(-\Delta E/kT)$.

Conformational Search

Conformational search is a process of finding low-energy conformations of molecular systems by varying user-specified dihedral angles. The method involved variation of dihedral angles to generate new structures and then energy minimizing each of these angles. Low-energy unique conformations are stored while high-energy duplicate structures are discarded. Because molecular flexibility is usually due to rotation of unhindered bond dihedral with little change in bond lengths or bond angles, only dihedral angles are considered in the conformational search. Its goal is to determine the global minimum of the potential energy surface of a molecular system. Several approaches have been applied to the problem of determining low-energy conformations of molecules. These approaches generally consist of the following steps with differences in details:

1. **Selection of an initial structure:** The initial structure is the most recently accepted conformation (e.g., energy minimized structure) and remains unchanged during the search. This is often referred to as a *random walk* scheme in Monte Carlo searches. It is based on the observation that low-energy conformations tend to be similar, therefore starting from an accepted conformation tends to keep the search in a low-energy region of the potential surface. An alternative method, called the *usage-directed method,* seeks to uniformly sample a low-energy region by going through all previously accepted conformations while selecting each initial structures. Comparative studies have found the usage-directed scheme to be superior for quickly finding low-energy conformations.
2. **Modification of the initial structure by varying geometric parameters:** The variations can be either systematic or random. Systematic variations can search the conformational space exhaustively for low-energy conformations. However, the number of variations becomes prohibitive except for the simplest systems. One approach to reduce the dimensionality of systematic variation is to first exhaust variations at a low resolution, then exhaust the new variations allowed by successively doubling the resolution. Random variations choose a new value for one or more geometric parameters from a continuous range or from sets of discrete values. To reduce the number of recurring conformations, several random variations have some sort of quick comparison with the sets of previous structures prior to performing energy minimization of the new structure.
3. **Geometry optimization of the modified structure to energy-minimized conformations:** The structures generated by variations in dihedral angles are energy-minimized to find a local minimum on the potential surface. Although the choice of optimizer (minimizer) has a minor effect on the conformational search, it is preferable to employ an optimizer that converges quickly to a local minimum without crossing barrier on the potential surface.
4. **Comparison of the conformation with those found previously:** The conformation is accepted if it is unique and its energy satisfies a criterion. Two types of criteria are used to decide an acceptance of the conformation. Firstly, geometric comparisons are made with previously accepted conformations to avoid duplication. Conformations are often compared by the maximum deviation of torsions or RMS deviation for internal

coordinates, interatomic distances, or least-squares superposition of conformers. Because geometry optimization can invert chiral centers, the chiral centers of the modified structures should be checked after the energy minimization. Secondly, the energetic test for accepting a new conformer may be carried out by a simple cutoff relative to the best energy found so far or a Metropolis criterion where higher-energy structures are accepted with a probability determined by the energy difference and a temperature, for example, $\exp(-\Delta E//kT)$.

COMPUTATIONAL APPLICATION OF MOLECULAR MODELING PACKAGES

Published FF parameters for MM and software for molecular modeling have been compiled. Some of the MM programs applicable to molecular modeling of biomolecules are listed in Table 7.1.

All of these programs a e available for Unix operating system. Most of the Windows versions are incorporated into commercial molecular modeling packages such as MM in Chem3D of CambridgeSoft (http://www.camsoft.com), AMBER and CHARMm in HyperChem of HyperCube (http://www.hyper.com), and SYBYL in PC Spartan of WaveFunction (http://www.wavefun.com). Online servers such as SWEET (http://dkfz-heidelberg.de/spec/sweet2/doc/mainframe.html) performs MM2/3, AMBER server (http://www.ambcr.ucsf.edu/amber/amber html) conducts *in vacuo* minimization with AMBER 5.0 and then electrostatic solvation with AMBER 6.0. B server (http://www.scripps.edU/nwhite/B/indexFrames.html) implements AMBER, and Swiss-Pdb Viewer (http://www.expasy.ch/spdbv/mainpage. html) executes GROMOS.

TABLE 7.1. Molecular Modeling Programs of Biochemical Interest

MM/FF program	*Source*	*Reference*
AMBER	University of California, San Francisco/HyperChem	Weiner *et al.* (1984)
CHARMM	Harvard University/Accelrys, Inc.	Brooks *et al.* (1983)
ECEPP	Cornell University	Nemathy *et al.* (1983)
GROMOS	University of Groningen/Biomos	Herman *et al.* (1984)
SYBYL	Tripos, Inc.	Clark *et al.* (1989)
MM2/3	University of Georgia/Chem3D	Allinger *et al.* (1989)
MACROMODEL	Columbia University	Chemistry, Columbia University
OPLS	Yale University/HyperChem	Jorgensen and Tirado Rives (1988)

Carbohydrate Modeling at SWEET

The online Sweet is a program for constructing 3D models of saccharides (Bohne *et al.* 1998; Bohne *et al.* 1999) with an extension to peptides and other simple biomolecules. To perform energy minimization of monosaccharides such as α-D-mannose, α-L-fructose, β-D-galactose, α-D-glucose, β-D-glucosamine, and their oligomers:

- Select Work page of beginner version to open the request form.
- Select monosaccharide unit(s) and glycosidic linkage(s) from the pop-up lists.
- Click the Send button to open the Result table.
- Choose desired tools to view the saccharide structure and save the structure as PDB file.
- Select Method options with Full MM3(96) parameters and Gradient with 1.0 (default) of Optimize tool.
- Click the Optimize button.

To perform energy minimization of custom-made saccharides, peptides and other biomolecules (a list of templates is available from http://www.dkfz-heidleberg.de/spec/sweet2/doc/input/liblist.html, but not all templates have parameters for minimization):

- Select expert version to open the request.
- Enter the compound name (e.g., a-D-Glcp for α-D-glucopyranose, β-D-Fruf for β-D-fructofuranose, three-letter symbols for amino acids) and the linkage (e.g., 1-4 for 1,4-glycosidic and N-C for peptide linkages).
- Click the Send button to access the Result table.
- From the Optimize tool, select Full MM3(96) parameters from Method option and 1.0 (default) from Gradient option.
- Click the Optimize button to access the Result table from which you may view and save the structure.

Folding of Nucleic Acids at mfold

RNA molecules form secondary structure by folding their polynucleotide chains via hydrogen bond formations between AU pairs and GC pairs. The thermochemical stability of forming such hydrogen bonds provides useful criterion for deducing the cloverleaf secondary structure of tRNAs; that is, tRNA molecules are folded into DH arm, φTC arm, anticodon arm, and extra loop (for some) with the hydrogen bonded stems. Some of the unpaired bases in these arms are hydrogen-bonded in the tertiary structures. The nucleotide sequence of RNA in fasta format can be submitted to RNA mfold server with up to 500 bases for an immediate job and 3000 for batch submission.

Enter the sequence name, paste the sequence in the query box, select or accept default options, and click the Submit button. The results (thermochemical data in text and plot files, and structure files in various formats including PostScrip) are returned (Figure 7.1). Analogous folding prediction for DNA is available at http://bioinfo.math.rpi.edu/ ~ mfold/dna/forml.cgi

Application of Chem 3D

Chem3D is the molecular modeling software for desktop computer marketed by CambridgeSoft Corp. as a component of the ChemOffice suite (ChemDraw, Chem3D and ChemFinder). It converts 2D structures to 3D renderings and imports PDB, MSI ChemNote, Mopac, and Gaussian files. The program displays molecular surfaces, orbitals, electrostatic potential, and charge densities, and it performs energy calculation with MM2 force field. The *ultra* version includes MOPAC, which calculates transition state geometries and physical properties using AMI, PM3, and MNDO's. A conformational search program, Conformer (Princeton Simulations), can be integrated with Chem3D to perform conformational search and analysis. User's guides for ChemDraw and Chem3D should be consulted.

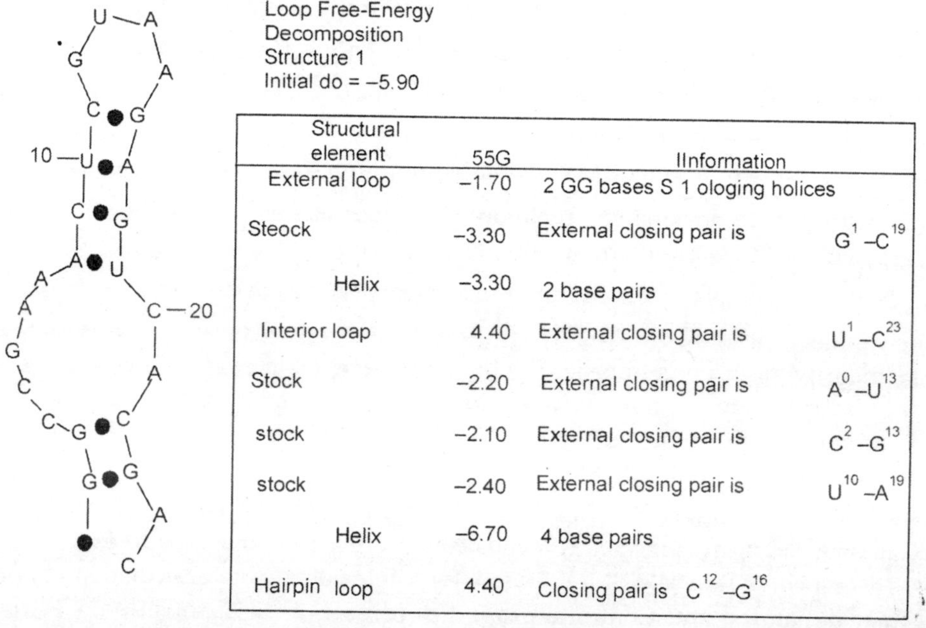

Loop Free-Energy Decomposition
Structure 1
Initial do = –5.90

Structural element	55G	IInformation	
External loop	–1.70	2 GG bases S 1 ologing holices	
Steock	–3.30	External closing pair is	$G^1 – C^{19}$
Helix	–3.30	2 base pairs	
Interior loap	4.40	External closing pair is	$U^1 – C^{23}$
Stock	–2.20	External closing pair is	$A^0 – U^{13}$
stock	–2.10	External closing pair is	$C^2 – G^{13}$
stock	–2.40	External closing pair is	$U^{10} – A^{19}$
Helix	–6.70	4 base pairs	
Hairpin loop	4.40	Closing pair is $C^{12} – G^{16}$	

Fig. 7.1. RNA fold to form secondary structure. The nucleotide sequence of ribozyme (B chain of URX057 from NDB) is submitted to RNA mfold server for fold analysis. The output includes computed structure (as shown) and thermodynamic data in text (as shown) and dot plot (not shown).

The opening window of Chem3D consists of the workspace (display window where 3D structures are displayed with rotation bar, slider knob, and action buttons), the menu bar (File, Edit, View, Tools, Object, Analyze, MM2, Gaussian, MOPAC, and Window menus), tool pallette (action icons for the cursor), and replacement text box (element, label, or structure name typed in this box is converted to chemical structure). Structure file in *.mol, *.pdb, or *.sml can be opened and saved from the File menu. (*Note:* PDB files saved from Chem3D do not contain residue IDs.) The accompanying program, ChemDraw, draws 2D structures (.cdx) that are converted into 3D models (.c3d) by Chem3D. The molecular sketches from ISIS Draw (*.skc) have to be converted to *.cdx with ChemDraw for the 3D conversion.

To draw biomolecules using ready-made substructures of Chem3D, go to the View menu (to view tables of topologies, parameters, force fields used in the program, and substructures for constructing 3D) and then to the substructure table (*.TBL) to select the desired substructures. Copy and paste the substructures on the display window for subsequent modeling. Alternatively, 3D models can be built by entering substructure names into the Replacement text box (upper left-hand corner of display window), such as HSerThreAlaAsnLeuGluTyrOH for heptapeptide, STANLEY. Invoke Tools → Clean up structure to quickly correct unrealistic bond lengths and angles.

The Tools menu contains commands for manipulating the displayed structures such as Fit, Move, Reflect, Invert, Dock, and Overlay. To dock two interacting structures,

- Copy the first structure into clipboard.

- Open the second structure on the display window and paste the first structure upon the second.
- Select two atoms (one each from the two structures) as the first interaction pair, and set the interaction distance between them (Open the measurement table via Object → Set distance, and enter the desired interaction distance in the Optimal cell).
- Repeat the procedure (setting interaction distance between the two interacting atoms) for, at least, four interaction pairs in order to achieve reasonable dock.
- Invoke Dock command from the Tools menu.
- Set the values of the minimum RMS error and minimum RMS gradient to 0.01.
- Click Start to initiate docking computation, which terminates when either RMS error or RMS gradient becomes less than the set values.
- Compute the MM2 energy (MM2 → Minimize energy) and save the file.

Follow the similar procedures to overlay two structures for structural comparison. Specify, at least, three similarity atom pairs in order to obtain the acceptable overlay. It is not necessary to give the bond distance between the two atoms in the pair (assume to be zero) for the overlay. Choose Overlay command from the Tools menu to set both RMS error and gradient to 0.01 and click start to initiate the overlay computation.

The Analyze menu provides commands for measuring geometries. The MM2, MOPAC, and Gaussian (if Gaussian program is installed) menus contain related commands to execute energy computations. These include Run MM2/MOPAC job (for single point energy calculation), Minimize energy, Molecular dynamics, and Compute properties (dipole, charge, solvation, electrostatic potential, polarizability, *etc.* by Mopac) commands. For energy minimization,

- Choose Minimize energy command from the MM2 menu to open the dialog box.
- With Job type tab active, select minimize energy, display options and set minimum RMS to 0.100.
- Click Run to initiate minimization.
- At the end of run, various energies (stretch, bend, torsion, van der Waals, and dipole-dipole) and total energy is displayed.
- Save the record and structure as emstruct.c3d.

For dynamic simulation,

- Choose Molecular dynamics command from the MM2 menu to open the dialog box.
- With Dynamics tab active, enter simulation parameters (default settings: step interval, 2.0 fs; frame interval, 10 fs; terminate after 1000 steps, heating/cooling rate, 1.0 kcal/atom/ps; and target temperature, 300 K).
- Click Run button to initiate the dynamic simulation.
- The simulation is terminated when the target temperature is reached. You can view the energetics of the simulation from the message window and the trajectories (structural frames) by moving the slider knob at the bottom of the model window.
- Save the simulation record and trajectories as dynamics.c3d.
- To save an individual trajectory, pick the desired frame, choose Edit → Select all, and save as (File → Save as) trajname.pdb.

Application of HyperChem

HyperChem is the PC-based molecular modeling and simulation software package marketed by Hypercube Inc. The program provides molecular mechanics (with MM + , AMBER, BIO + (CHARMm), and OPLS force fields), semiempirical (extented Huckel, CNDO, INDO, MINDO3, MNDO, AM1, PM3, and ZINDO), and *ab initio* quantum mechanics calculations. Computations are carried out for single-point energy, geometry optimization (energy minimization), molecular dynamics, Langevin dynamics, Monte Carlo simulation, and conforma-tional search. The accompanied manuals should be consulted for various applications of HyperChem. The HyperChem window displays the menu bar (File, Edit, Build, Select, Display, Databases, Setup, Compute, Annotation, Cancel, and Scipt menus), tool bar (draw, select, display, move and shortcut icons), Workspace, and status line.

The program supports various 2D and 3D files including HyperChem (*.hin), PDB (pdb files in .ent), MDL (*.mol), Tripos (*.mlz) files. ISISDraw (*.skc) can also be opened or saved from the File menu. The 2D sketch can be converted into a 3D structure (and calculation of atom types for MM) by selecting Model build from the Build menu. The Build menu also provides tools for building the structure. United atoms tool simplifies a molecular structure and calculations by including hydrogen atoms in the definition of carbon atoms. HyperChem uses atom types for molecular mechanical calculations. The atom types can be calculated (Calculate types) or changed (Set atom type).

The Edit menu contains items for manipulating displayed structure while the Display menu determines the appearance of molecules in the workspace (e.g., Show selection, Rendering, Overlay, Show isosurface, Show periodic box, Show multiple bonds, Show hydrogen bonds, Recompute H bonds, add Labels, and change Color). Rendering displays model rendering in sticks (stereo, ribbons, and wedges), balls, balls and cylinders, overlapping spheres, dots, and sticks and dots. Two selected molecules can be placed on top of one another by using Overlay tool of the Display menu. The Select menu enables the selection of Atoms, Residues, Molecules, and Spheres that encompasses all atoms within a 3D sphere plus all atoms with/without 2D rectangle.

The Setup and Compute menus contain tools for carrying out chemical calculations. For molecular mechanics energy computation,

- Select atoms or residues to be included in the calculation (default is the whole molecule).
- Choose Start Log command from the File menu if you want to store energy calculations in a log file (chem.log as default).
- Choose force field (MM + , AMBER, BIO+ (CHARMm), or OPLS) from the Setup menu.
- Click the Options button to open the option dialog box.
- For MM+ (energy calculations of small biomolecules or ligands): Choose either Bond dipoles or Atomic charges (assigned via Build → Set charge) for use in the calculations of nonbounded Electrostatic interactions. Select None (calculate all nonbonded interactions recommended for small molecules), Switched or Shifted for Cutoffs (for large molecules).
- For AMBER, BIO + , or OPLS (energy calculation of biomacromolecules): Choose Constant (for systems in a gas phase or in an explicit solvent) or Distance dependent (to approximate solvent effects in the absence of an explicit solvent) and set Scale factor for Dielectric permittivity (≥ 1.0 with the default of 1.0 being applicable to most systems). Select either Switched or Shifted for Cutoffs and set Electrostatic (the range is 0 to 1; use 0.5 for AMBER and OPLS, and use 0.4, 0.5, or 1.0 for BIO +) and van der Waals (the range is 0 to 1; use 0.5

for AMBER, 1.0 for BIO +, and 0.125 for OPLS). Different parameter sets are available for the AMBER force field (AMBER 2, 3, 94, or 96), which can be selected from Setup/Parameter of Force Field options.

- After choosing Setup menu options, execute chemical calculations from Compute menu commands to perform one of the fallowings.

(a) Single point: calculates the total energy and the RMS gradient.

(b) Geometry optimization: executes energy minimization, and calculates an optimum molecular structure with lowest energy and smallest RMS gradient.

(c) Molecular dynamics: calculates the motion of selected atoms over picosecond intervals to search for stable conformations.

(d) Langevin dynamics: calculates the motion of selected atoms over picosecond intervals using frictional effects to simulate the presence of a solvent.

(e) Monte Carlo: calculates ensemble averages for selected atoms.

Additional commands are available for semiempirical and *ab initio* calculations. These include:

(a) Vibration (calculates the vibrational motions of selected atoms)

(b) Transition state (searches for transition states of reactant or product atoms)

(c) Plot molecular properties (displays the electrostatic potential, total spin density or total charge density)

(d) Orbitals (analyzes and displays orbitals and their energy levels)

(e) Vibrational spectrum (analyzes and displays the vibrational frequencies)

(f) Electronic spectrum (analyzes and displays the ultraviolet-visible spectrum)

For energy minimization,

- Select Compute → Geometry optimization to open Molecular mechanics optimization dialog box.
- Choose algorithm such as Steepest descent, Fletcher-Reeves (conjugate gradient), or Polak-Ribiere (conjugate gradient, default of HyperChem), and choose options for termination condition such as RMS gradient (e.g., 0.1 kcal/mol Å) or number of maximum cycles.
- Click OK to close the dialog box and start optimization.

To apply periodic boundary conditions for solvation,

- Select Setup → Periodic box to open Periodic box options box.
- Referring to the dimension given for the smallest box enclosing solute, assign the dimension for the Periodic box size (e.g., twice of the largest dimension for the given smallest box or a cube of 18.70 Å on the side recommended by HyperChem but not exceeding 56.10 Å on the side), maximum number of water molecules (for information), and minimum distance between solvent and solute atoms (practical range: 1-5 with the default of 2.3 Å).
- Click OK to close the option box and start optimization.

For molecular dynamics calculation,

- Select Compute → Molecular dynamics to open Molecular dynamics options dialog box.

- Set heat time (e.g., 5 ps), run time (at equilibrium, e.g. 5 ps), step size (e.g., 0.005 ps), starting temperature (e.g., 0 K), simulation or final temperature (e.g., 300 K), and temperature step (e.g., 30 K).
- Choose *In vacuo* (for the system not in a periodic box) or Periodic boundary conditions (for the system in a defined periodic box).
- Set Bath relaxation time (suggested range: step size to the default of 0.1 ps) and assign any number between – 32,768 and 32,768 for Random seed as the starting point for the random number generator used for the simulations. Friction coefficient (any positive value) is needed only for the Langevin dynamics.
- Click the Snapshots button if you want playback of MolD trajectories (saved in a movie file .avi).
- Click the Average button to select average values of kinetic energy (EKIN), potential energy (EPOT), total energy (ETOT), and their RMS deviations and named selections (user selected interatomic distances, angles or torsion angles) to save (default, chem.csv) and plot after the simulation.
- Analogous procedures are applied to the Langevin dynamics (via Compute → Langevin dynamics) and Monte Carlo simulation (via Compute → Monte Carlo).

The biopolymer modeling of HyperChem includes Building polynucleotides, polypeptides and polysaccharides, Amino acid sequence (fasta format) editing, Mutations, Overlapping by RMS fit, and Merging structures. To facilitate manipulation of protein structures, there is often a need to display the protein backbone only as follows.

- Open PDB file (*.ent) from the File menu.
- Set the select level to Molecule and use selection tool to click on the protein molecule.
- Change the selection to water molecules by choosing Complement selection on the Select menu.
- Choose Clear command on the Edit menu to remove water molecules and to display the protein structure.
- Turn off Show Hydrogens on the Display menu.
- Choose Select Backbone on the Select menu and the Show Selection Only on the Display menu to display the backbone of the polypeptide chain.

The Databases menu provides tools for building polypeptides (Amino Acids, Make Zwitterion, Sequence Editor), polynucleotides (Nucleic Acids), polysacchar-ides (Saccharides), and organic polymers (Polymers) from residues (monomer units) as exemplified for DNA.

To build protein structure,

- Select Amino Acids from the Databases menu to open the amino acid dialog box.
- Choose chain conformation (Alpha Helix or Beta Sheet or Other) and isomer (L or D), and pick amino acids from the N-terminus.
- Close the dialog box to complete the chain.

To construct nucleic acids,

- Choose Nucleic Acids on the Databases menu.

- From the dialog box, choose the helical conformation of nucleic acid (A, B, Z, or other form) and add nucleotides (dA, dT, dG, and dC for DNA; rA, rU, rG, and rC for RNA) in the direction of 5' to 3' (default) or a choice of Backward and Double stranded. Both termini can be capped (5' Cap and 3' Cap).

To construct polysaccharides,

- Select Saccharide from the Display menu to open the Sugar builder window.
- Choose Add menu from the Sugar builder window to open the dialog box for linking monosaccharides (Aldoses, Ketoses, Derivatives, or End groups).
- Each selection opens another dialog box with options for choosing specific saccharide residues in pyranose or furanose forms, anomer (α, β, or acyclic), isomer (D or L), and type of link (linkage type).
- Add (pick) saccharide residues from the list of aldoses (hexoses in aldopyranose form, pentoses in aldofuranose form, and tetraoses in open-chain form), ketoses (hexoses in ketofuranose form, pentoses and tetraose in open-chain form), derivatives (glucosamine, galactosamine, N-acetylnuraminic acid, N-acetyl muramic acid, inositol, 2-deoxyribose, rhamnose, fucose, and apiose), and blocking groups (H, NH_2, = O, COO –, methyl, lactyl, *O*-methyl, *N*-methyl, *O*-acetyl, *N*-acetyl, phosphoric acid, sulfate, N-sulfonic acid) to build polysaccharides.

The Databases menu also permits the mutation of the selected amino acid residue(s) of a protein molecule. Choose Mutate from the Database menu to open a listing of amino acids. Highlighting the candidate amino acid effects the mutation.

To compare structures of two molecules by overlapping,

- Open the first molecule.
- Choose Merge command from the File menu to open the second molecule.
- Set different colors for the two structures (Display → Color after selecting the molecule).
- Select matching residue (s) from each of the two molecules via Select → Residue → Select to open a dialog box.
- Enter residue number under By number/Residue number. (Repeat the process in order to use more than one residue for the overlaying.)
- Choose RMS Fit and Overlay on the Display menu and pick the desired overlaying option (Molecular numbering or Selection order). This places one molecule on top of another.

WORKSHOPS

1. Aldopentose and aldohexose may exist in furanose and pyranose forms. The former favors the furanose form while the latter prefers the pyranose form. Would the energetic differences between the two forms as exemplified by the following two pairs of monosaccharides be sufficient to rationalize the preferences.

   (a) β-D-Ribofuranos versus β-D-ribopyranose

   (b) α–D–Glucofuranose versus α–D-glucopyranose

2. Both purine and pyrimidine bases tautomerize between enol form and keto form. Calculate the energy of tautomerization for thymidine.

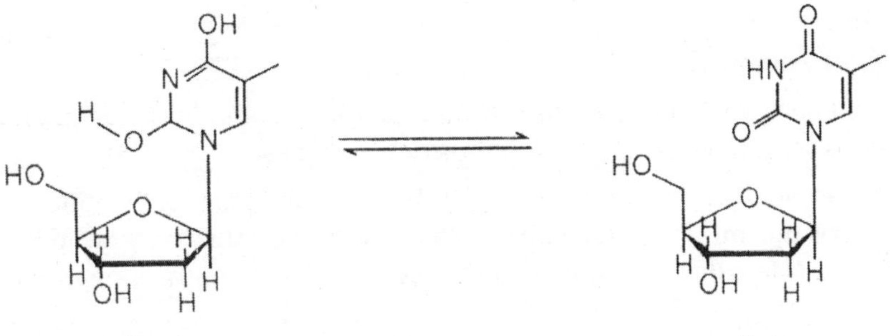

Enol Keto

3. Arachidonic acid (5*E*, 8*E*, 11*E*, 14*E*-eicosatetraenoic acid) is a precursor for the biosynthesis of prostaglandins. Apply torsion angle rotations involving bond C9-C10 and C10-C11 to search for the prostaglandin-like conformation.

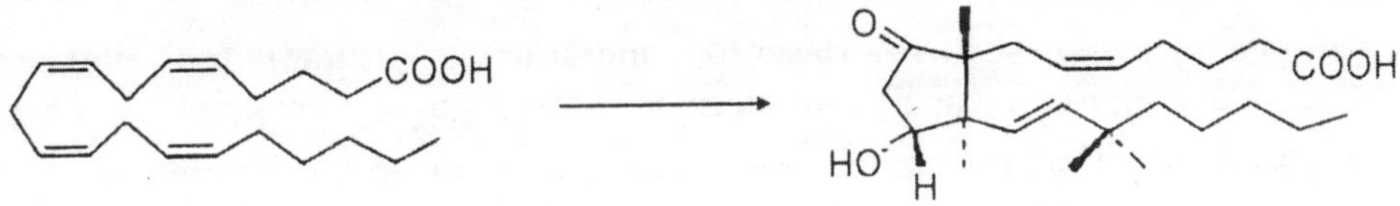

4. The nicotinamide adenosine dinucleotide (NAD^+) exits in *anti*- and *syn*-conformations: Compare their conformational energies as affected by solvation under periodic boundary conditions.

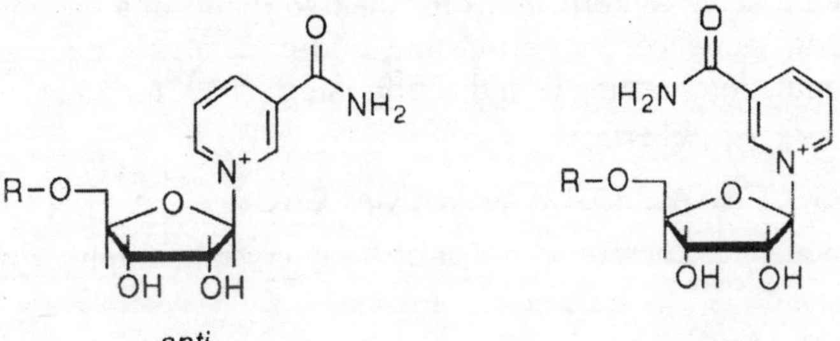

anti *syn*

5. Monosaccharide units are linked together by different glycosidic linkages to form oligosaccharides of diverse structures and conformations. Explore this diversity by performing geometry optimization of the following groups of oligosaccharides of D-glucopyranose (Glc*p*):
   (a) α,α-Trehalose [α-Glc*p*-(1 → 1) αGlc*p*], maltose [αGlc*p*-(1 → 4) αGlc*p*], and isomaltose [αGlcp-(1 → 6) αGlc*p*]
   (b) Hexasaccharides of Glcp linked by α-1,4 linkages, $[\alpha Glcp\text{-}(1 \rightarrow 4)\ \alpha Glcp]_3$, versus those linked by β-1,4 linkages, $[\beta Glcp\text{-}(1 \rightarrow 4)\ \beta Glcp]_3$

6. The double-stranded structure of DNA is stabilized by the hydrogen bonds formed between A and T pairs and between G and C pairs. Model the hydrogen bond interactions between = O... HN < of purine and pyrimidine bases for the AT pair at 2.70 ± 0.05 Å (between N and N or O at positions 1 and 6) and the GC pair at 2.95 ± 0.05 Å (between N and N or O at positions 1, 2 and 6).

(a) dAMPand TMP

(b) dGMP and dCMP

Perform geometry optimization until RMS gradient is equal to or less than 0.100. Formulate your conclusions with reference to their interaction energies.

7. Compare energies and conformation of the following pairs of biomolecules by performing geometry optimization (steepest descent then conjugated gradient until the energy difference is less than 1 kcal/mol) and molecular dynamics (heating to 300 K to be followed by equilibration for 5 ps at 300 K).

   (a) Met-enkephalin (TyrGlyGlyPheMet) versus Leu-enkephalin (TyrGlyGly-PheLeu)

   (b) Double-stranded hexadeoxyribonucleotides of AT chains versus GC chains

8. Retrieve nucleotide sequences (fasta files) of yeast cytosolic and mitochon-drial Gly-tRNA and submit them to RNA folding to obtain their secondary (cloverleaf) structures and thermochemical data of foldings.

9. Retrieve nucleotide sequence (fasta file) and atomic coordinates (pdb file) of yeast Asp-tRNA. Perform folding analysis/molecular modeling to display graphics of the following:

   (a) Secondary structure showing the maximum hydrogen bond formations

   (b) Tertiary structure with highlighted anticodon

   (c) Tertiary structure with molecular surface plot

   (d) Tertiary structure with electrostatic potential plot

10. Deduce the probable conformations for the two chains of a hammerhead ribozyme with the following sequences by performing folding analysis, geometry optimization and dynamic simulation (heating to and equilibrate at 300 K for 5 ps).

    Chain A: GUGGUCUGAUGAGGCC.

    Chain B: GGCCGAAACUCGUAAGAGUCACCAC.

Compare the modeled structure with that of X-ray crystallographic structure (299D.pdb).

MOLECULAR BIOMECHANICS

8

As an example of the use of the numerical algorithm developed, we first consider a model for switching in the bacterial flagellar motor proposed by Scharf *et al.* (1998). Some bacteria, such as *Escherichia coli,* swim by spinning long helical flagella. Each cell has multiple flagella, all of which have the same handedness. When the flagella are spun in the counterclockwise (CCW) direction, they come together to from a bundle that propels the cell through the fluid. The motor that is responsible for flagella rotation is reversible. When spun in the clockwise (CW) direction, the flagella fly apart and the cell undergoes a tumbling motion. Addition of a chemical attractant causes the cell to suppress tumbling when moving toward this food source.

One of the proteins in the signaling pathway is CheY. The binding of phosphorylated CheY to the portion of the motor located within the cytoplasm promotes CW rotation. To model motor reversals, the protein complex that forms the rotor is assumed to exist in two distinct conformational states that correspond to CW and CCW rotation. The binding of CheY decreases the free energy of the CW state, while at the same time increasing the free energy of the CCW by an equivalent amount. To capture this effect, the free energy of the rotor is assumed to have the following form :

$$G(x) = 4\Delta G_{nb}\left(\frac{x^4}{4} - \frac{x^2}{2}\right) - \frac{1}{2}\Delta G x, \tag{8.1}$$

where x is an appropriate reaction coordinate, $\Delta G = G(-1) - G(1)$ is the free energy difference between the CW and CCW states, and $\Delta G_{nb} = G(0) - G(-1)$ is the free energy difference between the transition state and either CW or CCW state when the CheY concentration is such that *(i.e.,* there is no bias toward CCW or CW rotation.

To model the chemical kinetics of CheY, we assume that binding of CheY to the motor is a two-rate processes with a single binding site being either empty or occupied. Let $P_E(t)$ be the probability that the site is empty at time t and $P_o(t)$ the probability that the site is occupied. The probabilities satisfy the following set of coupled equations:

$$\frac{dp_E(t)}{dt} = -k_{on}p_E(t) + k_{off}p_O(t), \tag{8.2}$$

$$\frac{dp_O(t)}{dt} = -k_{off}p_O(t) + k_{on}p_E(t), \tag{8.3}$$

where k_{off} is the dissociation rate constant and k_{on} is the rate at which CheY binds to the motor. From the law of mass action, k_{on} should be proportional to the CheY concentration. That is, $k_{on} = k'_{on}$[CheY], where k'_{on} is a second-order rate constant and the brackets stand for concentration.

If the concentration of CheY is held constant, then p_E and p_O will relax to their equilibrium values. These are found by solving (8.2) and (8.3) with the time derivatives set equal to zero. Doing this yields p_O = [CheY]/(K_d + [CheY]) where $K_d = k_{off}/k'_{on}$ = 9.1 μM is the dissociation constant.

There are approximately 26 binding sites on the motor, and the average number of occupied sites is $26p_O$. In the absence of CheY, $\Delta G = \Delta G_0 = 14k_BT$ and at saturating concentrations of CheY, $\Delta G = \Delta G_\infty = -9k_BT$. Therefore, the change in ΔG from low to high CheY concentrations is $\Delta\Delta G = \Delta G_0 - \Delta G_\infty = 23k_BT$, where the symbol ΔG indicates that we are talking about a change in the value of ΔG. Thus each CheY contributes roughly $0.88k_BT$ toward changing the relative free energy of the CW and CCW states. These considerations lead to the following expression for ΔG:

$$\Delta G = \Delta G_0 - \Delta\Delta G \frac{[\text{CheY}]}{K_d + [\text{CheY}]}. \tag{8.4}$$

Graphs of the free energy at various CheY concentrations are shown in Fig. 8.1A, with two minima located roughly at $x = \pm 1$. Generally, we shall measure distance x in nanometers (nm) and force in piconewtons (pN). In these units k_BT = 4.1 pN-nm at room temperature (T = 298 K). The force that arises from changes in free energy is $-\partial G/\partial x$. Therefore, the force vanishes at the minima.

Additionally, if the conformation of the rotor is slightly displaced from either minimum, it experiences a force that moves it back toward that minimum. The force also vanishes at the local maximum located near $x = 0$. However, when the conformation of the rotor is slightly displaced from the origin, the force acts to move away from the rotor at $x = 0$ and toward one of the two minima. Thus, we expect the rotor to spend most of its time near the minima. To surmount the energy barrier between the minima requires a substantial thermal fluctuation. The reaction coordinate $x(t)$, which determines the state of the rotor, can take on values anywhere between ± ∞. Clearly, we cannot use an infinite interval in our numerical algorithm.

However, since $G(x) \to \infty$ as $x \to \pm\infty$, there is a strong restoring force that drives the reaction coordinate back toward the origin when $|x|$ is large. This means that the probability of finding $x(t)$ at distances far from the origin is small, and ignoring large values of $|x|$ will not significantly affect our numerical solutions. For the parameters we shall consider, the interval (– 2, 2) is wide enough to ensure numerical accuracy. In practice, an appropriate interval can be determined by successively enlarging the length until the numerical results no longer change appreciably. At $x = \pm 2$ we enforce reflecting boundary conditions as described. The diffusion equation for this process is

$$\frac{\partial p}{\partial t} = D\frac{\partial}{\partial x}\left(\frac{\partial G(x)}{\partial x}\frac{p}{k_BT} + \frac{\partial p}{\partial x}\right) \tag{8.5}$$

To find the equilibrium distribution for $p(x, t)$, the above equation is solved with $\partial p/\partial t = 0$. This yields

$$p_{eq}(x) = \frac{\exp\left(-\dfrac{G(x)}{k_BT}\right)}{\int_{-\infty}^{\infty}\exp\left(-\dfrac{G(y)}{k_BT}\right)dy}. \tag{8.6}$$

The diffusion coefficient D in (8.5) represents an effective diffusion coefficient for the reaction coordinate that includes many microscopic effects. For all the results presented below, D = 70 nm^2/

s and $\Delta G_{nb} = 5k_BT$. These values were chosen to be consistent with experimental observation that at 14μM of CheY, motor reversals occur at an average rate of 2/s (we expand on this point below). Fig. 8.1B shows time series generated by the three potentials shown in Fig. 8.1A. The bistable nature of the system is clearly evident. The time series can be used to produce histograms of the reaction coordinate.

An approximation for the distribution $p_n^{(s)}$ is constructed by dividing the number of points in each bin of the histogram by the total number of points in the time series. Then we estimate $p(x_n) \approx p_n^{(s)}/\Delta x$. Fig. 8.1C shows distributions generated in this fashion. The solid lines are the exact results given by (8.6). As is clearly seen from the figure, the numerical algorithm accurately reproduces the equilibrium distribution. In the discussion of Markov chains and diffusion in the previous chapter, we encountered the idea of a waiting or first passage time. This is a very important mathematical concept that comes up in many different biological contexts. At a CheY concentration of 14 μM the motor reverses roughly twice per second, and there is no bias toward CW or CCW rotation.

To compute the switching rate, we must compute the average time for the system located at the reaction coordinate x to surmount the energy barrier at $x = 0$. To this end, the reaction coordinate is started at $x = -1$ at $t = 0$ with an absorbing boundary at $x = 0$. Fig. 8.2A shows the numerically generated probability density at various times. To generate this figure, 61 grid points were used. Note that the probability of finding the particle in the interval $(-\infty, 0)$ is continuously decreasing, due to the absorbing boundary. This probability can be used to determine the first passage time density $f(t)$ through the relation

$$\begin{aligned} f(t) &= -\frac{d}{dt}\text{Prob} = [-\infty < x(t) < 0] = -\frac{d}{dt}\int_{-\infty}^{0} p(x,t)dx \\ &\approx -\frac{d}{dt}\sum^{M} p_n(t) = p_M F_{M+1/2}^{abs} \end{aligned} \tag{8.7}$$

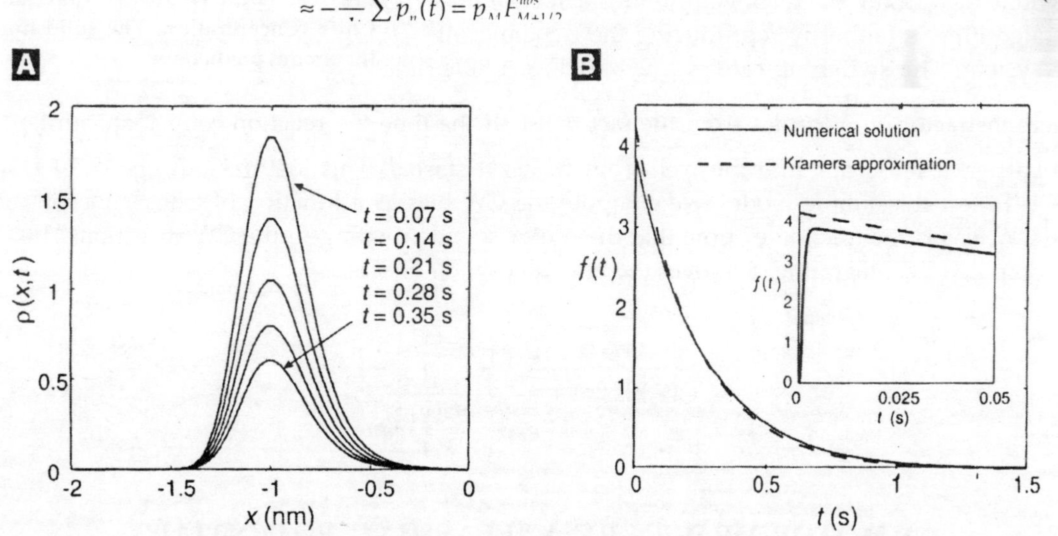

Fig. 8.2 Numerical results of the first passage time problem. The concentration of CheY has been chosen such that $\Delta G = 0$. At f = 0, the reaction coordinate is placed at $x = -1$. An absorbing boundary is placed at $x = 0$ and a reflecting barrier at $x = -2$. The first passage time is the time for the reaction coordinate to reach the origin. (A) Time evolution of the probability density. As time increases, the probability that the reaction coordinate remains inside the region decreases. (B) Probability density of the first passage time. Inset: an expanded view of the probability density near the origin showing the nonexponential nature of the distribution.

where the last equality follows from (12.42) and the absorbing boundary condition. Therefore, the numerical algorithm is well suited for computing first passage time densities. Fig. 8.2B shows the first passage time density. The solid line is the numerical result. The dashed line is the Kramers approximation, which assumes that the process has an exponential distribution with mean first passage time

$$MFPT \approx \frac{k_B T \pi}{D\sqrt{G''(-1)\,|\,G''(0)\,|}} \exp\left(\frac{G(0)-G(-1)}{k_B T}\right)$$

$$= \frac{k_B T \pi}{D\sqrt{32\Delta G_{nb}}} \exp\left(\frac{\Delta G_{nb}}{k_B T}\right). \quad ...(8.8)$$

A derivation of this result can be found in. Note that the most significant factor in determining the mean first passage time is ΔG_{nb}. The validity of the Kramers approximation depends on $\Delta G_{nb} >> k_B T$. As shown in the inset of Fig. 8.2B, the first passage time distribution is not exponentially distributed, since it must be equal to zero at $t = 0$. However, if we ignore this very short initial time interval, the distribution is approximated reasonably well with an exponential. Using the numerical distribution to compute the mean first passage time, we obtain MFPT = 0.253 s, and using Kramers approximation (8.8) we find MFPT = 0.236s. An exact expression gives MFPT = 0.259 s. The agreement between this value and the numerical result given above provides evidence that the algorithm is faithfully reproducing the dynamics of the system. The switching rate is l/(2 MFPT) = 1.98/s,

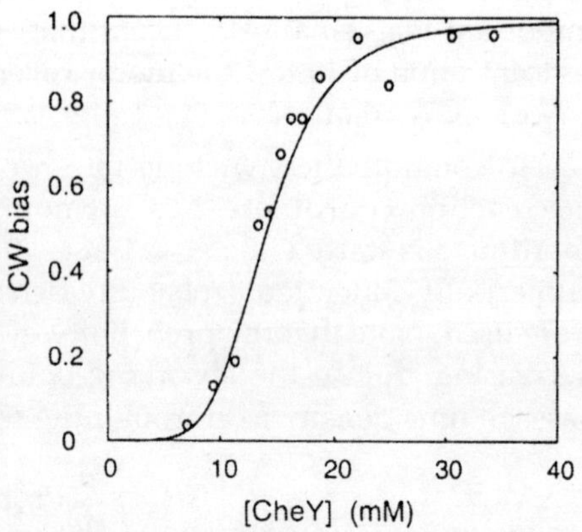

Fig. 8.3. The CW bias as function of CheY concentration. The solid line is the model prediction.

where the factor of $\frac{1}{2}$ comes from the fact that half the time the reaction coordinate surmounts the barrier, it falls back into the well from which it started. This justifies our choice of D and AG_nb-To test this simple model, we compute the CW bias as a function of CheY concentration. The CW bias is the fraction of time that the motor spends rotating in the CW direction. This can be computed by integrating (8.6) over the interval $(-\infty, 0)$. That is,

$$CW \text{ bias} = \frac{\int_{-\infty}^{0} \exp\left(-\frac{G(x)}{k_B T}\right) dx}{\int_{-\infty}^{\infty} \exp\left(-\frac{G(y)}{k_B T}\right) dy} \quad (8.9)$$

A MOTOR DRIVEN BY A "FLASHING POTENTIAL"

The following process, called the *flashing ratchet*, is a paradigm for molecular motors. It is also a good application for the methods developed. Imagine a protein driven by alternating its exposure to two potential energy profiles: V_1 (solid line) and V^2 (dashed line), as shown in Fig. 8.4. The first potential is a piece wise linear asymmetric sawtooth potential, while the second potential is a constant. Thus, in the first potential, the protein is localized near a local minimum, while in the second potential the protein diffuses freely. While in either potential, the motion of the particle is given simply by $\zeta dx/dt = -dV_i/dx + f_B(t), i = 1,2$. Switching between the potentials is governed by a chemical reaction (vertical arrows), which occurs with rate k.

Clearly, if the sawtooth potential is symmetric, the average displacement of the protein must be zero. However, in the case of the asymmetric potential shown in Fig. 8.4, the protein moves on the average to the right, although all steps are reversible. This phenomenon can be easily understood when the following inequalities are valid:

$$\frac{k_B TL}{FD} \ll \frac{1}{k} \ll \frac{L^2}{D} \tag{8.10}$$

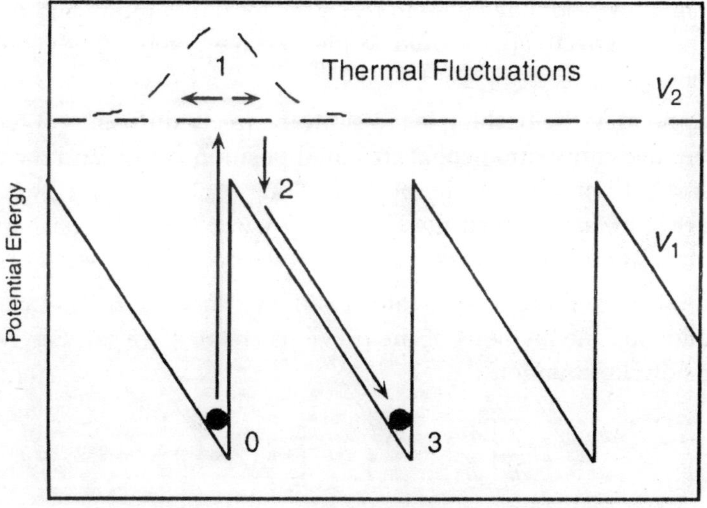

Fig. 8.4. The flashing ratchet. In the first potential, the protein is localized near a local minimum. Alternatively, the protein is free to diffuse. When the first potential is switched back on, the protein settles into the nearest local minimum.

Here L is the wavelength of the sawtooth potential, and $F = -dV_1/dx$ is the slope of the first potential, i.e., the force driving the protein to the right in the first potential. The corresponding drift rate of the protein is $F/\zeta = FD/k_BT$. The order of magnitude of the time for the protein to drift into a local minimum is k_BTL/FD. The diffusion in the first potential and the protein's deviations from a local minimum can be neglected if the slope of the sawtooth potential is very steep.

Quantitatively, this means that $k_BT << FL$ or $k_BTL << FL^2$; compare with (8.10). The first inequality (8.10)) means that the protein reaches a local minimum of the sawtooth potential well before this potential switches off. The second inequality (8.10) indicates that when the protein diffuses freely, it rarely can move farther than distance L before the sawtooth potential is switched back on: the mean time between "flashes" of the potential, l/k, is much less than the characteristic diffusion time L^2/D. Thus by the time the sawtooth potential is switched off, the protein is in a local minimum. When the sawtooth potential is switched off, the protein diffuses with equal probability to the left and to the right, and will not diffuse very far compared to the period of the potential. If the protein diffuses to the left, then by the time the sawtooth potential is switched back on, the protein is in the *basin of attraction* of the same local minimum it started from.

When the potential is on, the protein returns to the starting point of the cycle. However, if the diffusion to the right took place, the protein is in the basin of attraction of the next local minimum to the right. Thus, the protein either does not move, or moves the distance L to the right, with equal probability. Said another way, the diffusion in the flat potential can be viewed as a spreading Gaussian distribution. The asymmetry of the sawtooth potential, when it is switched on, cuts a larger portion of the distribution into the next domain of attraction.

The corresponding average rate of motion is $\langle V \rangle = Lk/4$ (the mean duration of a cycle is $2/k$, and on average, the system steps a distance L per two cycles). We considered the highly peculiar sawtooth potential with one of the slopes being infinitely steep. For more regular smooth potentials, the velocity of the flashing ratchet is computed analytically in the so-called fast and slow flashing limits. In both of these limits, the protein advances very slowly. If the flashing is too fast, the protein does not have time to reach the local minimum, and the effect of asymmetry is lost. The protein is effectively exposed to the average potential, which does not support any steady movement.

On the other hand, if the flashing is too slow, the freely diffusing protein moves too far from a local minimum, the information about its initial position is lost, and the average displacement becomes very small. The mean velocity of the flashing ratchet reaches a maximum $\approx Lk/4$ (for the smooth asymmetric potential, such that $\delta V \approx k_B T$) when $k \approx D/L^2$ (flashing frequency is of the same order of magnitude as the inverse time to diffuse over the potential's period).

In the general case, the average velocity of the flashing ratchet can be computed only numerically. Following the methods of the previous chapter, we can describe the ratchet by two coupled Smoluchowsky equations:

$$\frac{\partial p_i}{\partial t} = D\frac{\partial}{\partial x}\left[\frac{\partial p_i}{\partial x} + \frac{\partial V_i/\partial x}{k_B T}p_i\right] + k(-p_i + p_j), i = 1,2, j \neq i. \tag{8.11}$$

These equations must be solved numerically (see Exercise 2) on the finite domain $[0, L]$ with periodic boundary conditions and normalization condition

$$\int_0^L (p_1(x,t=0) + p_2(x,t=0))dx = 1.$$

When the probability distributions achieve steady state, the net current is

$$J = -D\left[\frac{\partial^2(p_1 + p_2)}{\partial x^2} + \frac{\partial V_1/\partial x}{k_B T}p_1\right],$$

from which the average velocity can be found: $\langle V \rangle = LJ$.

It is important to realize that in the process energy is consumed from the chemical reaction that drives the switching between the two potentials. The motion down the slope of the sawtooth potential 2 $\rightarrow$ 3 generates heat by frictional dissipation. Finally, if a small load force directed to the left is applied to the protein, the movement slows down. The load force is equivalent to tilting the potential to the left. Thus, the flashing ratchet is able to generate force, and has all characteristics of a molecular motor. However, there is no direct correspondence of the flashing ratchet mechanism to a real motor. In what follows, we consider two simple models of actual molecular motors.

THE POLYMERIZATION RATCHET

Perhaps the simplest way to convert chemical energy into a mechanical force is by polymerizing a filament against a load force. Here the energy source is the free energy of binding of a monomer onto the tip of the polymer, ΔG_b. If a polymer assembles against no resistance, the polymerization velocity (elongation rate) is simply

$$V_p = L\,(k_{on}M - k_{off}) \tag{8.12}$$

where L (nm) is the size of the monomer, M (μM) the monomer concentration, and k_{on} (1/μM·s), k_{off} (1/s) are the polymerization and depolymerization rate constants, respectively.

If an object is placed ahead of the growing polymer, there are two mechanisms by which the polymer can "push" the object: (i) by rectifying the Brownian motion of the object; (ii) by actively "pushing" against the object, i.e., a power stroke. First, we discuss the Brownian ratchet. We assume that the polymer is anchored at the left end and is perfectly rigid. The object has a diffusion coefficient $D = k_BT/\zeta$. For the moment, we neglect depolymerization ($k_{off} = 0$). In order for a monomer to bind to the end of the filament the object must open up a gap of size L by diffusing away from the tip, and remaining there for a time $\approx (k_{on}M)^{-1}$ to allow a polymerization event to take place. In the limiting case when polymerization.is much faster than diffusion, i.e., $k_{on}M >> D/L^2$, we can consider the polymerization to happen instantaneously once a gap of size L appears. Then the time for the load to diffuse a distance L is simply the mean first passage time $\langle T\rangle = L^2/2D$. To cover N such intervals takes $N \cdot \tau$ time units, so the average velocity is simply $\langle V_p\rangle = NL/(N\cdot\langle T\rangle) = 2D/L$. This is the speed of an ideal Brownian ratchet. Note that by reducing the size of the monomer L, the speed of the ratchet increases, since the likelihood of a thermal fluctuation of size L increases exponentially as L decreases. However, this is true only as long as our approximation holds: $k_{on}M >> D/L^2$, or $L >> \sqrt{D/(k_{on}M)}$. For smaller values of L the polymerization reaction becomes the limiting factor, so that $V_p \approx L \cdot k_{on}M$ (cf. (8.12) with $k_{Off} = 0$).

We can picture the situation as shown in Fig. 8.5B: the object diffuses on a "staircase" sequence of identical free energy functions $\phi(x)$, each with a step height of the monomer binding free energy $\Delta G = -\,k_BT\,\ln(k_{on}M/k_{off})$. If a load force F_L opposes the diffusive motion of the object, the potential becomes $\phi(x) - F_Lx$. This corresponds to tilting the potential so that the object must diffuse "uphill," as shown in Fig. 8.5B.

Including the depolymerization rate complicates the analysis considerably. However, a diffusion equation can be formulated that can be solved exactly when $k_{on}ML$, $k_{off}L << 2D/L$. In this regime, the approximate load-velocity relationship is given by the simple formula (Peskin *et al.* 1993)

$$V_p = L(k_{on}Me^{-f_L L/k_BT} - k_{off}). \tag{8.13}$$

That is, the polymerization rate in (8.13) is weighted by a Boltzmann factor where the exponent F_LL/k_BT is the work done by the load in moving the object one step distance L. The stall load F_s is reached when the work done in moving the object a distance L *is* just equal to the free energy from the binding reaction, so that $V_p = 0$:

$$F_s = \frac{k_BT}{L}\ln\left[\frac{k_{on}M}{k_{off}}\right] \tag{8.14}$$

Note that without depolymerization, $k_{off} \to 0$, there is no finite stall load.

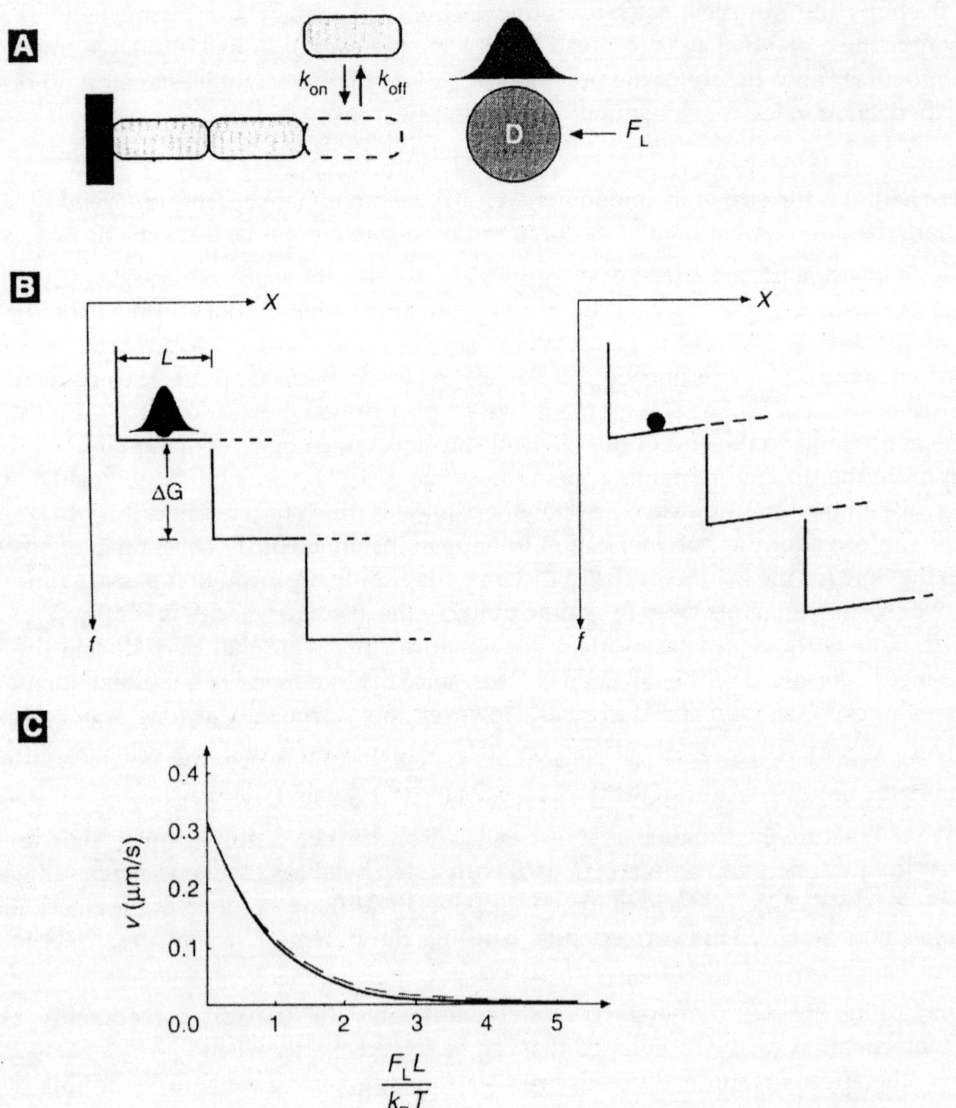

Fig. 8.5. The polymerization ratchet, (a) Monomers of length L polymerize onto the end of a growing filament with rate constants k_{on}, k_{off}. An object with diffusion coefficient D has its thermal motions rectified by the insertion of each new monomer. A load force F_L opposes the the motion of the object to the right, (b) Free energy diagram of the polymerization process. The total free energy $\Delta G = DG_b + F_L x$ where the binding free energy satisfies $\Delta G_b \gg k_B T$. Left panel: $F_L = 0$; right panel: $F_L > 0$ tilts the potential so that the object must diffuse uphill. (c) The load-velocity curve for the polymerization ratchet given by the approximation (8.13). The exact solution is shown by the dashed line.

Variations and elaborations on the polymerization ratchet have been used to model a variety of cellular processes, including lamellipodial protrusion, the polymerization of microtubules, the propulsion of intracellular pathogens and the translocation of proteins.

SIMPLIFIED MODEL OF THE F_0 MOTOR

To further illustrate the numerical formalism developed, we shall examine in detail a simplified model based on the principle of the ion-driven motor of ATP synthase. This enzyme uses electrochemical energy stored in a proton motive force across the inner membrane of mitochondria to produce ATP. This will illustrate many of the principles of mechanochemical energy conversion by proteins, but is sufficiently simple to analyze analytically and numerically. The motor is sketched schematically in Fig. 8.6. It consists of two reservoirs separated by an ion-impermeable membrane.

The reservoir on the left is acidic (high proton concentration) with concentration c^{acid}, and the reservoir on the right is basic (low concentration) with concentration c^{base}. The motor itself consists of two "parts": (i) a "rotor" carrying negatively charged sites spaced a distance L apart that can be protonated and deprotonated; (ii) a "stator" consisting of a hydrophobic barrier that is penetrated by an apolar strip that can allow a protonated site to pass through the membrane, but will block the passage of an unprotonated site. (The height of the energy barrier blocking passage of a charge between two media with different dielectric constants ϵ_1 and ϵ_1 is $\Delta G \approx 200[(1/\epsilon_1) - (1/\epsilon_2)] \approx 45k_BT$. This energy penalty arises from the necessity of stripping hydrogen-bonded water molecules from the rotor sites.)

Qualitatively, the motor works like this. Rotor sites on the acidic side of the membrane are frequently protonated. In this state (a nearly neutral dipole) the rotor can diffuse to the right, allowing the protonated site to pass through the membrane-stator interface to the basic reservoir. Once exposed to the low proton concentration in the basic reservoir, the proton quickly dissociates from the rotor site. In its charged state, the rotor site cannot diffuse backwards across the interface: Its diffusion is "ratcheted." We will show that thermal fluctuations will consistently drive the rotor to the right in Fig. 8.6.

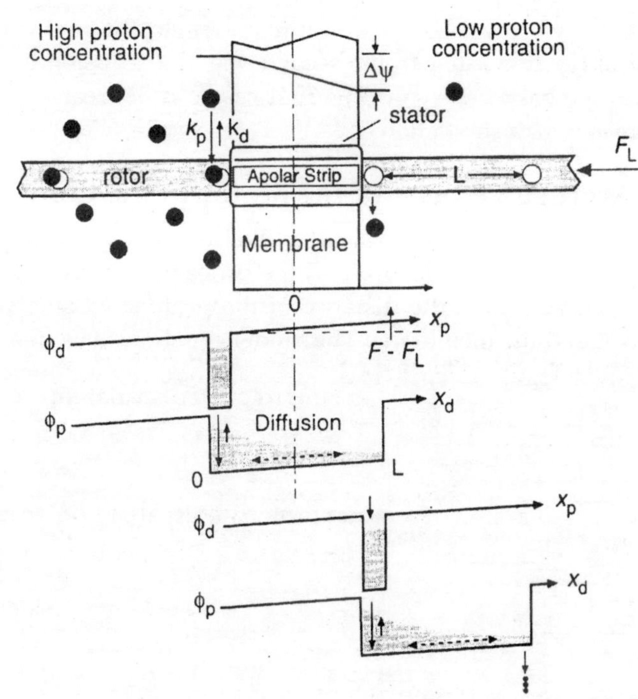

Fig. 8.6. Simplified model illustrating the principle of the F_0 motor. The state of the rotor can be pictured as a probability cloud (shown shaded) that drains from one potential to the next. ϕ_P is the potential seen by the site adjacent to the left side of the membrane in the protonated state, and ϕ_d is the potential in the deprotonated state. The potentials are tilted by an amount equal to the difference between the driving force ($F_I = e \cdot d\psi/dx$), and the load force, F_L. In the deprotonated state the site is immobilized, and in the protonated (almost neutral) state it can diffuse in the potential well ϕ_p. In the fast diffusion limit, the probability cloud quickly settles into the exponential Boltzmann distribution inside ϕ_p, which determines the probability of draining into the next deprotonated well, thus completing one step to the right.

Thus a rotor site can exist in two states: unprotonated and protonated. To specify the mobility state of the rotor, we need to keep track only of the site immediately adjacent to the membrane on the acidic side. In its unprotonated state, the site adjacent to the membrane is immobilized, since it cannot pass into the stator, nor can it diffuse to the left, since the next rotor site on the basic side of the membrane is almost always deprotonated, and cannot diffuse to the left. (Of course, this depends on the thickness of the membrane being the same as the rotor spacing; this is unrealistic, but the full model treated in the references does not have this constraint.)

Thus the progress of the model can be pictured as a sequence of transitions between two potentials. When deprotonated, the rotor is immobilized in potential ϕ_d, and when protonated, it can move in potential ϕ_p. The effect of the load force F_L is to tilt the potentials upward, so the motion in potential ϕ_p is "uphill" (i.e., the total potential when protonated can be written as $\phi_p(x) - F_L(x)$.

Below, we consider two limiting cases. In the first one, diffusion is much faster than the chemical reaction rates. In the second one, the diffusion time scale is comparable to the reaction rates in the basic reservoir. The first case can be treated *analytically,* while the second one will require *numerical* simulation.

The Average Velocity of the Motor in the Limit of Fast Diffusion

The model can be formulated mathematically in terms of the probability of the deprotonated state $p_d(t)$ (non-dimensional), and the probability density of the protonated state $p_p(x, t)$ (1/nm). Here x, $0 \le x \le L$, is the distance of the protonated site from the interface between the acidic reservoir and the membrane. The model equations have the form

$$\frac{dp_d}{dt} = \text{net deprotonated spatial flux} + \text{net reaction flux}$$

$$= J_{x_d} + J_\xi \qquad \text{...(8.5)}$$

$$\frac{\partial p_p}{\partial t} = \text{net protonated spatial flux} - \text{net reaction flux}$$

$$= J_{x_p} - J_\xi \qquad \text{...(8.6)}$$

where

$$J_x = 0,$$

$$J_{x_p} = D\frac{\partial}{\partial x}\left(\frac{\partial p_p}{\partial x} - \frac{F_1 - F_L}{k_B T} p_p\right),\ J_{x_p}(0) = J_{xp}(L) = 0,$$

$$J_\xi = \text{deprotonation at acidic reservoir}$$
$$+ \text{deprotonation at basic reservoir}$$
$$- \text{net protonation at both reservoirs}$$
$$= k_d p_p(0) + k_d p_p(L) - \bar{k}_p p_d.$$

Protonation rates are proportional to hydrogen ion concentration on either side of the membrane, and the net protonation at both reserviors is $\bar{k}_p = k_p c^{acid} + k_p c^{base}$. Note that the rates of protonation and deprotonation have the dimensions $\bar{k}_p$ (1/s) and k_d (nm/s), respectively. We assume that the deprotonation rates are the same at both reservoirs.

First we nondimensionalize the model equations using the rescaled coordinate $(x/L) \to x$, the rescaled time $(k_d t/L) \to t$, and the following dimensionless parameters:

- Ratio of reaction to diffusion time scales: $\Lambda = D/k_d L$.
- Net work done in moving the rotor a distance $L : w = (F_I - F_L)L/k_B T$.
- Equilibrium constant: $\bar{k} = \bar{k}_p L/k_d$.

Here $F_I = e\Delta\psi/L$ is the electrical driving force, which is assumed to be constant. Substituting these variables and parameters into equations (8.15) and (8.16) gives

$$\frac{dp_d}{dt} = -\bar{k}p_d + p_p(0) + p_p(1), \tag{8.17}$$

$$\frac{\partial p_p}{\partial t} = \bar{k}p_d - p_p(0) - p_p(1) + \Lambda\frac{\partial}{\partial x}\left(\frac{\partial p_p}{\partial x} - wp_p\right), \tag{8.18}$$

where x, t are now the nondimensional coordinate and time, respectively.

In many situations it turns out that diffusion is much faster than the chemical reaction rates: $\Lambda >> 1$. (At $D \approx 10^7$ nm^2/s, L $\approx$ 10 nm, $k_d \approx 10^3$ nm/s, the order of magnitude of the parameter Λ is 10^3.) This means that the time between dissociation events is much longer than the time to diffuse a distance L, so the process is limited by the speed of the reactions, not by the diffusion of the rotor. In other words, the diffusive motion of the rotor is so fast that it achieves thermodynamic equilibrium, and so its displacement can be described by a Boltzmann distribution. In this case we can express the probability distribution as $p_p(x, t) = p_p(t) \cdot P(x)$, where $p_p(t)$ is the probability of the site being in the protonated state and $P(x)$ is the *equilibrium* spatial probability density describing the rotor's position relative to the stator.

We can obtain the steady-state Boltzmann distribution $P(x)$ from (8.18). First, we divide through by Λ and take advantage of the fact that $\Lambda >> 1$: All terms but the last are rendered negligible, so that the distribution of rotor positions in the protonated state potential well ϕ_p $(0 \le x \le 1)$ in Fig. 18.6 is governed by

$$\frac{dP}{dx} - wP = 0.$$

The solution must be normalized to 1, since it represents a probability density. The result is:

$$P(x) = \left(\frac{w}{e^w - 1}\right)e^{wx}, 0 \le x \le 1, \tag{8.19}$$

where the quantity in parentheses is the normalization factor. Thus, the rates of protonation at time t are $p_p(0, t) = p_p(t) \cdot P(0)$ and $p_p(1, t) = p_p(t) \cdot P(1)$, respectively. Substituting this into (8.17) and imposing the conservation of probability, $p_d(t) + p_p(t) = 1$, we reduce the problem to the two-state Markov chain described by:

$$\frac{dp_d}{dt} = -\frac{dp_p}{dt} = -\kappa p_d + (P_+ + P_-)p_p, \tag{8.20}$$

where $P_+(w) = w/(1 - e^{-w})$, $P_-(w) = w/(e^w - 1)$. Therefore, the stationary probabilities are obtained directly by setting the time derivatives equal to zero and solving for p_p:

$$p_d(w,\kappa) = \frac{P_+ + P_-}{\kappa + P_+ + P_-}, \; p_p(w,\kappa) = \frac{\kappa}{\kappa + P_+ + P_-}, \tag{8.21}$$

or, in the dimensional variables:

$$\overline{\kappa}_p \tag{8.22}$$

$$P_p(w,k) = \frac{k_p(c^{\text{acid}} + c^{base})L}{k_p(c^{\text{acid}} + c^{base})L + k_d P_+ + k_d P_-'} \tag{8.23}$$

The average velocity of the motor can be found using the following heuristic argument. The rotor effectively moves to the right either from the protonated state, when the proton is released to the basic reservoir with the effective rate $k_d P_+$, or from the deprotonated state, when the protonation takes place at the acidic reservoir with the effective rate $k_p c^{\text{acid}} L$. The corresponding effective rate of movement to the right is the sum of the corresponding rates weighted by the respective state probabilities: $\langle V_r \rangle = k_p c^{acid} Lpd + k_d P_{+p} p$. Similarly, the rotor effectively moves to the left either from the protonated state, when the proton is released to the acidic reservoir with the effective rate $k_d P_-$, or from the deprotonated state, when the protonation takes place at the basic reservoir with the effective rate $k_p c^{\text{base}} L$. The corresponding effective rate of movement to the left is the sum of the corresponding rates weighted by the respective state probabilities: $\langle V_1 \rangle = k_p c^{\text{base}} L p_d + k_d P_{-pp}$. The net average velocity $\langle V \rangle = \langle V_r \rangle - \langle V_l \rangle$, can be obtained by using the expressions for the state probabilities (8.22), (8.23) and some algebra:

$$\langle V \rangle (F_L) = \frac{k_p k_d L w (c^{\text{acid}} e^{w} - c^{\text{base}})}{(e^{w} - 1) k_p L (c^{\text{acid}} + c^{\text{base}}) + k_d w (e^{w} + 1)}, \tag{8.24}$$

$$w = \frac{(F_1 - F_L)L}{k_B T}$$

As a check, note that when there is no load ($F_L = 0$), no membrane potential ($F_l = 0$), and no proton gradient ($c^{\text{acid}} = c^{\text{base}}$), the velocity vanishes, as it should. The load-velocity relationship given by (8.24) looks very similar to the one in the next limiting case, and is plotted in Fig. 8.8C.

The *stall force* F_s is reached when the load force just brings the motor to a halt ($c^{acid} e^{w} - c^{\text{base}} = 0$):

$$F_s = F_l + \frac{k_B T}{L} \ln\left[\frac{c^{\text{acid}}}{c^{\text{base}}}\right] \tag{8.25}$$

Since the electrical driving force satisfies $F_l = e\Delta\psi/L$, (8.25) can be written as an equilibrium thermodynamic relation in terms of the energy:

$$F_l = e\Delta\psi - 2.3 k_B T \Delta \text{pH} \tag{8.26}$$

This says that the reversible work done to move a rotor site across the membrane is equal to the work done by the electrical field plus the "entropic work" done by the Brownian ratchet. (The term "reversible" in this context means that the velocity is so slow (near stall) that we can neglect the viscous dissipation.) Dividing through by the unit charge *e* gives the work per unit charge, which is just the protonmotive force discussed previously. One point about (8.26) is worth noting. Since the motor is working against a conservative load force, as the motor approaches stall its efficiency approaches 100%. For a motor working against a viscous load, a more sophisticated treatment is required.

It may seem from (8.24) that there is a definite average velocity of the motor in the limit $T \to \infty$. In other words, the motor continues to move in the absence of thermal fluctuation. The reason is that the solution of the Lengevin and Smoluchowski equations cannot be treated as a regular perturbation problem in the limit of low temperatures. This is a singular perturbation problem, and the protein behavior at absolute zero temperature cannot be quantified as a simple limit of such behavior at low temperatures.

Brownian Ratchet vs. Power Stroke

In the last example the motion of the rotor was driven by a combination of Brownian motion and the membrane potential. The rotor diffusion is biased by the electrostatic forces that are switched off and on by the binding and dissociation of protons to the rotor sites. The membrane potential appears to drive the rotor unidirectionally without the aid of Brownian motion. However, this cannot happen without the binding and dissociation of protons, a stochastic process driven by thermally excited transitions.

Thus even the "power stroke" depends on Brownian motion, so that setting $k_B T = 0$ in the model equations arrests the rotor motion. This is a fundamental distinction between molecular and macroscopic motors. The distinction between a motor driven by a "Brownian ratchet" and one driven by a "power stroke" may not be so clear in other systems. In the polymerization ratchet model described above, the movement of the *load* is driven entirely by its Brownian motion.

The chemical step simply rectifies, or biases, this motion (Peskin *et al.* 1993). By comparison, the FI motor in ATP synthase is driven by the hydrolysis of ATP. The conformational change in the protein that constitutes the power stroke is known: Binding of ATP to the catalytic site drives the change in protein shape that drives rotation. That is, the load is not driven significantly by Brownian motion; it sees the proteins conformational change as a "power stroke." However, a closer look at how ATP binds to the catalytic site reveals that it is a multistep process involving the sequential annealing of hydrogen bonds between the protein and the nucleotide.

Each step in this process is driven by Brownian motion, i.e., a thermally activated process as illustrated in Fig. 8.5. Therefore, the power stroke itself can be viewed as a kind of Brownian ratchet, one that proceeds at a smaller length scale than the protein (ATP synthase is $\approx$ 10 nm in diameter, while the catalytic site is $\approx$ 1 nm). Thus the distinction between a process driven by a Brownian ratchet and by a power stroke can be largely a matter of size scale; a fuzzy boundary separates the two notions. In the extreme case where the motion of the *load* is due only to its diffusion, and the role of the chemical reaction is only to inhibit diffusion in one direction, we can say that the motor is a Brownian ratchet.

The Average Velocity of the Motor When Chemical Reactions Are as Fast as Diffusion

Next we consider a different limiting case: where the diffusion time scale is comparable to the reaction rates in the basic reservoir. In this case we have to change the mathematical formulation of the model. We make the simplifying assumption that the proton concentration in the acidic reservoir is so high that the binding sites on that side are always protonated. Now we define the right boundary of the membrane as the origin, and the distance between the membrane and the binding site nearest the membrane in the basic reservoir x is always between 0 and L.

The chemical state of the rotor is determined by the state of all the binding sites in the acidic reservoir. In general, if there are N binding sites on this side, the total number of chemical states is 2^N. However, for now we focus on the binding site nearest the membrane. In this case there are just two states: "off" if this site is unprotonated and "on" if it is protonated. The mechanochemistry of the motor was described by the following set of coupled diffusion equations:

$$\frac{\partial p_d}{\partial t} = D\frac{\partial}{\partial x}\left(\frac{F_L - F_I}{k_B T} p_d + \frac{\partial p_d}{\partial x}\right) - \bar{k}_p p_d + k_d p_p, \quad ...(8.27)$$

$$\frac{\partial p_d}{\partial t} = D\frac{\partial}{\partial x}\left(\frac{F_L - F_I}{k_B T} p_p + \frac{\partial p_p}{\partial x}\right) + \bar{k}_p p_d - k_d p_p, \quad ...(8.28)$$

where $p_p(x, t)$ and $p_d(x,t)$ (1/nm) are the probability densities for being at position x and in the protonated and deprotonated states, respectively, at time t. The proton association and dissociation rates in basic reservoir are $\bar{k}_p$ (1/s) and k_d (1/s), respectively. Note that dimensions of some of the model parameters and variables are different in this limit.

We can nondimensionalize these equations using the rescaled coordinate $(x/L) \rightarrow x$, the rescaled time $k_d t \rightarrow t$, and the dimensionless parameters $\Lambda = (D/k_d L^2)$, $w = (F_I - F_L)\, L/k_B T$, and $\kappa = k_p/k_d$. The nondimensional equations have the form

$$\frac{\partial p_d}{\partial t} = \Lambda\frac{\partial}{\partial x}\left(w p_d + \frac{\partial p_d}{\partial x}\right) - \kappa p_d + p_p, \quad (8.29)$$

$$\frac{\partial p_p}{\partial t} = \Lambda\frac{\partial}{\partial x}\left(w p_d + \frac{\partial p_p}{\partial x}\right) + \kappa p_d - p_p. \quad (8.30)$$

Equations (8.29) and (8.30) are second-order partial differential equations. This means that four boundary conditions are required in order to have a mathematically complete description of the problem. One boundary condition is that $x = 0$ is reflecting:

$$\left[w p_d + \frac{\partial p_d}{\partial x}\right]_{x=0} = 0 \quad (8.31)$$

This takes into account the fact that an unprotonated site cannot pass back through the membrane. The remaining three boundary condtions require knowing the state of all the binding sites in the basic reservoir, which would necessitate solving a large number of coupled diffusion equations (one for each possible chemical state of the rotor). However, to keep things simple, we construct a reflecting boundary condition at $x = 1$ (x is now measured in units of L) when the rotor is in the protonated state. That is,

$$\left[w p_p + \frac{\partial p_p}{\partial x}\right]_{x=1} = 0 \quad (8.32)$$

If the proton dissociation rate is fast and the proton concentration is low in the basic reservoir, we don't expect this artificial boundary condition to have much of an effect, since in this limit the probability of the first site being occupied when $x = 1$ is very small, so that this boundary condition is rarely encountered.

When an unprotonated site moves to the right of $x = 1$, *it* brings a protonated site out of the membrane channel and into the region $0 < x < 1$. This protonated site becomes the new site that we follow. The state of the motor goes from unprotonated to protonated, and the coordinate of the motor goes from $x = 1$ to $x = 0$. Conversely, when a protonated site moves into the membrane, it brings an unprotonated site into the region $0 < x < 1$. This unprotonated site becomes the new site we follow. The state and the coordinate of the motor change accordingly. These considerations are illustrated in Fig. 8.7 by the arrows connecting the right end of p_d to the left end of p_p. The boundary conditions that model this situation are:

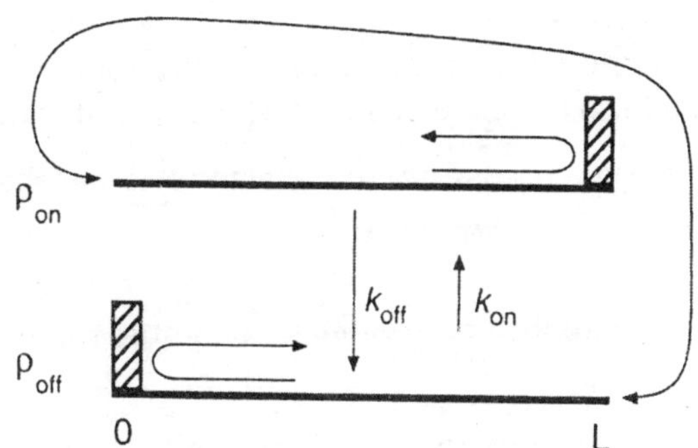

Fig. 8.7. In general, the chemical state of the motor is determined by all the binding sites in the right chamber. In the simplified version of the model only the binding site nearest the membrane is considered. The reflecting boundary condition in the off state at $x = 0$ is due to the fact that an unprotonated site cannot pass back through the membrane. The reflecting boundary condition in the on state at $x = L$ is artificial and is used to simplify the problem.

$$\left(wp_d + \frac{\partial p_d}{\partial x}\right)_{x=1} = \left(wp_p + \frac{\partial p_p}{\partial x}\right)_{x=0}, \; p_d(1,t) = p_p(0,t), \tag{8.33}$$

which is the mathematical statement of the fact that the rotor in the off state at $x = 1$ is equivalent to the rotor in the on state at $x = 0$. Therefore, to implement these boundary conditions numerically, we make use of the periodic boundary condition discussed in the previous chapter. Note that there are two mechanisms for changing the chemical state of the rotor: movement of the rotor and chemical kinetics.

We are now in a position to use the numerical algorithm developed to approximate (8.29)-(8.33). The interval (0,1) is divided into M segments. For each of the M grid points there are two possible states of the rotor, off and on. Therefore, there are 2M possible states in the discrete approximation of the process. Let the first M states correspond to the offstate, and the states M +1 to $2M$ correspond to the onstate. The equations used in the numerical scheme for $1 < n < M$ are

$$\begin{aligned} \frac{dp_n}{dt} &= (F_{n-1/2}p_{n-1} - B_{n-1/2}p_n) - (F_{n+1/2}p_n - B_{n+1/2}p_{n+1}) - \kappa p_n + p_{n+M}, \\ \frac{dp_n + M}{dt} &= (F_{n-1/2}p_{,,-1+M} - B_{n-1/2}p_{n+M}) - (F_{n+1/2}p_{n+M} - B_{n+1/2}p_n + 1 + M) - p_{n+m} + \kappa p_n \end{aligned} \tag{8.34}$$

and the equations used to implement the boundary conditions are

$$\frac{dp_1}{dt} = -(F_{3/2}p_1 - B_{3/2}p_2) - \kappa p_1 + p_{M+1}, \tag{8.35}$$

$$\frac{dpM}{dt} = (F_{M-1/2}p_{M-1} - B_{M-1/2}pM) - (F_{1/2}pM - B_{1/2}pM + 1) - \kappa p_M + p_{2M}, \tag{8.36}$$

$$\frac{dp_{M+1}}{dt} = (F_{1/2}p_M - B_{1/2}p_{M+1}) - (F_{3/2}p_{M+1} - B_{3/2}p_{M+2}) - p_{M+1} + kp_1, \tag{8.37}$$

$$\frac{dp_{2M}}{dt} = (F_{M-1/2}p_{2M} - 1 - B_{M-1/2}p_{2M}) - p_{2M} + \kappa p_M \tag{8.38}$$

The potential used in F_n and B_n is $\phi(x) = (F_L - F_1)\,x$. Note that because of the chemical kinetics, the matrices **A** and **C** required for the numerical scheme are no longer tridiagonal. However, they are still sparse, so that solving (8.39) is not computationally expensive.

Let us discuss briefly how to calculate the protonation and deprotonation rates $\bar{\kappa}_p$ and k_d. The chemical reaction is

$$\text{site}^- + \text{H}^+ \leftrightarrow \text{site . H}. \tag{8.39}$$

At equilibrium, protonation and deprotonation balance. That is,

$$\bar{\kappa}_p\,[\text{site}^-] = k_d[\text{site}\cdot\text{H}]. \tag{8.40}$$

Proton concentrations are generally reported as a pH value:

$$\text{pH} = -\log_{10}[\text{H}^+], \quad [\text{H}^+] = 10^{-\text{pH}} \tag{8.41}$$

The higher the pH value, the lower the proton concentration. The pK_a value of the binding site is calculated from the measured concentration values of [site · H] and [site$^-$] as

$$\text{pK}_a = \text{pH} + \log_{10}\frac{[\text{site}\cdot\text{H}]}{[\text{site}^{-1}]} \tag{8.42}$$

Combining (8.40) and (8.42), we see that the rates $\bar{\kappa}_p$ and k_d are related to pH and pK_a by

$$\frac{\bar{k}_p}{k_d} = 10^{pK_a - pH} \tag{8.43}$$

Generally, $\bar{\kappa}_p$ is limited by the rate at which protons diffuse to the binding site. In this limit, the association rate can be computed from the Smoluchowski formula (this rate is proportional to the proton concentration, c^{base}):

$$\bar{k}_p = \begin{pmatrix}\text{proton}\\ \text{concentration}\end{pmatrix} \cdot \begin{pmatrix}\text{absorption rate to a perfectly}\\ \text{absorbing disk of radius } r\end{pmatrix}$$
$$= \underbrace{0.6\ \text{nm}^{-3} \cdot 10^{-\text{pH}}}_{\text{protons/nm}^3} \cdot \underbrace{4r\,D_{\text{proton}}}_{\text{absorption rate}} \tag{8.44}$$

Here D_{proton} is the diffusion coefficient of protons. If we know $\bar{k}_p$, k_d can be determined by (8.43). The table shown below lists typical parameter values for ATP synthase.

Parameter Name	**Parameter Value**
Diffusion coefficient of the rod	$D = 10^4\ \text{nm}^2/\text{s}$
pK_a value of binding site	$\text{pK}_a = 6.0$
pH of the right compartment	pH = 6.0 to 8.0 (variable)
External loadforce on the rod	$F_L = 0$ to 3 pN (variable)
Distance between binding sites	$L = 8$ nm
Diffusion coefficient of proton	$D_{\text{proton}} = 10^{10}\text{nm}^2/\text{s}$
Absorbing radius of binding site	$r = 0.5$ nm

Note that at these values of the model parameters, at pH = 7, the values of the nondimensional parameters are $\Lambda \approx 0.1$, $\kappa \approx 0.1$, $w \approx 1$.

We now have all the necessary information to use the numerical scheme. Fig. 8.8A shows the relaxation of the marginal density $p(x, t) = p_p(x, t) + p_d(x, t)$ to steady state. The (dimensional) average velocity is a steady-state property of the system and is related to the total flux by the relations

$$\langle V\rangle = LJ = -LD\left[\frac{F_L - F_I}{k_B T}\left(p_p^{(s)} + p_d^{(s)}\right) + \frac{\partial}{\partial x}\left(p_p^{(s)} + p_d^{(s)}\right)\right], \tag{8.45}$$

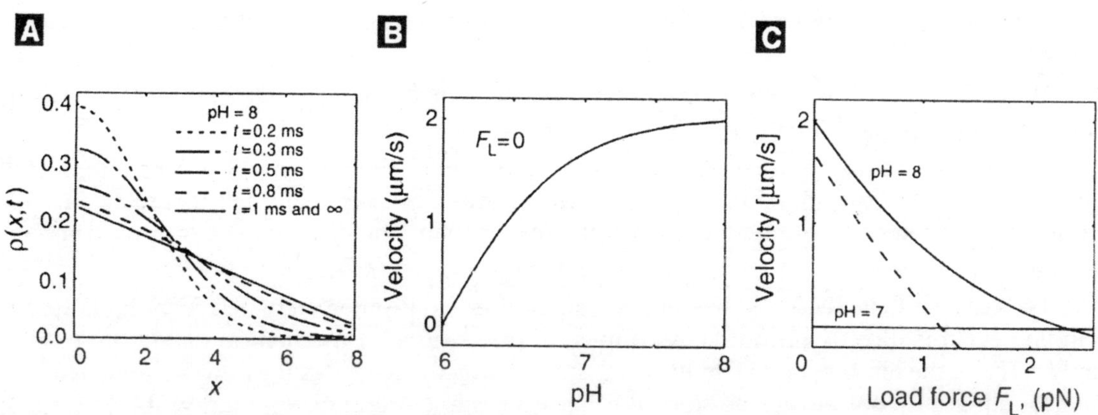

Fig. 8.8. Numerical results of the simplified F_0 motor. Here pH is the pH value of the right compartment. The external load force is F_L- In all the figures the number of grid points used in the simulations was $M = 32$. (A) Relaxation of the marginal density $\rho = \rho_{rmon} + \rho_{off}$ to steady state for pH = 8 and no external load. (B) Motor velocity as a function of pH with no external load. (C) Motor velocity as a function of the load force for pH = 7 and pH = 8.

where the superscripts in the above equation stands for steady state. Once steady state is achieved, the average velocity *(V)* can then be computed from the relation

$$\langle V\rangle = L(F_{n-1/2}pn - 1 - B_{n-1/2}p_n) + L(F_{n-1/2}p_{n+M}) \tag{8.46}$$

for $1 < n < M$. Typical results for the mean velocity are shown in Fig. 8.8B and Fig. 8.8C.

OTHER MOTOR PROTEINS

As we discussed there is such a variety of protein motors that no classification scheme can do justice to their diversity. However, for the purposes of discussion we can identify several physical properties that delineate classes of motors. The literature on molecular motors is vast, so we shall restrict our discussion here to those for which reasonably complete mathematical models exist as outlined in Table 8.5. Some of the most important characteristics of molecular motors are the following:

Fuel. The two most common energy sources for molecular motors are nucleotide hydrolysis (e.g., ATP, GTP) and transmembrane ion gradients. Certain specialized motors depend on stored elastic energy that has been captured during the assembly of the motor (e.g., spasmoneme and the acrosome of *Limulus*.

Mechanical escapement. Three common motor types are (i) rotary (e.g., the bacterial flagellar motor F_0 ATPase); (ii) linear motors that run along a "track," usually actin, microtubules, or nucleic acid polymers (e.g., myosin, kinesin, dynein, RNA poly-merase); (iii) polymerization or depolymerization motors that directly push or pull against a load (e.g., the acrosome, cellular lamellipodia, the propulsive tail of *Listeria).* The latter category suggests a subclassification into those that operate in a continuous cycle (e.g., myosin, FI), and motors that are "one-shot" processes; i.e., they function for but a single episode of polymerization, after which they are usually disassembled.

Cooperative vs. "loners." Because of its small "duty cycle" (i.e., attachment time to the load), myosin II must act in concert with many other partners to produce a continuous force on an actin filament. Myosin V and kinesin, however, have longer duty cycles, and so they can transport a vesicle without the cooperation of other motors. These categories do not begin to delineate the variety of possibilities.

However, one common event generally commences the transduction process between chemical energy and mechanical force.

Because molecular motors can be viewed as enzymes, the binding of a substrate onto the motor initiates the transduction process. However, this does not really tell us much, since it is simply a restatement that chemical reactions (other than isomerizations) begin by combining substrates.

The feature that distinguishes molecular motors from other enzyme reactions is that the binding event is directly or indirectly coupled to the creation of mechanical forces. For example, in FI ATPase, the binding of ATP to the catalytic site directly generates the power stroke. However, the coupling may not always be so direct. For example, binding of a proton to the F_0 rotor site switches off the local electrostatic field surrounding the rotor, permitting bidirectional diffusion.

The binding of monomers to polymerizing actin captures thermal fluctuations in elastic strain, which is subsequently released to power protrusion. Several points are important. First, the role of thermal fluctuations in all these processes is central, so that none could operate when $k_B T = 0$. Second, energy captured by binding or dissociation events can be stored and released later, and in a different location, to produce mechanical work. Third, the operation of every molecular motor depends on its specialized protein geometry, so that models of motors that ignore geometry are generally not useful to biologists. Finally, we do not believe that the operation of molecular motors involves any novel physics or chemistry.

However, the amazing variety of protein shapes requires that we treat each motor individually. General principles are not likely to provide more than philosophical comfort in understanding any particular motor. In the words of Katchalsky:

"It is easier to make a theory of everything, than a theory of something" —*Aharon Katchalsky*

Some Molecular Motors that Have Been Modeled Mathematically

Motor	References
Acrosome	(Mahadevan and Matsudaira 2000; Oster et al. 1982)
F_1 ATPase	(Oster and Wang 2000a; Wang and Oster 1998; Oosawa and Hayashi 1986)
F_0 ATPase	(Elston *et al.* 1998; Dimroth *et al.* 1999; Lauger 1991; Stein and Lauger 1990)
Bacterial flagellar motor	(Elston and Oster 1997; Lauger 1990; Berry 1993)
HSP70	(Peskin *et al.* 1993; Simon *et al.* 1992; Elston 2000; Chauwin *et al.* 1998)
Kinesin	(Peskin and Oster 1995; Fox and Choi 2001; Derenyi and Vicsek 1996; Keller and Bustamante 2000)
Myosin	(Huxley and Simmons 1971; Huxley 1957; McMahon 1984; Smith and Geeves 1995)
Polymerization	(Peskin *et al.* 1993; Mogilner and Oster 1996; Mogilner and Oster 1999)
RNA polymerase	(Wang et al. 1998; Julicher and Bruinsma 1998)

DYNAMICS PROTEOMICS

9

This chapter focuses on the application of molecular modeling to calculate, manipulate, and predict protein structures and functions. Concepts of structure similarity/overlap, homology modeling, and molecular docking, which are special concerns of protein biochemists, are considered. Approaches to protein modeling by the use of two programs (Swiss-Pdb Viewer and KineMage) and two online servers (B and CE) are described.

STRUCTURE SIMILARITY AND OVERLAP

General Consideration

With protein structures, we can observe similarities in topology or even in structural details. Comparison of protein structures can reveal distant evolutionary relationships that would not be detected by sequence alignment alone. In devising measures of similarity and difference between two proteins, it is sometimes clearer to note how to proceed in comparing sequences than to do so in comparing structures. If the amino acid sequences of two proteins can be aligned, then we can either count the number of identical residues or use a similarity index between amino acids. Such an index would take the form of a 20 × 20 matrix, *M*, such that each entry corresponds to a pair of amino acids and M_{ij} gives a measure of the similarity between any pair of amino acids.

The similarity between two sequences is then the sum of values for each pair of aligned amino acids taken from the matrix, plus a correction to account for the gaps found in the sequences at sites of insertions or deletions of amino acids. Indeed, it is by maximizing such a similarity score that the optimal alignment of two sequences is conventionally calculated. In three dimensions the problem can be more complex. If two protein structures are very closely related and we align the residues, it is then easy to superpose the corresponding residues and to measure and analyze the nature of the deviations in position of corresponding atoms.

Structure tends to change more conservatively than sequence, and it is not uncommon to be able to recognize a relationship between proteins from their structures, when no evidence of homology appears in the sequences. Using computer graphics, it is possible to superpose two (or more) structures in one picture, and this can reveal, immediately and obviously, which features are conserved and which are different between two related proteins or in the conformational changes during allosteric transition/upon complexation.

It is this facility in which lies the real advantage of computer graphics. There are two ways to superpose structures. One may display pictures of each structure, and use interactive graphics to provide the facility to rotate and translate one with respect to the other, superposing the two "by eye." The second method is numerical. By selecting corresponding sets of atoms from two structures, a program can calculate the best "least-square fit" of one set of atoms to another. Both methods are useful: Fitting by eye permits rapid experimentation to assess the goodness of fit of different portions of the molecules. Numerical fitting permits quantitative comparisons of the relative goodness of fit of different structures and substructures.

Suppose we are dealing with two or more protein structures that contain regions in which the backbone atoms are almost congruent. We wish to superpose the two structures by moving one with respect to the other so that the corresponding atoms in the well-fitting regions are optimally matched. There are two aspects to this problem:

1. Choosing the corresponding sets of atoms from the two structures.
2. Finding the best match of the corresponding atoms.

Rigid-Body Motions and Least-Squares Fitting

The general numerical approach employs rigid body motions and least-squares fitting. Given two sets of points : x_i, $i = 1, 2 ..., N$ and y_i, $i = 1, 2, ..., N$ (here x_i and y_i are vectors specifying atomic coordinates), find the best rigid motion of the points $y_i \rightarrow Y_i$ such that the sum of the squares of the deviations $\Sigma|x_i - Y_i|^2$ is a minimum.

Two mathematical facts that we apply without proof are :

1. The most general motion of a rigid body is a combination of a rotation and a translation.
2. At the minimum, the mean positions (colloquially, the centers of gravity) of the two sets of points coincide.

It follows that Y_i must be equal to $Ry_i + t$, where R is a rotation matrix and t is a translation vector, and that to minimize

$$\Sigma|x_i - (Ry_i + t)|^2$$

A program must choose t to move the set y_t so that its center of gravity coincides with that of the x_t, and determine the optimal rotation matrix R. These solutions provide the following :

1. The minimum value of the root-mean-square (RMS) deviation between corresponding atoms :

   $$\Delta = [\Sigma\{x_i - (Ry_i + t)\}^2/N]^{1/2}$$

   This quantity is a useful measure of how similar the structures are.
2. A translation vector t and rotation matrix R: From the rotation matrix, one can derive the angle of rotation θ.
3. Assessment of the fit: By calculating the transformed points $Y_t = Ry_i + t$, one can report the deviation in individual atomic positions $|x_i - Y_i|$. Scanning such a list can reveal that certain portions of the selected regions fit well and that others do not.

Finding which regions fit well and which do not is an important step in the analysis of the structural similarities/differences. The repeated least-squares fitting of different set of atoms called "sieving" procedure is carried out to extract a well-fitting subset.

Identification of Similar Substructure

One difficulty in identifying similar substructures is that it is hard to formulate the problem in such a way to yield a unique answer. For this reason the use of interactive graphics as a guide to formulating the appropriate question is generally adapted. The basic idea is that if some set of corresponding residues from each of two proteins fits well, then any subsets of corresponding residues will also fit well though the converse is not necessarily true. For two related structures, the main question is finding the best *geometric transformation* in order to superimpose their frameworks together as closely as possible and evaluate their degree of similarity.

Generally, the superimposition is carried out by finding the rigid-body translation and minimizing the root-mean-square (RMS) distance between corresponding atoms of the two molecules; that is, two structures are superimposed and the square root is calculated from the sum of the squares of the distances between corresponding atoms:

$$\text{RMS (rms)} = \left\{ \sum_{i}^{N} (| u_i - v_i |)^2 \right\}^{1/2}$$

where u_i and v_i are corresponding vector distances of the ith atom in the two structures containing N atoms. The result is a measure of how each atom in the structure deviates from each other, and an RMS value of 0–3 Å signifies strong structural similarity. It is useful to set a threshold for goodness of fit—for example, 0.5 Å. By recording only those pairs of substructures for which RMS deviation, $\Delta\text{RMS} \leq 0.5$, one has generated a list of pairs of well-fitting substructures.

The next step is to work toward larger substructures by merging entries in this list. The procedure can be iterative; given a list of pairs of well-fitting substructures, form the unions of pairs of entries, fit them, and append to the list any new well-fitting substructures discovered. Another approach is to characterize the conformation by the sequence of main-chain conformational angles ϕ and ψ, and extract sets of consecutive residues with similar conformations.

Structural Relationships Among Related Molecules

Families of related proteins tend to retain similar folding patterns. If one examines sets of related proteins, it is clear that general folding pattern of the structural core is preserved. But there are distortions that increase in magnitude as the amino acid sequences diverge. In proteins, the common core generally contains the major elements of secondary structures and segments flanking them, including active site peptides. Large structural changes in the regions outside the core make it difficult to measure structural change quantitatively by straightforward application of simple least-squares superposition techniques.

To define a useful measure of structural divergence, it is necessary first to extract the core and then carry out the least-squares superposition on the one alone. For each major element of secondary structure of given two related proteins, perform a succession of superposition calculations, which include the main-chain atoms (N, $C\alpha$, C, O) of corresponding secondary structural elements plus additional residues extending from either end. Include more and more additional residues now identified as well-fitting contiguous region containing an element of secondary structure plus flanking segments. After finding such pieces corresponding to all common major elements of secondary structure, do a joint superposition of the main chain of all of them.

Homology Modeling

One of the ultimate goals of protein modeling is the prediction of 3D structures of proteins from their amino acid sequences. The prediction of protein structures rely on two approaches that are complementary and can be used in conjunction with each other :

1. Knowledge-based model combining sequence data to other information, such as homology modeling.
2. Energy-based calculations through theoretical models and energy minimization, such as, *ab initio* prediction.

The most promising methodology relies on modification of a closely related (homologous sequence), functionally analogous molecule whose 3D structure has been elucidated. This is the basis of homology modeling for deriving a putative 3D structure of a protein from a known 3D structure. Residues are changed in the sequence with minimal disturbance to the geometry, and energy minimization optimizes the altered structure.

The understanding is that functionally analogous proteins with homologous sequences will have closely related structures with common tertiary folding patterns. When sequence homology with known protein is high, modeling of an unknown structure by comparison can be carried out with reasonable success. However, it is noted that structure homology may remain significant even if sequence homology is low; that is, 3D structure seems better conserved than the residue sequence. It is shown that protein pairs with a sequence homology greater than 50% have 90% or more of the residues within a structurally conserved common fold. The homology modeling normally consists of four steps:

Table 9.1. Some Web Sites for Protein Structure Alignment

Program	*URL Address*	*Reference*
CE	http://cl.sdsc.edu	Shindyalov and Bourne (1998)
Dali	http://www2.ebiac.uk/deli	Holm and Sander (1993)
K2	http://sullivan.bu.edu/k2	Szustakowaski and Weng (2000)
ProSup	http://anna.came.sbg.ac.at/prosup/main.html	Feng and Sippl (1996)
TOP	http://bioinfol.mbfys.luse/TOP	Lu (2000)
VAST	http://www.ncbinlm.nih.gov:80/structure/VAST/vast.shtml	Gibrat *et. al.*, (1996)

1. Start from the known sequences.
2. Assemble fragments/substructures from different, known homologous structures.
3. Carry out limited structural changes from a known neighboring protein.
4. Optimize the structure by energy minimization.

In principle, predicting structure from the sequence by comparison to a known homologous structure is satisfactory when sequence homology is greater than 50%. Part of the problem of homology modeling at lower levels of similarity is to correctly align unknown and target proteins. Sequence alignments are more or less straightforward for levels of above 30% pairwise sequence identity. The region between 20% and 30% sequence identity (the twilight zone) is less certain. A means to automatically intrude into the twilight zone by detecting remote homologues (sequence identity <25%) are threading techniques.

A sequence of unknown structure is threaded into a sequence of known structure, and the fitness of the sequences for that structure is assessed. Comparing and overlapping two protein structures quantitatively remain an active area of development in structural biochemistry.

Methods for protein comparison generally rely on a fast full search of protein structure database. Some of these methods that are available over the Internet are listed in Table 9.1.

STRUCTURE PREDICTION AND MOLECULAR DOCKING

Ab initio Prediction of Protein Structure

It is well established that the sequence of the amino acid constituting a protein is of prime importance in the determination of its 3D structure and functional properties. Because sequence assignment is much faster and easier than 3D structure determination, the prediction of the 3D structure from the sequence of amino acids has been a great challenge for protein modeling. Energy-based structure prediction relies on energy minimization and molecular dynamics.

The method is faced with the problem of a large number of possible multiple minima, making the traversal of the conformational space difficult and making the detection of the real energy minimum or native conformation uncertain. The genetic algorithm is applied to the problem of protein structure prediction with a simple force field as the fitness function to generate a set of suboptimal/native-like conformations. Because the number of probable conformations is so large (for the main chain conformations with 2 torsion angles per residue and assuming 5 likely values per torsion angle, a protein of medium size with 100 residues will have $(2 \times 5)^{100} = 10^{100}$ conformations even if we further assume optimal (constant) bond lengths, bond angles, and torsion angles for the side chains), it is computationally impossible to evaluate all the conformations to find the global optimum.

Various constraints and approximations have to be introduced. For example, an assumption of constant bond lengths and bond angles by carrying out folding simulation in vacuum simplifies the total energy expression to

$$E = E_{\text{tor}} + E_{vdW} + E_{\text{elec}} + E_{pe}$$

where E_{tor}, E_{evW}, E_{elec}, and E_{pe} are torsion angle potential, van der Waals interaction, electrostatic potential, and pseudoentropic term that drive the protein to a globular state ($E_{pe} \approx 4^{\Delta d}$ kcal/mol, where Δd = largest distance between any Cα atoms in one conformation). The combination of heuristic criteria with force field components may alleviate the inadequacy in the simplified fitness functions. The secondary structure prediction may be performed to reduce the search space. Thus, either idealized torsion angles or boundaries for torsion angles according to the predicted secondary structures can be used to constrain main-chain torsion angles.

It is shown that the incorrect structures have less stabilizing hydrogen bonding, electrostatic, and van der Waals interactions. The incorrect structures also have a larger solvent accessible surface and a greater fraction of hydrophobic side-chain atoms exposed to the solvent.

Molecular Docking

Molecular docking explores the binding modes of two interacting molecules, depending upon their topographic features or energy-based consideration, and aims to fit them into conformations that lead to favorable interactions. Thus, one molecule (*e.g.*, ligand) is brought into the vicinity of another (*e.g.*, receptor) while calculating the interaction energies of the many mutual orientations of the two interacting molecules. In docking, the interacting energy is generally calculated by computing the van der Waals and the Coulombic energy contributions between all atoms of the two molecules.

Ligand-receptor interaction is an important initial step in protein function. The structure of ligand-receptor complex profoundly affects the specificity and efficiency of protein action. The molecular docking performs the computational prediction of the ligand-receptor interaction and the structures of ligand-receptor complexes. In the complex, ligand and receptor molecules are presumed to adopt the energetically most favorable docking structures. Thus, the goal of molecular docking is to search for the structure and stability of the complex with the global minimum energy. There are two classes of strategies for docking a ligand to a receptor.

The first class uses a whole ligand molecule as a starting point and employs a search algorithm to explore the energy profile of the ligand at the binding site, searching for optimal solutions for a specific scoring function. The search algorithms include geometric complementary match, simulated annealing, molecular dynamics, and genetic algorithms. Representative examples are DOCK3.5, AutoDock, and GOLD. The second class starts by placing one or several fragments (substructures) of a ligand into a binding pocket, and then it constructs the rest of the molecule in the site. Representative examples are DOCK4.0, FlexX, LUDI, GROWMOL, and HOOK.

In an interactive docking, an initial knowledge of the binding site is normally required.

The ligand is interactively placed onto the binding site. Geometric restraints such as distances, angles, and dihedrals between bonded or nonbonded atoms may facilitate the docking process. Two parameters, the equilibrium value of the internal coordinate and the force constant for the harmonic potential, need to be specified. For interatomic distance, the equilibrium restraint may be the initial length of the bond if you would like a particular bond length to remain constant during a simulation. If you want to force a bond coordinate to a new value, the equilibrium internal coordinate is the new value. Normally, the force constants are 7.0 kcal/mol Å^2 for an interatomic distance (larger for nonbonded distances), 12.5 kcal/mo Å^2 for an angle, and 16.0 kcal/mol Å^2 for a dihedral angle. Such constraints also ensure the confinement of the ligand molecule at proximity of the binding site of the receptor during energy minimization.

One may freeze most of the receptor molecule while allowing the ligand and contact residues to move in the field of the frozen atoms. Thus only the selected atoms (*e.g.*, ligand molecule and contact residues) move while other (frozen) atoms influence the calculation. In an automatic docking for example, a ligand is allowed to fit into potential binding cleft of the receptor of known crystal structure. Initially, the surface complementarity between the ligand and the receptor is determined by searching for a geometrical fit using the molecular surface as starting images (spheres).

From each surface points, a set of spheres that fill all pockets and grooves on the surface of the receptor is generated, and various criteria are introduced to reduce their number to one sphere per atom. Within this approximation of both ligand and receptor shapes by sets of spheres, the search is made to fit the set of ligand spheres within the set of receptor spheres. The matching algorithm collects all fits that are possible by comparing internal distances in both ligand and receptor and lists pairs of ligands and receptor spheres having all internal distances matching within a tolerance value.

Ligand atom coordinates are calculated, and the locations of the ligand atoms are optimized to improve the fit. There are numerous cases of proteins for which structures have been determined in more than one state of ligation. In some cases, the structures undergo little change, except perhaps for specific and localized changes associated with particular functional residues — for example, triose phosphate isomerase. Other cases, such as citrate synthase, show conformational change in which binding of ligand in a site between two domains leads to a closure of the interdomain cleft. Finally, there are the long-range integrated conformational changes associated with allosteric transitions of proteins.

Solvation

Solvation can have a profound effect on the results of molecular calculation. Solvent can strongly affect the energies of different conformations. It influences the hydrogen bonding pattern, solute surface area, and hydrophilic/hydrophobic group exposures of protein molecules. Solvation is an important, ongoing problem in protein modeling. At present, a protein molecule is placed in an imaginary box of water molecules, and periodic boundary conditions are set at constant dielectric of 1.0. The dielectric constant defines the screening effect of solvent molecules on nonbonded (electrostatic) interactions and can vary from 1 (*in vacua*) to 80 (in water). For the binding site, a constant dielectric of 4.0 is often used where no measured data are available. Alternatively, a distance-dependent dielectric constant is used to mimic the effect of solvent in molecular mechanics calculations in the absence of explicit water molecules.

APPLICATIONS OF PROTEIN MODELING

Most comprehensive software programs suitable for protein modeling are commercial packages, some of which are listed in Table 9.2. The application of HyperChem in the biomolecular modeling has been described and can be extended to the modeling of protein structures. The aspects of protein modeling will be illustrated with two freeware programs and two online servers.

Protein Modeling with Swiss PDB Viewer

Swiss-PDB Viewer is an application program that provides a user interface for visualization and analysis of bio-molecules in particular proteins. The program (Spdbv) can be downloaded from http://expasy.ch/spdbv/text/getpc.htm, and the user guide is available from http://www.expasy.ch/spdbv/mainpage.html. Spdbv implements GROMOS96 force field to compute

Table 9.2. Some Commercial Suppliers for Modeling Packages

Supplier	*URL Address*	*Modeling Package*
Accelrys, Inc.	http://www.accelrys.com	Cerius², Insigh II, QUANTA
CambridgeSoft Corp.	http://www.camsoft.com	Chem3D
Hypercube, Inc.	http://www.hyper.com	HyperChem
Tripos, Inc.	http://www.tripos.com	Alchemy, Biopolymer, SYBYL
Wavefunction, Inc.	http://www.wavefun.com	Spartan, PC Spartan

energy and to execute energy minimization by steep descent and conjugate gradient methods. For standalone modeling, an accompanying Spdbv Loop Database should be downloaded and placed into the "_stuff_" directory. Typically, the workspace consists of a menu bar, tool icons, and three windows; Main (Display) windows, Control panel, and Align window. The Control panel and Align window can be turned on and off from the Window menu. The Control panel provides a convenient way to select and manipulate the attributes of individual residues. The first column shows names of amino acid residues/nucleotides with check mark (✔) toggle between display/hide. The subsequent columns are for the display of side chains, residue labels, dot surfaces (VDW or accessibility), and secondary structures if the check marks (✔) are present. The last column with small square boxes is used to highlight the residue (s) with color.

The Align window permits an easy means of threading sequence in the homology modeling. The common tools are located at the top of the display window under the menu bar. These include move molecule (translate, enlarge, and rotate) tools and general usage (bond distance, bond angle, torsion angle, label, range, center, fit, mutation, and torsion) tools. Clicking the icon displays an instruction on the text space under the tool icons. For example, upon clicking the label icon, you will be asked to pick an atom. Upon picking an atom, the label (Residue ID and atom class) is displayed. The mutation icon provides a convenient approach to mutate selected residue(s) of the picked atom(s). To perform amino acid mutation of the displayed molecule,

- Color the residue (s) to be mutated from the Control panel.
- Click the mutation icon to open a pop-up list of amino acids.
- Pick the mutating amino acid and hit Enter key to automatically selecting the rotamer.
- Revisualize the molecule (check on the Control panel) if necessary.
- Click the mutation icon again and answer Yes to accept the mutation.
- Repeat the procedures for further mutations.
- Perform energy minimization to release local constraints of the mutant protein.
- Refine the structure by Selecting amino acids making clashes (Select → aa Making clashes) and initiate fixation (Tool → Fix selected sidechains).
- Save (File → Save → Layer) the mutant molecule as mutmolec.pdb.

The Preferences menu provides tools for defining and selecting options such as Loading protein molecule (Cα-trace, backbone, backbone + side chains or ribbon), Labels, Colors, Ribbon, Surface, and Electrostatic potential representations. If hydrogen bonds are to be calculated, the H-Bond detection threshold option should be checked. Hydrogen bonds are detected if an H is within 1.20–2.76 Å of an acceptor in the presence of explicit hydrogen or 2.35–3.20 Å between the prospective donor and acceptor in the absence of explicit hydrogen. If energy minimization is to be performed, the Energy minimization option should be checked.

The set preferences can be saved and opened for later operations as filename.prf. The PDB file can be opened and saved from the File menu. The Edit menu enables an online access to PROSITE and BLAST searches. The Edit menu also offers utilities for assigning secondary structure types to the selected amino acids. The groups/residues can be selected for visualization and manipulation from the Select menu or the Control panel. The Select features includes group kind (amino acids and nucleotides), group property (basic, acidic, polar or nonpolar), secondary structure, accessible amino acids (dialog box for defining % accessible surface) and neighbors of selected amino acids (dialog box for defining the distance from the selected molecule/residue).

The selected residues become highlighted (red) in the Control panel except that the neighbors of selected amino acids appear with side chains on the displayed molecule and as check marks in the side-chain column of the Control panel. After selection, the selected groups/residues can be highlighted from the Control panel by hitting the small square box to open a color palette or saved as newfile.pdb that contains the atomic coordinates of the selected molecule/residues. The

Tools menu provides tools for computing hydrogen bonds, molecular surface, electrostatic potential, threading energy, and force field energy.

If the appropriate Show option in the Display menu is checked, hydrogen bonds (with/without distance), surface dots, or electrostatic potential grids (automatic) are displayed on completion of the computation. The threading energies are plotted in the Align window (by clicking the little arrow located at the top of the window).

To perform energy minimization,

- Select Compute Energy (Force field) tool to open a dialog box for specifying angles/bonds to be computed. Be sure to check Show energy report, which is saved in the "temp" directory as molecule.En (*n* according to the order of computations). Each residue needs a topology for the force field to work. The force field calculation of the current version (version 3.7) of Spdbv only treats proteins (*i.e.*, energies of heteroatoms are not calculated).
- Before performing energy minimization, check the Energy Minimization option of the Preference menu.
- On the Energy minimization preferences box, specify the number of steps/cycles and method of minimization (Steepest descent or Conjugate gradients), angles/bonds to be minimized, and conditions for terminating the minimization. Be sure to check the radio button for Lock nonselected residues and the box for Show energy report.
- Choose the residues to be energy-minimized.
- Invoke the Energy Minimization tool of the Tools menu to initiate minimization.
- The results of each cycle are displayed and saved in the "temp" directory as molecule.En (*n* = order of executing minimization).

The structural similarity of two molecules can be compared by superimposition with Spdbv:

- Open the reference molecule and color the whole molecule with reference color (shift-click the square box in the Control panel to open color palette, and select the desired color).
- Open the trial molecule and color the whole molecule with another color.
- Invoke the Magic Fit tool from the Fit menu. The two molecules become superimposed and their sequences are displayed in the Align window.
- As you move the cursor pointing to amino acid residues of the aligned sequences in the Align window, the corresponding residues of the superimposed molecules in the Display window blink from blue to yellow, allowing an easy visualization of the overlap. In addition, the rms distance of the superimposed residues is displayed.
- Activate the trial molecule by clicking its name.
- Select the Generate Structural Alignment tool of the Fit menu to initiate the sequence alignment which can be viewed by clicking the little text icon.
- Save the file as alignefile.txt.
- To improve the superimposition (lowering rms distance), select the Iterative Magic Fit tool from the Fit menu.
- You can save each of the overlapped molecules separately (File → Save → Layer) as individual.pdb or save the superimposed molecules together (File → Save → Project) as overlap.pdb.

The modeling of a protein structure from its sequence against the known 3D structure of homologous protein (s) in the homology modeling can be attempted with Spdbv as follows :

- Download the trial sequence in fasta format (start with >) and save it as trialseq.txt.
- Open the template structure of the reference molecule (refmol.pdb) and color it with reference color.
- Choose the Load Raw Sequence to Model tool from the SwissModel menu and open trialseq.txt. The trial sequence appears on the Display window as α helix (color it other than the reference color if desired).
- Select the Magic Fit tool from the Fit menu. The α helix changes to the structure overlapping the reference structure.
- Make sure that the option "Update Threading Display Automatically" via SwissModel → Update threading display now is not checked.
- Invoke the Generate Structure Alignment tool of the Fit menu.
- Click the little arrow beside the question mark of the Alignment window to view a plot of threading energies.
- Click smooth to set smooth = 1 and check Update threading display now tool.
- Select Color → By Threading Energy to display threading energy profile of the structure (The mean force potential energy of the polypeptide chain increases with color varying from blue to green, yellow, and then red).
- Activate the trial structure from the Control panel for structural refinement.
- Perform two operations: Select → aa Making Clashes, and Tools → Fix Selected Side Chains.
- Repeat the process to observe a decrease in the number of amino acid residues which make clashes.
- Save the trial structure (File → Save → Layer) as trialmol.pdb, which can be submitted to Swiss-Model (http://www.expasy.ch/swissmod/) using Optimise (project) mode for optimization.

The two structures (or part of structures) from two opened pdb files can be merged. The merge can be applied to dock a ligand from one file onto the receptor site of the other. For example,

- Activate the first file (ligand file) and select the ligand molecule.
- Activate the second file (receptor file) and select residues of the receptor site.
- Select Create Merged Layer from the Selection tool of the Edit pull-down menu. The selected structures are merged.
- Saved (File → Save → Layer) as pdb file.

PROSITE is a database of biologically significant sites and patterns (amino acid residues) formulated from a known family of proteins as the characteristic familial identifiers or signatures that can be used to detected such family from the query sequences of proteins. If the PROSITE database (prosite.dat downloaded from http://expasy.hcuge.ch/sprot/prosite.html) is installed in the "usrstuff" directory, selecting Search for PROSITE tool from the Edit menu returns a list of ProSite for the protein molecule. Picking the desired ProSite highlights amino acid residues of the ProSite of the displayed molecule as well as the amino acid residues in the Control panel. This provides a convenient way of preparing *pdb* files of selected amino acid residues comprising ProSites of proteins (*e.g.*, pdb file of the active site residues).

Representation and Animation with KineMage

KineMage (kinetic images) is an interactive 3D structure illustration software that can be downloaded from http://orca.st.usm.edu/~rbateman/kinemage. It is adapted for the structure representation of biochemical molecules by many biochemistry textbooks. The program consists of two components; PREKIN and MAGE. The PREKIN program interprets/converts pdb file (molecule.pdb) to kinemage file (molecule.kin), which is then displayed and manipulated with the MAGE program. Launch the PREKIN program :

- Click Proceed to input the pdb file and to assign the output file (*e.g.*, molecule.kin). This opens the Starting Ranges dialog box, which offers options (radio buttons) for "Backbone browsing script," "Selection of built-in scripts," "New ranges," "Focus only," and Range list and Reset options.
- Choose Selection of built-in scripts to open the Built-in Scripts dialog box, which offers various scripts for illustrating macromolecules. "CaSS" is the default Browsing backbone script connecting all Cα of the backbone plus disulfides, "mcHb" gives main chain with hydrogen bonds, "aasc" offers Cα backbone with all the amino acid side chains colored by amino acid types, and "lots" includes me, sc, ca, and heteroatoms of the bound ligands and water. Two scripts for DNA/RNA are "naba" and "separate bases." The "ribbon" representations include (*a*) a thin ribbon with variable width dependent on the curvature of the backbone (an option for adding an arrow head on β strand) and (*b*) "ribbon HELIX_SHEET," which offers further options for specifying different widths for a helix, β strand, and β coil and for specifying the number of subunits to be transcribed.

For customerized structural representation (*e.g.*, main chain with side chains of active site residues and bound ligands) :

- Select radio button of New Ranges of the Starting Ranges dialog box to open the Range Controls box.
- Enter residue numbers for the start and end residues.
- Check me (main chain), sc (side chain), or ht (heteroatoms of ligands), respectively, and choose either "OK accept and. comes back for more" for multiple selections or "OK accept and end ranges" to terminate the selection.
- Accept defaults from the Focus Options box and specify the subunits to be transcribed in the Run Conditions box.
- Go to File then Quit.

PREKIN compiles molecule.kin file (ascii text file), which can be read and edited with Microsoft Word or Corel WordPerfect. The kinemage file (*.kin) has the following general organization:

```
@text: Description of kinemage to enter into the text window.
@kinemage i: Starting a new kinemage with numbers (where i = 1,2...n)
@caption: List for the caption window
@onewidth: All lines 2 pixels wide
 optional @?viewid {} to @?matrix for displaying different views
                                            derived from *.vw files
```

```
@group {identifier} [parameter] : High level display object
@subgroup {identifier} [parameter] : Mid level display object
@vectorlist {identifier} [parameter]: Low level display object,
        [color = ] is the most common
        {atomid} [-/P/L] x,y,z: Individual vectorlist item e.g.
        coordinates of atoms
```

The color of structural elements can be changed by changing color description of color = {color} on the appropriate @subgroup or @vectorlist line. The control checkboxes corresponding to @groups are displayed automatically. However, the control checkboxes relating to @subgroup or @vectorlist can be added/deleted by adding/deleting master = {label} on the desired lines where the checkboxes applied. Similarly, the label of the control checkbox can be changed as well. The molecu^i.kin files can be merged in a usual manner. The merged files can remain as independent kinemages (as ©kinemage *i*) or edited to become @group {identifier} within the merged kinemage.

The former is displayed under the control of Kinemage menu and is useful if each kinemage consists of different views. The latter offers mechanisms for superimposing structures (if multiple control checkboxes for @group are checked) and animation. To create animation (displaying structures in successive frames), merge molecu_ i.kin files either by text editor or from the Edit menu of MAGE program. Edit the merged file such that the merged kinemages become ©groups. Set parameter of @group to animate (adding an animate statement on @group line, *i.e.*, @group {mergedkinemage} animate).

For different views of structures, write molecule.vw (from @viewid to @matrix) from the MAGE program and insert molecule.vw files into the molecu_i.kin file. After editing, save the file as newname.kin or molecule.kin. The following lysozyme.kin file exemplifies the animated structures of lysozyme and its binary complexes showing different colors for the main chain Cα (white), contact side chains (cyan), hidden (off) catalytic side chains (blue), and bound ligand (orange, NAGs) with multiple control checkboxes (main chain, contact, catalytic and NAGs). @groups are derived from merged molecu_i.kin files which can be animated, ©subgroups designate different structural components with different colors assigned by @vectorlists.

The statement "@vectorlist off {catalytic} color = blue master = {catalytic}" provides a mechanism for highlighting the side chains of two catalytic residues (Glu35 and Asp52) to blue, which is shown initially in cyan via the checkbox.

```
@kinemage 1
@caption
 Lysozyme and its binary complexes.
@onewidth
@zoom 1.00
@zslab 200
@matrix
1.000000 -0.000000 0.000000 0.000000 1.000000 0.000000 0.000000 -
0.000000 1.000000 . . . . . . . . . . .
@group dominant {Lysozyme} animate
```

```
@subgroup {mainchain} master = {main chain}
@vectorlist {ca-ca} color = white
{ ca lys a 1}  P2.420, 10.480, 9.130
{ ca lys a 1 } L 2.420, 10.480, 9.130
{ ca val a 2 } L 2.390, 13.800, 7.270
{ ca phe a 3 } L-1.140, 15.180, 7.230
{ ca gly a 4 } L-2.630, 17.250, 4.400
{ ca arg a 5 } L -4.240, 20.510, 5.660
. . . .           . . . . . . . .
@subgroup {sidechain}
@vectorlist {ss} color = yellow
{ca cys a 64} P 4.450, 13.410, 28.940
(cb cys a 64} L4.340, 13.240, 27.440
(sg cys a 64} L5.670, 14.060, 26.450
{sg cys a 80}, L7.200, 12.800, 26.570
{cb sys a 80} L5.820, 10.990, 24.930
. . . .   . . . . .
@vectorlist (contact} colour = cyan master = {contact}
{ ca glu a 35 } P 4.850, 23.610, 15.140
{ cb glu a 35 } L3.960, 23.120, 16.260
{ eg glu a 35 } L 3.380, 24.160, 17.230
{ cdglu a 35 } L 4.450, 24.740, 18.110
{ oel glu a 35 } L 5.560, 24.230, 18.140
{ cd glu a 35 } P 4.450, 24.740, 18.110
{ oe2 glu a 35 } L 4.160, 25.770, 18.750
{ ca asn a 46 } P8.750, 18.590, 22.160
{ cb asn a 46 }L 8.590, 19.930, 21.500
{ eg asn a 46 } L 8 .940, 21.040, 22.460
{ odl asn a 46 } L 8.740, 21.050, 23.650
{ eg asn a 46 } P 8.940, 21.040, 22.460
{ nd2 asn a 46 } L970, 22.010, 21.910
. . . . . . . . . . . . . .
@group dominant {E-NAG4} animate
@subgroup (mainchain}
@vectorlist {ca-ca} colour = white master = {main chain}
. . . . . . . . . . . . . . . . . . . . .
```

```
@subgroup {NAG4} master = {NAGs}
@vectorlist {sc} colour = orange
{ cl nag b 1 } P 0.740, 27.090, 34.670
{ cl nag b 1 } 0.740, 27.090, 34.670
{ c2 nag b 1 } 0.280, 26.450, 35.950
{ c3 nag b 1} ▮1.070, 26.970, 36.340
{ c4 nag b 1} ▮1.150, 28.470, 36.320
{ c5 nag ba 1 } ▮0.520, 29.110, 35.070
. . . . . . . . . . . . .
```

Launch MAGE and click Proceed to display the desktop window, which consists of three component windows: the Caption window, which provides information about the file, the Text window, which describes the features of .kin file; and the Graphic window for interactive viewing of the structure. The structural features corresponding to the checkboxes are displayed if they are checked. More than one molecule can be displayed (overlapped) if multiple checkboxes corresponding to the group name are checked. Click the checkbox labeled Animate to display structures (with asterisk in front of the group name) in successive frames . To measure geometry, successively click two atoms for the distance and three atoms for the angle. The information appears at the lower left corner of the Graphic window.

To merge *.kin files, select Append File tool of the File menu and open the file to merge. To change default color, invoke Change Color tool of the Edit menu and click the structure element to open Color selection box from which the desired color can be assigned. Different structure views are created/added to the *.kin file by selecting Keep Current View tool of the Edit menu and entering View number and View ID to the dialog box. This adds @nviewid {viewid}, @nzoom, @nzslb, @ncenter, and @nmatrix (n is the view number) lines (ahead of @group) to the .kin file.

Biopolymer Modeling at *B* (Biopolymer) Server

The *B* server offers an online biomolecular modelling program that builds biomolecular models, measures their geometry, and implements an AMBER force field in the geometry optimization as well as in molecular dynamics. The Biomer authentification certificate is required to enable opening/saving files (though it is not needed to run the *B* program).

- Click the Start *B* button to open the *B* window.
- Hit the Grant button to the Biomer certificate to open the file browser box.
- Select Polynucleotide/Polypeptide/Polysaccharide from the Build menu to open a dialog box.
- For polysaccharide, choose connectivity (O1–C[1 – 6]), anomer (alpha or beta), isomer (*L* or *D*), and conformation (define phi and psi angle with omega = 180). Add sugars (alsohexoses or aldopentoses) to build polysaccharide chain.
- For polynucleotide, choose various options such as either DNA or RNA, Form (*a, b, c, d, e, t* or *z* forms), Build (5′ to 3′ or 3′ to 5′), single strand/double strand, and conformation of pentofuranose ring. Add bases (pairs) in the specified direction to build polynucleotide chain.

- For polypeptide, the *B* program provides options for building various protein conformations including 3 – 10 helix, alpha helix, alpha helix (*L-H*), beta sheet (anti-prl), beta sheet (parallel), various beta turns, extended, gamma turns, omega helix, pi helix, polyglycine, and polyroline. Choose the desired conformation and isomer (*L* or *D*) and then add amino acids from *N*-terminus to construct polypeptide chain.
- Cap the termini if desired for polypeptide or polysaccharide. Click Done button to view the molecule on the display window.

The Tools menu provides utilities to calculate net mass, net charge, energy and to invert chirality. For energy minimization.

- Select the Molecular mechanics menu and choose Steepest descent as the initial Minimizer that can be changed to Conjugated gradient later.
- Click Minimize to start energy minimization. The gradient and energy are displayed on the lower left corner of the display window.

For dynamic, simulation,

- Select the Molecular dynamics menu to open the dialog box. Specify Time, Temperature, and Step size for Heating cycle, Equilibrium period, and Cooling cycle.
- Click Start trajectory to initiate simulation. The lower left corner shows the progress of the simulation.
- Click the Grant button to the certificate to save the molecule.pdb.

Structure Comparison and Alignment at *CE* (Combinatorial Extension) Server

The 3D protein structure comparison and alignment can be performed online using the combinatorial extension algorithm at CE server. On the CE home page (http ://cl.sdsc.edu/ce :html), click All or Representatives PDB, then enter PDB ID into the Specify protein chain box to search for structural alignments from the CE database. From the hit list, select desired entries from the check boxes for sequence alignment and select Neighbors to display the structure neighbors superimposing with the protein specified by the query PDB ID. To perform sequence alignment and structure comparison,

- Click Two Chains to upload two PDB IDs (e.g., user's query pdb file against PDB file supplied by the user or retrieved form the CE database) on the CE home page.
- On the query form, enter the template PDB ID (either from the CE database or from user's directory) for Chain 1 and enter/browse the query pdb file for Chain 2.
- Click the Calculate Alignment button to receive Structure Alignment result showing sequence alignment of USR1 (template PDB) and USR2 (query PDB), which can be downloaded as a pdb file.
- Click the Press to Start Compare 3D button to open the display window showing the overlapped structures. The menu bar consists of Command (Reset, Exit), Feature table (Browse features...), Align_menu (Control penal), and 3D_menu (rotate, translate, zoom, switch C_{α}/all atoms).
- Select Feature table → Browse features → Structure or Sequence to analyze the superimposed structures. For example, choosing Structure (Phys)-Secondary structure illustrate secondary structural features of the superimposed structures and choosing

Structure (Comp)-Difference displays the structural differences between the superimposed structures.

To perform sequence alignment/structure comparison of a query pdb file against the CE database,

- Click *Uploaded by the user against the PDB* to upload the query pdb file.
- Enter the e-mail address and click the Search Database button. The search and analysis results are returned by the e-mail if the submission is followed by a message : "Your query [ce no.] has been successfully submitted. Results will be sent to [your e-mail address]."

WORKSHOPS

All software programs discussed whichever are the most appropriate, are to be applied for this workshop.

1. Deduce the probable conformation of porcine glucagon with the following sequence (construct initial structure in α helix from amino acids with HyperChem which lists residue ID in PDB file) by geometry optimization until the difference in energies between the successive steps is less than 1.0 kcal/mol or rms gradient of 0.1 kcal/(mol-Å) :

HSQGTFTSDYSKYLDSRRAQDFVQWLMNT

Compare the optimized structure with that of the X-ray crystallographic structure (1GCN.pdb).

2. Retrieve pdb files of lysozyme from different species including chicken (1LYZ.pdb), swan (IGBS.pdb), pheasant (1GHL.pdb), and turkey (1LZY.pdb), and compare their main chain structures. Write a new pdb file for the superimposed structures of chicken and turkey lysozymes.

3. Bovine chymotrypsinogen A is activated by excision of two dipeptides, Ser14-Arg15 and Thr147-Asn148, to form γ-chymotrypsin A and α-chymotrypsin A. The activation of inactive zymogen to active γ-chymotrypsin and α-chymotrypsin can be explained by comparing X-ray crystallographic structures of zymogen versus the active enzyme, which shows the formation of the properly oriented substrate binding pocket. The substrate binding pocket of chymotrypsin can be inferred from ProSite around Ser195. Retrieve pdb files of chymotrypsinogen (*Note:* The atomic coordinates of several residues in chymotrypsinogen are not shown in the pdb files due to poor resolution) and γ-chymotrypsin or α-chymotrypsin, and examine the structural change induced by the activation of zymogen in the movement of Glyl93. Write the new pdb file illustrating the change. Compare the strain energies and geometries of the catalytic triad – His57, Asp102, and Ser195 – of the optimized chymotrypsinogen and chymotrypsin.

4. The allosteric enzyme, aspartate transcarbamylase, exists in different confor-mational R-state and T-state. Retrieve the pdb files (2AT1.pdb and 3AT1.pdb) of both states that give the tetrameric structures, each consisting of two larger catalytic (A and C) and smaller regulatory (B and D) subunits. Write the new pdb file for the superimposed catalytic subunit structures and identify major conformational changes accompanied the allosteric transition with reference to

(*a*) Surface plots

(*b*) Conformational (Ramachandran) plots

(*c*) Solvent accessible amino acid residues (*e.g.*, 30% accessible surface)

(*d*) Amino acid residues interacting (*e.g.*, within 4.0 Å) with the effector, pal-mitoleic acid (PAM)

(*e*) Aspartate carbamoyltransferase signature (ProSite) amino acid residues

5. The induced fit model (Koshland, 1958) provides good explanation for enzyme specificity and catalysis. The binding of a substrate induces conformational changes that align the catalytic groups in the correct orientations for catalysis. The 3D structures of *E. coli* dihydrofolate reductase and its binary (with $NADP^+$ or NADPH) as well as ternary (with $NADP^+$ and folate.) complexes have been determined. Retrieve these pdb files (5DFR for apoenzyme, 6DFR and 1DRH for binary complexes, and 7DFR for ternary complex) and examine conformational differences upon complexations. Construct animated kinemage files showing induced conformational changes accompanying formation of the binary and ternary complexes (*Note:* The coordinates for the residues 16-20 are missing for some pdb files, and the coordinates for nicotinamide mononucleotide moiety of $NADP^+$ in 6DRF.pdb are not shown.)

6. Retrieve pdb file of ribonuclease A and use it as the reference coordinates to perform homology modeling of sheep ribonuclease A with the following amino acid sequence :

KESAAAKFERQHMDSSTSSASSNYCNQMMKSRNLTQDRCKPVNTFVHESLADVQAVCSQKNVACKNGQT

NCYQSYSTMSITDCRETGSSKYPNCAYKTTQAEKHIIVACEGNPYVPVHFDASV

Apply ProSite to identify pancreatic ribonuclease family signature.

7. The three disulfide bonds linking Cysteines 5 & 55 [5-55], 14 & 38 [14-38], and 30 & 51 [30-51] stabilize the tertiary fold (comprising two antiparallel β strands and two short α helices) of a 58-amino-acid protein, bovine pancreatic trypsin inhibitor (BPTI). The main folding pathway of BPTI has been elucidated. The disulfide bond [30-51] is formed as an initial intermediate. This is followed by the formation of disulfide bonds, [30-51] and [5-55], which adopt a native-like fold to bring Cys14 and Cys38 into correct proximity to form the last disulfide bond and the native structure (1BPI.pdb). Perform mutations by replacing cysteines with serines successively to produce mutants (C5S, CMS, C30S, C38S, C51S, and/or C55S) that mimic the unfolded polypeptide chain and folding intermediates. Use KineMage animation to simulate structural changes accompanied the folding pathway of BPTI at 300 K.

8. The active site of hen's egg-white lysozyme (1LYZ.pdb) is a cleft consisting of six subsites, *A* to *F*, which accommodate hexasaccharide with alternative N-acetylglucosamine (NAG) and W-acetylmuramie acid (NAM). Two catalytic residues, Glu35 and Asp52, are situated on the opposite sides of the active cleft between subsites *D* and *E*. Attempt to dock trisaccharide, NAM-NAG-NAM (9LYZ.pdb), to the active site (subsites B-D) of lysozyme interactively by placing Cl' of the reducing NAM at approximately equal distance from the carboxyl oxygens of Glu35 and Asp52 (Tsai, 1997). Perform energy minimization and molecular dynamic with respect to the trisaccharide and the two catalytic residues while freezing the rest of the lysozyme molecule. Deduce the structural changes upon complexation between lysozyme and trisaceharide.

9. Protein engineering has been carried out to redesign substrate specificity of lactate dehydrogenase from *Bacillus stearothermophilus* :

(*a*) Mutant (Q102M, K103V, P104S, A235G, A236G) tolerates substrates with large hydrophobic side chains.

(*b*) Mutant (Q102R) shifts substrate specificity from pyruvate to oxaloacetate.

The amino acid sequences of corresponding regions from dogfish M_4 and *Bacillus stearothermophilus* lactate dehydrogenases are shown below :

```
dogfish    VITAGARQQEGESRLNLV  ...   ...  KDVVDSAYEV
B.Stear.   VICAGANQKPGETRLDLV  ...   ...  VNVRDAAYQI
```

Simulate these mutants for dogfish M_4 lactate dehydrogeanse (6LDH.pdb). After energy minimization, superimpose the wild-type and mutant enzymes and construct their animated kinemage files including views for the complete molecular structures and close-ups of the *L*-lactate dehydrogenase active site as identified by the ProSite.

10. The two nicotinamide coenzymes, NAD(H) and NADP(H), differ in that the 2′-hydroxy group of the adenylyl moiety of NAD(H) is phosphorylated in NADP(H). Site-directed mutagenesis has been used to identify amino acid residues that are responsible for the coenzyme specificity in oxidoreductases. Glutathione reductase and dihydrolipoamide dehydrogenase catalyze related chemical reactions and bear a striking similarity to one another, both structurally and mechanistically. Both enzymes are dimeric and contain functionally reactive disulfide in addition to two molecules of FAD at the active site. Glutathione reductase catalyzes the electron transfer between NADPH and glutathione, whereas dihydrolipoamide dehydrogenase catalyzes NAD^+-dependent oxidation of dihydrolipoamide. A comparison of amino acid sequences of the βαβ dinucleotide binding fold of several glutathione reductases and dihydrolipoamide dehydrogenases as shown below suggests candidate residues (bold blue Glu for 2′-hydroxy of NAD(H) and bold red Arg for 2′-phosphate of NADP(K)) for investigating the coenzyme specificity by directed mutagenesis.

Glutathione reductase :

```
Human           194  GAGYIAVEMAGILSALGSKTSLMIRHDKVLRSFD  227
Drosophila      155  GAGYIGVEMAGILSALGYGT-VMVR-SIVLRGFD  186
Yeast           206  GAGYIGIELAGVFHGLGSETHLVIRGETVLRKFD  239
P. aeruginosa   188  GGGYIAVEFASIFNGLGAETTLLYRRDLFLRGFD  221
E. coli         174  GAGYIAVELAGVINGLGAKTHLFVRKHAPLRSFD  207
```

Dihydrolipoamide dehydrogenase :

```
Human           220  GAGVIGVELGSWQRLGADVTAVEFLGHVGGVGID  254
Pig             220  GAGVIGVELGSVWARLGADVTAVELLGHVGGIGID  254
Yeast           209  GGGIIGLEMGSVYSRLGSKVTVVEFQPQIGGPM-D  242
P. putida       178  GGGYIGLGLGIAYRKLGQVSTVVEARERILP-TYD  211
E. coli         180  GGGILGLEMGTVYHALGSQIDVVEMFDQVIP-AAD  213
```

Retrieve human erythrocyte glutathione reductase (3GRS.pdb) and perform the triple mutations by replacing Ile217 with Glu (I217E), Arg218 with Phe (R218F), and Arg224 with Gly (R224G). Save the energy-minimized mutant enzyme as 1GRM.pdb. Dock NAD^+ and $NADP^+$, respectively, to the mutant enzyme (save the complexes as 1GRM.pdb and 1GRN.pdb, respectively). Evaluate the structural and thermodynamic contributions of these three residues by comparing the mutant enzyme and complexes (1GRM.pdb, 1GRN.pdb and 1GRP.pdb) against the similarly energy-minimized wild-type enzyme and complexes (3GRS.pdb, 1GRA.pdb, and 1GRB.pdb). Construct animated kinemage files to display their structural differences with views of the complete molecular structures and close-ups of FAD, NAD(P) bound to nicotinamide nucleotide-disulfide oxidoreductase active site (identified by ProSite).

BIOSTATICS IN PROTEIN STRUCTURE

10

The aim of comparative or homology protein structure modeling is to build a three-dimensional (3D) model for a protein of unknown structure (the target) based on one or more related proteins of known structure (the templates). The necessary conditions for getting a useful model are that the similarity between the target sequence and the template structures is detectable and that the correct alignment between them can be constructed. This approach to structure prediction is possible because a small change in the protein sequence usually results in a small change in its 3D structure. Although considerable progress has been made in the ab initio protein structure prediction, comparative protein structure modeling remains the most accurate prediction method.

The overall accuracy of comparative models spans a wide range. At the low end of the spectrum are the low resolution models whose only essentially correct feature is their fold. At the high end of the spectrum are the models with an accuracy comparable to medium resolution crystallographic structures. Even low resolution models are often useful for addressing biological questions, because function can often be predicted from only coarse structural features of a model. At this time, approximately one-half of all sequences are detectably related to at least one protein of known structure. Because the number of known protein sequences is approximately 500,000, comparative modeling could in principle be applied to over 200,000 proteins. This is an order of magnitude more proteins than the number of experimentally determined protein structures (~13,000).

Furthermore, the usefulness of comparative modeling is steadily increasing, because the number of different structural folds that proteins adopt is limited and because the number of experimentally determined structures is increasing exponentially. It is predicted that in less than 10 years at least one example of most structural folds will be known, making comparative modeling applicable to most protein sequences.

All current comparative modeling methods consist of four sequential steps. The first step is to identify the proteins with known 3D structures that are related to the target sequence. The second step is to align them with the target sequence and pick those known structures that will be used as templates. The third step is to build the model for the target sequence given its alignment with the template structures. In the fourth step, the model is evaluated using a variety of criteria. If necessary, template selection, alignment, and model building are repeated until a satisfactory model is obtained.

The main difference among the comparative modeling methods is in how the 3D model is calculated from a given alignment (the third step). For each of the steps in the modeling process, there are programs and servers available on the World Wide Web.

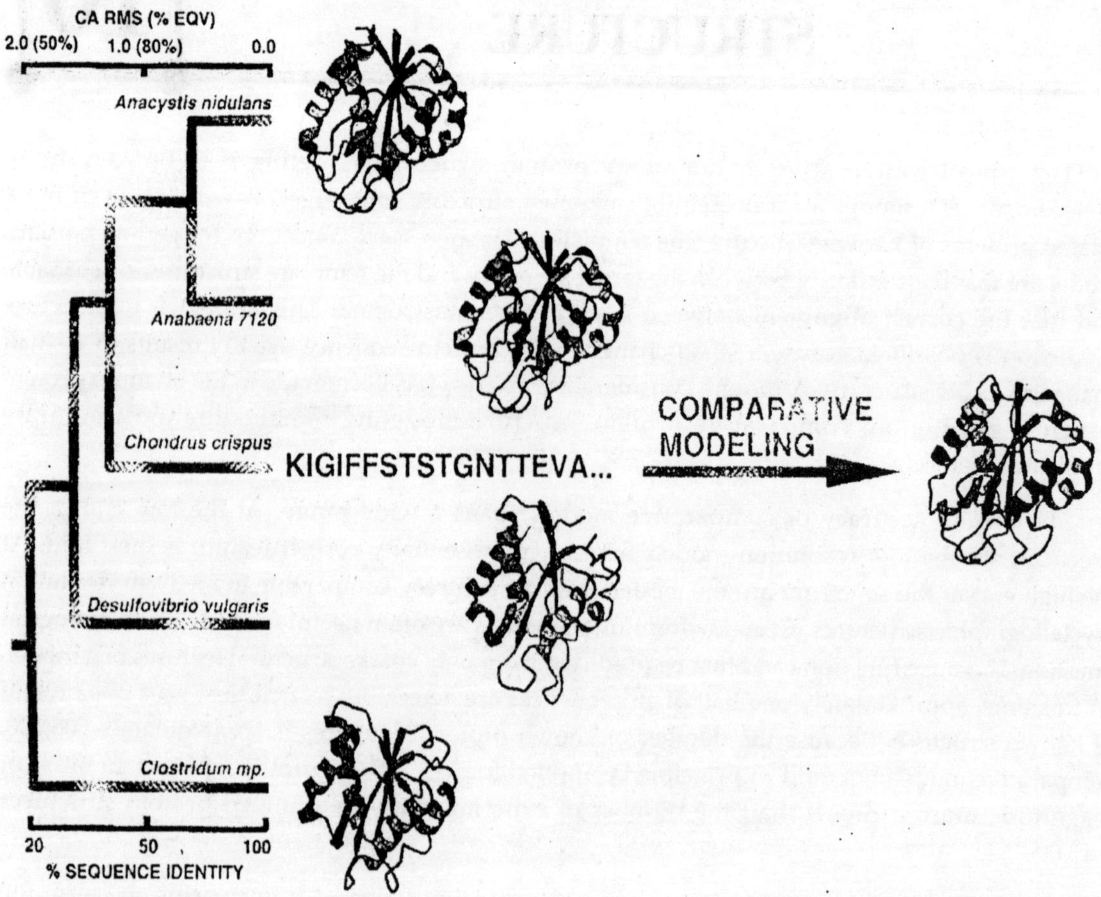

Fig. 10. 1. The basis of comparative protein structure modeling. Comparative modeling is possible because evolution resulted in families of proteins, such as the flavodoxin family, modeled here, which share both similar sequences and 3D structures. In this illustration, the 3D structure of the flavodoxin sequence from *C. crispus* (target) can be modeled using other structures in the same family (templates). The tree shows the sequence similarity (percent sequence identity) and structural similarity (the percentage of the C_α atoms that superpose within 3.8 Å of each other and the RMS difference between them) among the members of the family.

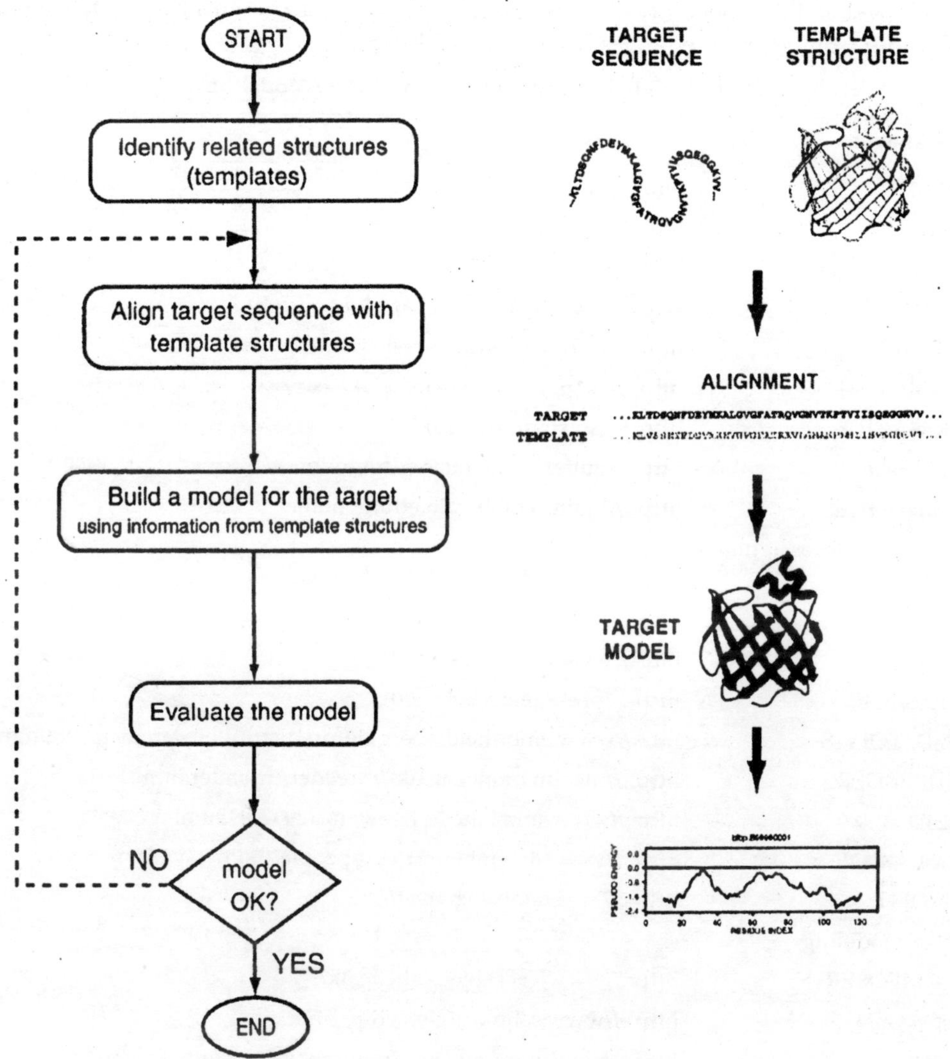

Fig. 10.2. The flowchart for comparative protein structure modeling.

STEPS IN COMPARATIVE MODELING

Identifying Known Protein Structures Related to the Target Sequence

The first task in comparative modeling is to identify all protein structures related to the target sequence, some of which will be used as templates. This is greatly facilitated by databases of protein sequences and structures and by software for scanning those databases. The target sequence can be searched against sequence databases such as PIR, GenBank, or TrEMBL/SWISS-PROT and/or structure databases such as the Protein Data Bank, SCOP, DALI, and CATH. Searching against sequence databases can be useful even if it identifies only proteins of unknown structure, because such sequences can be used to increase the sensitivity of the search for the template structures. At present, the probability of finding related proteins of known structure for a sequence picked randomly from a genome ranges from 20% to 70% [8-11].

Table 10.1. Web Sites Useful for Comparative Modeling

Databases	
NCBI	http ://www.ncbi.nlm.nih.gov/
PDB	http ://www.rcsb.org/pdb/
MSD	http ://msd.ebi.ac.uk/
CATH	http ://www.biochem.ucl.ac.uk/bsm/cath/
TrEMBL	http ://www.expasy.ch/sprst/sprst-top.html
SCOP	http ://scop.mrc-lmb.cam.ac.uk/scop/
PRESAGE	http ://csb.stanford.edu/
ModBase	http ://guitar.rockefeller.edu/modbase/
GeneCensus	http ://bioinfo.mbb.yale.edu/genome
Template search, fold assignment	
BLAST	http ://www.ncbi.nlm.nih.gov/BLAST/
FastA	http ://fasta.bioch.virginia.edu/
DALI	http ://www2.ebi.ac.uk/dali/
PRESAGE	http ://presage.berkeley.edu
PhD, TOPITS	http:/www.embl-heidelberg.de/predictprotein/predictprotein.html
THREADER	http ://insulin.bmnel.ac.uk//threader/threader.html
123D	http ://www-lmmb.ncifcrf.gov/~nicka/123D.html
UCLA-DOE	http ://www.doe-mbi.ucla.edu/people/frsvr/frsvr.html
PROFIT	http ://lore.came.sbg.ac.at/
Comparative modeling	
COMPOSER	http ://www-cryst.bioc.cam.ac.uk/
CONGEN	http ://www.cabm.rutgers.edu/~bruc
DRAGON	http ://www.nimr.mrc.ac.uk/~mathbio/a-aszodi/dragon.html
MODELLER	http ://guitar.rockefeller.edu/modeller/modeller.html
PrISM	http ://honiglab.cpmc.columbia.edu/
SWISS-MODEL	http ://www.expasy.ch/swissmod/SWISS-MODEL.html
WHAT IF	http ://www.cmbi.kun.nl/whatif/
ICM	http ://www.molsoft.com/
SCRWL	http ://www.cmpharm.ucsf.edu/~dunbrack
InsightII	http ://www.msi.com/
GENEMINE	http ://www.bioinformatics.ucla.edu/genemine
SYBYL	http ://www.tripos.com/
Model evaluation	
PROCHECK	http ://www.biochem.ucl.ac.uk/~roman/procheck/procheck.html
WHATCHECK	http ://www.sander.embl-heidelberg.de/whatcheck/

ProsaII	http ://www.came.sbg.ac.at
ProCyon	http ://www.horus.com/sippl/
BIOTECH	http ://biotech.embl-ebi.ac.uk :8400/
VERIFY3D	http ://www.doe-mbi.ucla.edu/Services/Verify3D.html
ERRAT	http ://www.doe-mbi.ucla.edu/Services/Errat.html
ANOLEA	http ://www.fundp.ac.be/pub/ANOLEA.html
AQUA	http ://www-nmr.chem.ruu.nl/users/rull/aqua.html
SQUID	http ://www.yorvic.york.ac.uk/~oldfield/squid
PROVE	http ://www.ucmb.ulb.ac.be/UCMB/PROVE/

There are three main classes of protein comparison methods that are useful in fold identification. The first class compares the target sequence with each of the database sequences independently, using pairwise sequence-sequence comparison. The performance of these methods in sequence searching and fold assignments has been evaluated exhaustively. The most popular programs in the class include Fasta and BLAST. Program MODELLER, which implements all the stages in comparative modeling, can also automatically search for proteins with known 3D structure that are related to a given sequence. It is based on the local dynamic programming method for pairwise sequence comparison. The second class of methods rely on multiple sequence comparison to improve greatly the sensitivity of the search.

The best-known program in this class is PSI-BLAST. Another similar approach that appears to perform even slightly better than PSI-BLAST has been described. It begins by finding all sequences in a sequence database that are clearly related to the target and easily aligned with it. The multiple alignment of these sequences is the target sequence profile. Similar profiles are also constructed for all potential template structures. The templates are then found by comparing the target sequence profile with each of the template sequence profiles, using a local dynamic programming method that relies on the common BLOSUM62 residue substitution matrix. These more sensitive fold identification techniques are especially useful for finding structural relationships when sequence identity between the target and the template drops below 25%.

In fact, methods of this class, which rely on multiple sequence information, appear to be currently the most sensitive fully automated approach to detecting remote sequence-structure relationships. The third class of methods rely on pairwise comparison of a protein sequence and a protein structure; that is, structural information is used for one of the two proteins that are being compared, and the target sequence is matched against a library of 3D profiles or threaded through a library of 3D folds. These methods are also called fold assignment, threading, or 3D template matching. These methods are especially useful when it is not possible to construct sequence profiles because there are not enough known sequences that are clearly related to the target or potential templates. What similarity between the target and template sequences is needed to have a chance of obtaining a useful comparative model? This depends on the question that is asked of a model. When only the lowest resolution model is required, it is tempting to use one of the statistical significance scores for a given match that is reported by virtually any sequence comparison program to select the best template.

However, it is better to proceed with modeling even when there is only a remote chance that the best template is suitable for deriving a model with at least a correct fold. The usefulness of the template should be assessed by the evaluation of the calculated 3D model. This is the best approach, because the evaluation of a 3D model is generally more sensitive and robust than the evaluation of an alignment.

Aligning the Target Sequence with the Template Structures

Once all the structures related to the target sequence are identified, the second task is to prepare a multiple alignment of the target sequence with all the potential template structures. When the sequence identity between the target and the template is higher than approximately 40%, this is straightforward. The gaps and errors in the alignments are rare, whether they are prepared automatically or manually. However, at 30% sequence identity, the fraction of residues that are correctly aligned by pairwise sequence-sequence comparison methods is only 80% on average, and this number drops sharply with further decrease in sequence similarity.

Thus, an additional effort in obtaining a more accurate alignment is needed because comparative modeling cannot, at present, recover from an incorrect alignment; the quality of the alignment is the single most important factor determining the accuracy of the 3D model. In the more difficult alignment problems, it is frequently beneficial to rely on the multiple structure and sequence information as follows. First, the alignment of the potential templates is prepared by superposing their structures. Typically, all residues whose C_α atoms are within 3.5 Å of each other upon least-squares superposition are aligned. Next, the sequences that are clearly related to the templates and easy to align with them are added to the alignment. The same is done for the target sequence. And finally, the two profiles are aligned with each other, taking structural information into account as much as possible. In principle, most sequence alignment and structure comparison methods can be used for these tasks.

In practice, it is frequently necessary to edit manually the positions of insertions and deletions to ensure that they occur in a reasonable structural context. For example, gaps are favored outside secondary structure segments, in exposed regions, and between residues that are far apart in space. Secondary structure prediction for the target sequence or its profile is also frequently useful in obtaining a more accurate alignment to the template structures. Although 3D profile matching and threading techniques are relatively successful in identifying related folds, they appear to be somewhat less successful in generating correct alignments. When there is an uncertainty about a region in the alignment, the best way to proceed is to generate 3D models for all alternative alignments, evaluate the corresponding models, and pick the best model according to the 3D model evaluation rather than the alignment score. Once a multiple alignment is constructed, matrices of pairwise sequence similarities are usually calculated and employed to construct a phylogenetic tree that expresses the relationships among the proteins in the family.

All significantly different structures in the cluster that contains the target sequence are usually used as templates in the subsequent model building, although other considerations should also enter into the template selection. For example, if the model is prepared to study the liganded state of a protein, then a template in the liganded state is preferred over a template without a ligand. Some methods allow short segments of known structure, such as loops, to be added to the alignment at this stage.

Model Building

1. Modeling by Assembly of Rigid Bodies

The first approach and one still widely used in comparative modeling is to assemble a model from a small number of rigid bodies obtained from the aligned protein structures. This approach is based on the natural dissection of the protein structure into conserved core regions, variable loops that connect them, and side chains that decorate the backbone. For example, the following semiautomated procedure is implemented in the computer program COMPOSER. First, the template structures are selected and superposed. Second, the "framework" is calculated by averaging the coordinates of the C_α atoms of structurally conserved regions in the template structures. Third, the core main chain atoms of each core region in the target model are obtained by superposing on the framework the core segment from the template whose sequence is closest to that of the target. Fourth, the loops are generated by scanning a database of all known protein structures to identify the structurally variable regions that fit the anchor core regions and have a compatible sequence. Fifth, the side chains are modeled based on their intrinsic conformational preferences and on the conformation of the equivalent side chains in the template structures. And finally, the stereochemistry of the model is improved either by a restrained energy minimization or a molecular dynamics refinement.

The accuracy of a model can be somewhat increased when more than one template structure is used to construct the framework and when the templates are averaged into the framework using weights corresponding to their sequence similarities to the target sequence. For example, differences between the model and *X*-ray structures may be slightly smaller than the differences between the *X*-ray structures of the modeled protein and the homologs used to build the model. Possible future improvements of modeling by rigid-body assembly include incorporation of rigid-body shifts such as the relative shifts in the packing of α-helices.

2. *Modeling by Segment Matching or Coordinate Reconstruction*

The basis of modeling by coordinate reconstruction is the finding that most hexapeptide segments of protein structure can be clustered into only 100 structurally different classes. Thus, comparative models can be constructed by suing a subset of atomic positions from template structures as "guiding" positions, then identifying and assembling short all-atom segments that fit these guiding positions. The guiding positions usually correspond to the C_α atoms of the segments that are conserved in the alignment between the template structure and the target sequence. The all-atom segments that fit the guiding positions can be obtained either by scanning all the known protein structures, including those that are not related to the sequence being modeled, or by conducting a conformational search restrained by an energy function.

For example, a general method for modeling by segment matching is guided by the positions of some atoms (usually C_α atoms) to find the matching segments in the representative databse of all known protein structures. This method can construct both main chain and side chain atoms and can also model gaps. It is implemented in the program SEGMOD which is part of the Genemine package. Even some side chain modeling methods and the class of loop construction methods based on finding suitable fragments in the databse of known structures can be seen as segment-matching or coordinate reconstruction methods.

3. *Modeling by Satisfaction of Spatial Restraints*

The methods in this class begin by generating many constraints or restraints on the structure of the target sequence, using its alignment to related protein structures as a guide. The restraints are generally obtained by assuming that the corresponding distances between aligned residues in the template and the target structures are similar. These homology-derived restraints are usually supplemented by stereochemical restraints on bond lengths, bond angles, non-bonded atom-atom contacts, etc., which are obtained from a molecular mechanics force field. The model is then derived by minimizing the violations of all the restraints. This can be achieved by either distance geometry or real-space optimization. For example, an elegant distance geometry approach constructs all-atom models from lower and upper bounds on distances and dihedral angles. Lower and upper bounds on $C_\alpha - C_\alpha$ and main chain-side chain distances, hydrogen bonds, and conserved the dihedral angles were derived for *E. coli* flavodoxin from four other flavodoxins; bounds were calculated for all distances and dihedral angles that had equivalent atoms in the template structures. The allowed range of values of a distance or a dihedral angle depended on the degree of structural variability at the corresponding position in the template structures. Distance geometry was used to obtain an ensemble of approximate 3D models, which were then exhaustively refined by restrained molecular dynamics with simulated annealing in water.

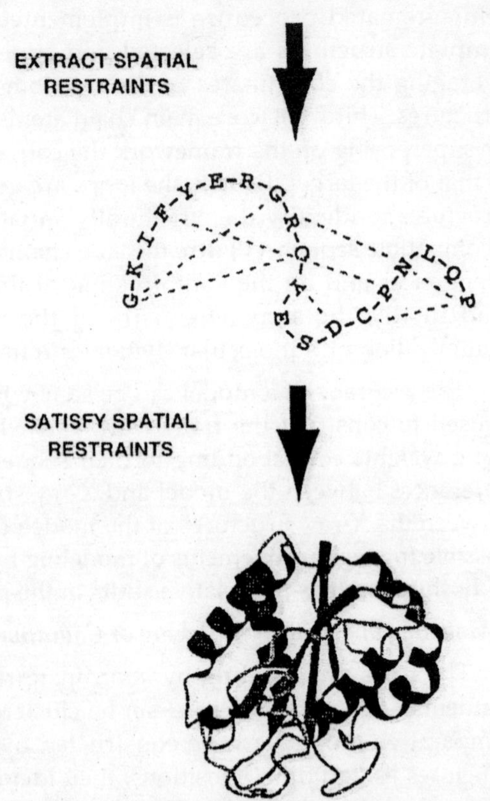

Fig. 10.3. Model building by Modeller. First, spatial restraints in the form of atomic distances and dihedral angles are extracted from the template structure(s). The alignment is used to determine equivalent residues between the target and the template. The restraints are combined into an objective function. Finally, the model for the target is optimized until a model that best satisfies the spatial restraints is obtained. This procedure is technically similar to the one used in structure determination by NMR.

We now describe our own approach in more detail. The question addressed is, What is the most probable structure for a certain sequence, given its alignment with related structures? The approach was developed to use as many different types of data about the target sequence as possible. It is implemented in the computer program MODELLER. The comparative modeling procedure begins with an alignment of the target sequence with related known 3D structures. The output, obtained without any user intervention, is a 3D model for the target sequence containing all main chain and side chain non-hydrogen atoms.

In the first step of model building, distance and dihedral angle restraints on the target sequence are derived from its alignment with template 3D structures. The form of these restraints was obtained from a statistical analysis of the relationships between similar protein structures. The analysis relied on a database of 105 family alignments that included 416 proteins of known 3D structure. By scanning the database of alignments, tables quantifying various correlations were obtained, such as the correlations between two equivalent $C_\alpha - C_\alpha$ distances or between equivalent main chain dihedral angles from two related proteins. These relationships are expressed as conditional probability density functions (pdf's) and can be used directly as spatial restraints. For example, probabilities for different values of the main chain dihedral angles are calculated from the type of residue considered, from main chain conformation of an equivalent residue, and from sequence similarity between the two proteins. Another example is the pdf for a certain $C_\alpha - C_\alpha$ distance given equivalent distances in two related protein structures. An important feature of the method is that the forms of spatial restraints were obtained empirically from a database of protein structure alignments.

In the second step, the spatial restraints and the CHARMM22 force field terms enforcing proper stereochemistry are combined into an objective function. The general form of the objective function is similar to that in molecular dynamics programs such as CHARMM22. The objective function depends on the Cartesian coordinates of ~ 10,000 atoms (3D points) that form a *system* (one or more molecules):

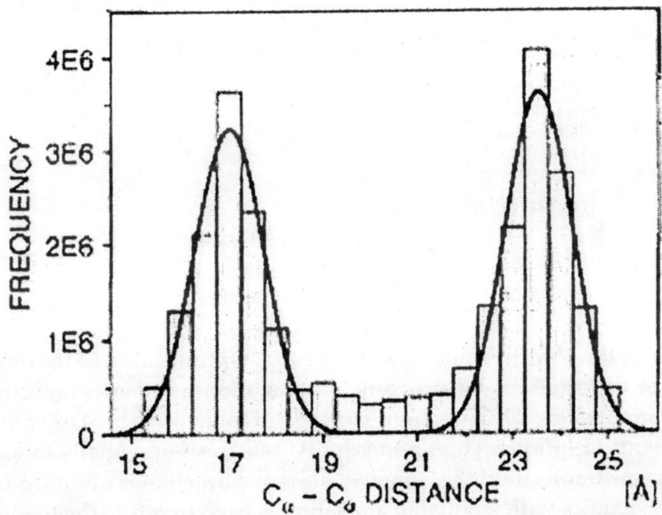

Fig. 10.4. Sample spatial restraint in MODELLER. A restraint on a given C_α–C_α distance, *d*, is expressed as a conditional probability density function that depends on two other equivalent distances (*d'* = 17.0 and *d''* = 23.5): *p*(*d*/*d'*, *d''*). The restraint (continuous line) is obtained by least-squares fitting a sum of two Gaussian functions to the histogram, which in turn is derived from many triple alignments of protein structures. In practice, more complicated restraints are used that depend on additional information such as similarity between the proteins, solvent accessibility, and distance from a gap in the alignment.

where F_{symm} is an optional symmetry term that restrains several parts of the structure to the same conformation. *R* are Cartesian coordinates of all atoms, *c* is a restraint term, *f* is a geometrical feature of a molecule, and p_i are parameters. For a 10,000 atom system there can be on the order of 200,000 restraints. The form of *c* is simple; it includes a quadratic function, harmonic lower and upper bounds, cosine, a weighted sum of a few Gaussian functions, Coulomb's law, Lennard-Jones potential, and cubic splines.

The geometrical features presently include a distance; an angle; a dihedral angle; a pair of dihedral angles between two, three, four atoms and eight atoms, respectively; the shortest distance in the set of distances; solvent accessibility in square angstroms; and atomic density, which is expressed as the number of atoms around the central atom. A pair of dihedral angles can be used to restrain strongly correlated features such as the main chain dihedral angles Φ and Ψ. Each of the restraints also depends on a few parameters p_i that generally vary from restraint to restraint. Some restraints can be used to restrain pseudo-atoms such as the gravity center of several atoms.

Finally, the model is obtained by optimizing the objective function in Cartesian space. The optimization is carried out by the use of the variable target function method, employing methods of conjugate gradients and molecular dynamics with simulated annealing. Several slightly different models can be calculated by varying the initial structure, and the variability among these models can be used to estimate the lower bound on the errors in the corresponding regions of the fold.

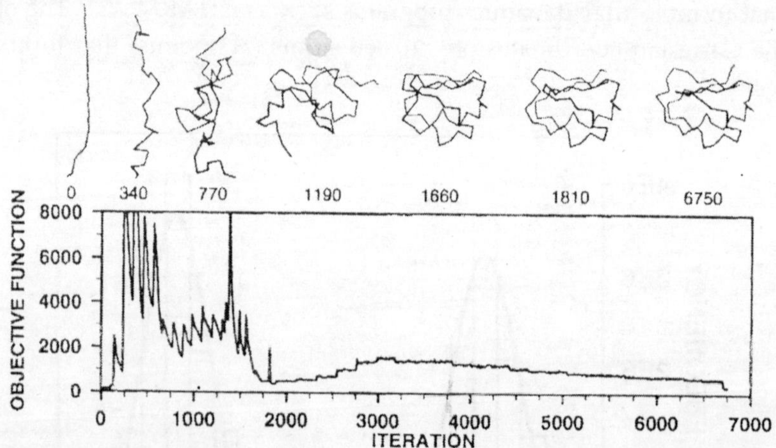

Fig. 10.5. Optimization of the objective function in MODELLER. Optimization of the objective function (curve) starts with a random or distorted model structure. The iteration number is indicated below each sample structure. The first approximately 2000 iterations correspond to the variable target function method relying on the conjugate gradients technique. This approach first satisfies sequentially local restraints, then slowly introduces longer range restraints until the complete objective function is optimized. In the remaining 4750 iterations, molecular dynamics with simulated annealing is used to refine the model. CPU time needed to generate one model is about 2 min for a 250 residue protein on a medium-sized workstation.

Because modeling by satisfaction of spatial restraints can use many different types of information about the target sequence, it is perhaps the most promising of all comparative modeling techniques. One of the strengths of modeling by satisfaction of spatial restraints is that constraints or restraints derived from a number of different sources can easily be added to the homology derived restraints. For example, restraints could be provided by rules for secondary structure packing, analyses of hydrophobicity and correlated mutations, empirical potentials of mean force, nuclear magnetic resonance (NMR) experiments, cross-linking experiments, fluorescence spectroscopy, image reconstruction in electron microscopy, site-directed mutagenesis, intuition, etc. In this way, a comparative model, especially in the difficult cases, could be improved by making it consistent with available experimental data and/or with more general knowledge about protein structure.

Loop Modeling

In comparative modeling, target sequences often have inserted residues relative to the template structures or have regions that are structurally different from the corresponding regions in the templates. Thus, no structural information about these inserted or conformationally variable segments can be extracted from the template structures. These regions frequently correspond to surface loops. Loops often play an important role in defining the functional specificity of a given protein framework, forming the active and binding sites. The accuracy of loop modeling is a major factor determining the usefulness of comparative models in applications such as ligand docking. Loop modeling can be seen as a mini protein folding problem. The correct conformation of a given segment of a polypeptide chain has to be calculated mainly from the sequence of the segment itself. However, loops are generally too short to provide sufficient information about their local fold. Even identical decapeptides do not always have the same conformation in different proteins. Some additional restraints are provided by the core anchor regions that span the loop and by the structure of the rest of a protein that cradles the loop.

Although many loop modeling methods have been described, it is still not possible to model correctly and with high confidence loops longer than approximately eight residues. There are two main classes of loop modeling methods : (1) the database search approaches, where a segment that fits on the anchor core regions is found in a database of all known protein structures, and (2) the conformational search approaches. There are also methods that combine these two approaches. The database search approach to loop modeling is accurate and efficient when a database of specific loops is created to address the modeling of the same class of loops, such as β-hairpins, or loops on a specific fold, such as the hypervariable regions in the immunoglobulin fold. For example, an analysis of the hypervariable immunoglobulin regions resulted in a series of rules that allowed a very high accuracy of loop prediction in other members of the family. These rules were based on the small number of conformations for each loop and the dependence of the loop conformation on its length and certain key residues.

There have been attempts to classify loop conformations into more general categories, thus extending the applicability of the database search approach to more cases. However, the database methods are limited by the fact that the number of possible conformations increases exponentially with the length of a loop. As a result, only loops up to four to seven residues long have most of their conceivable conformations present in the database of known protein structures. Even according to the more optimistic estimate, approximately 30% and 60% of all the possible eight- and nine-residue loop conformations, respectively, are missing from the database. This is made even worse by the requirement for an overlap of at least one residue between the database fragment and the anchor core regions, which means that the modeling of a five-residue insertion requires at least a seven-residue fragment from the database.

Despite the rapid growth of the database of known structures, there is no possibility of covering most of the conformations of a nine-residue segment in the foreseeable future. On the other hand, most of the insertions in a family of homologous proteins are shorter than nine residues. To overcome the limitations of the database search methods, conformational search methods were developed. There are many such methods, exploiting different protein representations, objective function terms, and optimization or enumeration algorithms. The search algorithms include the minimum perturbation method, molecular dynamics simulations, genetic algorithms, Monte Carlo and simulated annealing, multiple copy simultaneous search, self-consistent field optimization, and an enumeration based on the graph theory.

We now describe a new loop modeling protocol in the conformational search class. It is implemented in the program Modeller.

The modeling procedure consists of optimizing the positions of all non-hydrogen atoms of a loop with respect to an objective function that is a sum of many spatial restraints. Many different combinations of various restraints were explored. The best set of restraints includes the bond length, bond angle, and improper dihedral angle terms from the CHARMM22 force field, statistical preferences for the main chain and side chain dihedral angles, and statistical preferences for non-bonded contacts that depend on the two atom types, their distance through space, and separation in sequence. The objective function was optimized with the method of conjugate gradients combined with molecular dynamics and simulated annealing. Typically, the loop prediction corresponds to the lowest energy conformation out of the 500 independent optimizations. The algorithm allows straightforward incorporation of additional spatial restraints, including those provided by template fragments, disulfide bonds, and ligand binding sites.

To simulate comparative modeling problems, the loop modeling procedure was evaluated by predicting loops of known structure in only approximately correct environments. Such environments were obtained by distorting the anchor regions corresponding to the three residues at either end of the loop and all the atoms within 10 Å of the native loop conformation for up to 2-3 Å by molecular dynamics simulations. In the case of five-residue loops in the correct environments, the average error was 0.6 Å, as measured by local superposition of the loop main chain atoms alone (C, N, C_{α}, O). In the case of eight-residue loops in the correct environments, 90% of the loops had less than 2 Å main chain RMS error, with an average of less than 1.2 Å.

Side Chain Modeling

As for loops, side chain conformation is predicted from similar structures and from steric or energy considerations. The geometry of disulfide bridges is modeled from disulfide bridges in protein structures in general and from equivalent disulfide bridges in related structures. Modeling the stability and conformation of point mutations by free energy perturbation simulations is not discussed here.

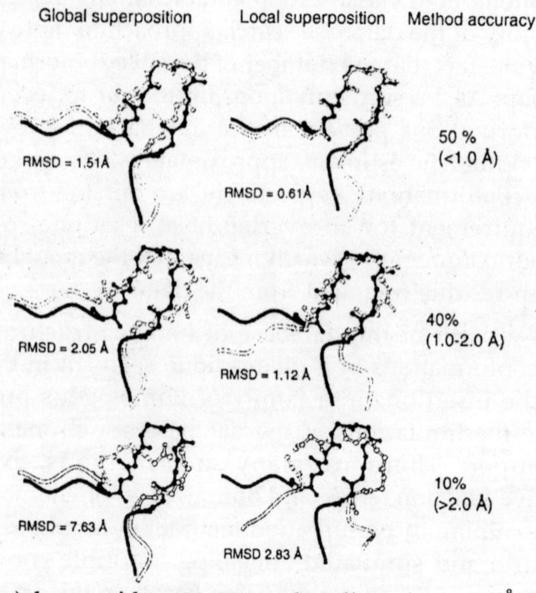

Fig. 10.6. Oxidoreductase (2nac), loop residues 28–35. Anchor distortion = 1.2 Å. Sample models of varying accuracy for an eight-residue loop in an approximately correct protein environment. The calculated loops (shaded) are compared with the *X*-ray structure (black). Three levels of accuracy are illustrated : High accuracy corresponding to the backbone RMSD < 1 Å (top), medium accuracy corresponding to the backbone RMSD < 2 Å (middle), and low accuracy corresponding to the backbone RMSD > 2 Å (bottom). The panels on the left compare the loop backbone conformations after least-squares superposition of the complete protein structure. The panels on the right compare the loop backbone conformations after local superposition of the loops. The RMSD values are quoted for the main chain atoms only. The fraction of the loops modeled at each accuracy level is given in the rightmost column. The figure was prepared using MOLSCRIPT.

Vasquez reviewed and commented on various approaches to side chain modeling. The importance of two effects on side chain conformation was emphasized. The first effect was the coupling between the main chain and side chains, and the second effect was the continuous nature of the distributions of side chain dihedral angles; for example, 5–30% of side chains in crystal structures are significantly different from their rotamer conformations and 6% of the χ_1 or χ_2 values are not within ±40° of any rotamer conformation. Both effects appear to be important when correlating packing energies and stability. The correct energetics may be obtained for the incorrect reasons; *i.e.*, the side chains adopt distorted conformations to compensate for the rigidity of the backbone.

Correspondingly, the backbone shifts may hinder the use of these methods when the template structures are related at less than 50% sequence identity. This is consistent with the *X*-ray structure of a variant of λ, represser, which reveals that the protein accommodates the potentially disruptive residues with shifts in its α-helical arrangement and with only limited changes in side chain orientations. Some attempts to include backbone flexibility in side chain modeling have been described, but the methods are not yet generally applicable. Significant correlations were found between side chain dihedral angle probabilities and backbone Φ, Ψ values. These correlations go beyond the dependence of side chain conformation on the secondary structure.

For example, the preferred rotamers can vary within the same secondary structure, with the changes in the Φ, Ψ dihedral angles as small as 20°. Since these changes are smaller than the differences between closely related homologs, the prediction of the side chain conformation generally cannot be uncoupled from backbone prediction. This partly explains why the conformation of equivalent side chains in homologous structures is useful in side chain modeling. A backbone-dependent rotamer library for amino acid side chains was developed and used to construct side chain conformations from main chain coordinates. This automated method first places the side chains according to the rotamer library and then removes steric clashes by combinatorial energy minimization. It was also demonstrated that simple arguments based on conformational analysis could account for many features of the observed dependence of the side chain rotamers on the backbone.

Recently, the main chain-dependent side chain rotamer library was recalculated and extensively evaluated. The accuracy of the method was 82% for the χ_1 dihedral angle and 72% for both χ_1 and χ_2 dihedral angles when the backbones of templates in the range from 30% to 90% sequence identity were used; a prediction was deemed correct when it was within 40° of the target crystal structure value. Chung and Subbiah gave an elegant structural explanation for the rapid decrease in the conservation of side chain packing as the sequence identity decreases below 30%. Although the fold is maintained, the pattern of side chain interactions is generally lost in this range of sequence similarity.

Two sets of computations were done for two sample protein sequences. The side chain conformation was predicted by maximizing packing on the fixed native backbone and on a fixed backbone with approximately 2 Å RMSD from the native backbone; the 2 Å RMSD generally corresponds to the differences between the conserved cores of two proteins related at 25–30% sequence identity. The side chain predictions based on the two kinds of backbone turned out to be unrelated. Thus, inasmuch as packing reflects the true laws determining side chain conformation, a backbone with less than 30% sequence identity to the sequence being modeled is no longer sufficiently restraining to result in the correct packing of the buried side chains. The solvation term is important for the modeling of exposed side chains. It was also demonstrated that treating hydrogen bonds explicitly could significantly improve side chain prediction.

Calculations that do not take into account the solvent, either implicitly or explicitly, introduce errors into the hydrogen-bonding patterns even in the core regions of a protein.

Residues with zero solvent accessibility area can still have a significant interaction energy with the solvent atoms. A recent survey analyzed the accuracy of three different side chain prediction methods. These methods were tested by predicting side chain conformations on near-native protein backbones with <4 Å RMSD to the native structures. The three methods included the packing of backbone-dependent rotamers, the self-consistent mean-field approach to positioning rotamers based on their van der Waals interactions, and the segment-matching method of Levitt. The accuracies of the methods were similar. They were able to predict correctly approximately 50% of χ_1 angles and 35% of both χ_1 and χ_2 angles. In typical comparative modeling applications where the backbone is closer to the native structures (<2 Å RMSD), these numbers increase by approximately 20%.

AB Initio Protein Structure Modeling Methods

This section briefly reviews prediction of the native structure of a protein from its sequence of amino acid residues alone. These methods can be contrasted to the threading methods for fold assignment, which detect remote relationships between sequences and folds of known structure, and to comparative modeling methods discussed in this review, which build a complete all-atom 3D model based on a related known structure. The methods for ab initio prediction include those that focus on the broad physical principles of the folding process and the methods that focus on predicting the actual native structures of specific proteins.

The former frequently rely on extremely simplified generic models of proteins, generally do not aim to predict native structures of specific proteins, and are not reviewed here. Although comparative modeling is the most accurate modeling approach, it is limited by its absolute need for a related template structure. For more than half of the proteins and two-thirds of domains, a suitable template structure cannot be detected or is not yet known. In those cases where no useful template is available, the ab initio methods are the only alternative. These methods are currently limited to small proteins and at best result only in coarse models with an RMSD error for the C_α atoms that is greater than 4 Å. However, one of the most impressive recent improvements in the field of protein structure modeling has occurred in ab initio prediction. *Ab initio* prediction relies on the thermodynamic hypothesis of protein folding.

The thermodynamic hypothesis suggests that the native structure of a protein sequence corresponds to its global free energy minimum state. Accordingly, *ab initio* prediction methods are generally formulated as optimizations. As such, they can be distinguished by the representation of a protein and its degrees of freedom, the function that defines the energy for each of the allowed conformations, and the optimization method that attempts to find the global minimum on a given energy surface.

Although the folding of short proteins has been simulated at the atomic level of detail, a simplified protein representation is often applied. Simplifications include using one or a few interaction centers per residue as well as a lattice representation of a protein. Some methods are hierarchical in that they begin with a simplified lattice representation and end up with an atomistic detailed molecular dynamics simulation. The energy functions for folding simulations include atom-based potentials from molecular mechanics packages such as CHARMM, AMBER, and ECEPP, the statistical potentials of mean force derived from many known protein structures, and simplified potentials based on chemical intuition.

Some methods also incorporate non-physical spatial restraints obtained from multiple sequence alignments and other considerations to reduce the size of the conformational space that needs to be explored. Many different optimization methods—even enumerations with some lattice models—have been applied to the protein folding problem. These methods include molecular dynamics simulations, Monte Carlo sampling, the diffusion equation method, and genetic algorithm optimization. A recent and particularly successful approach assembles the whole protein model from relatively short building blocks. Many candidate blocks are obtained from known protein structures by relying on energetic, geometrical, and sequence similarity filters.

The model of a whole protein is then assembled from such pieces by a Monte Carlo optimization of a statistical energy function. There is scope for combining the comparative modeling and *ab initio* methods. The modeling of inserted loops in comparative prediction is based primarily on the sequence information alone. In addition, the alignment errors as well as large distortions of the target relative to the template require that such regions be modeled *ab initio* without relying on the template structure. It is likely that the *ab initio* approaches will help reduce some of the limitations of comparative modeling.

Errors in Comparative Models

The errors in comparative models can be divided into five categories :

1. Errors in side chain packing.
2. Distortions or shifts of a region that is aligned correctly with the template structures.
3. Distortions or shifts of a region that does not have an equivalent segment in any of the template structures.
4. Distortions or shifts of a region that is aligned incorrectly with the template structures.
5. A misfolded structure resulting from using an incorrect template.

Significant methodological improvements are needed to address all of these errors.

Errors 3–5 are relatively infrequent when sequences with more than 40% identity to the emplates are modeled. For example, in such a case, approximately 90% of the main chain atoms re likely to be modeled with an RMS error of about 1 Å. In this range of sequence similarity, the lignment is mostly straightforward to construct, there are not many gaps, and structural ifferences between the proteins are usually limited to loops and side chains. When sequence dentity is between 30% and 40%, the structural differences become larger, and the gaps in the lignment are more frequent and longer. As a result, the main chain RMS error increases to about .5 Å for about 80% of the residues. The rest of the residues are modeled with large errors because ne methods generally fail to model structural distortions and rigid-body shifts and are unable recover from misalignments. Below 40% sequence identity, misalignments and insertions in ne target sequence become the major problems. Insertions longer than about eight residues cannot et be modeled accurately, but shorter loops can frequently be modeled successfully.

When sequence identity drops below 30%, the main problem becomes the identification of elated templates and their alignment with the sequence to be modeled. In general, it can be expected that about 20% of residues will be misaligned and consequently incorrectly modeled with an error greater than 3 Å at this level of sequence similarity. This is a serious impediment for comparative modeling because it appears that at least one-half of all related protein pairs are related at less than 30% sequence identity.

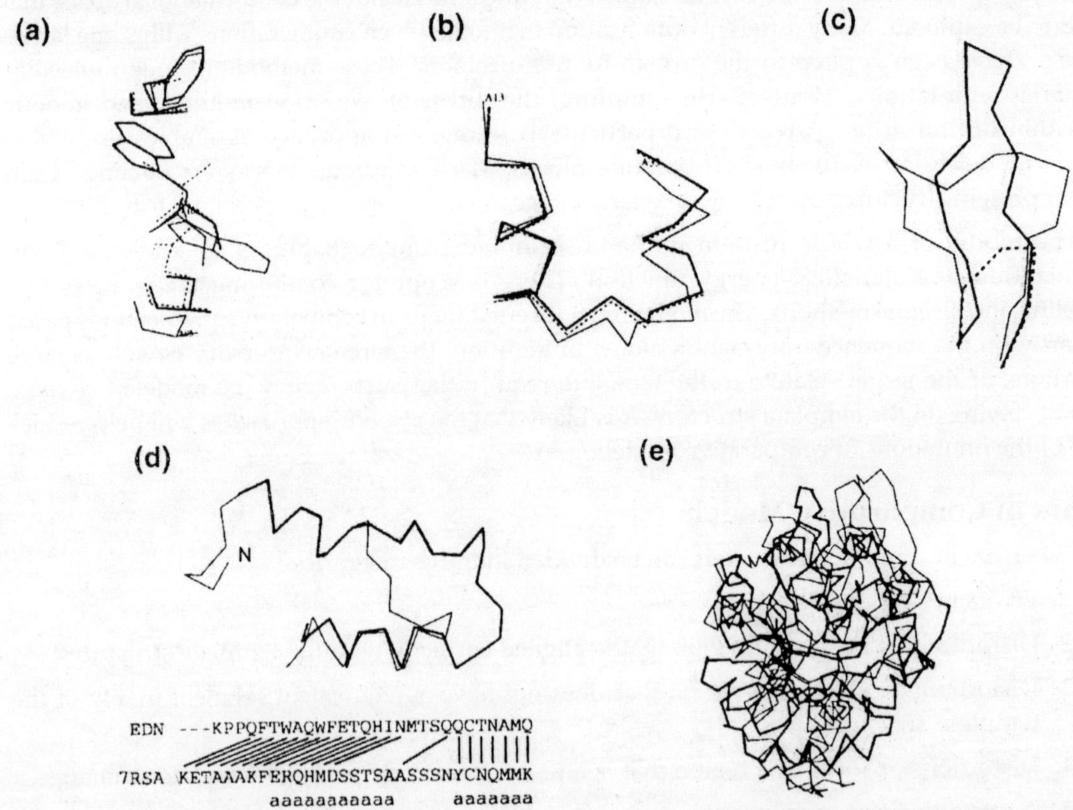

Fig. 10.7. Typical errors in comparative modeling, (*a*) Errors in side chain packing. The Trp 109 residue in the crystal structure of mouse cellular retinoic acid binding protein I (thin line) is compared with its model (thick line) and with the template mouse adipocyte lipid-binding protein (broken line), (*b*) Distortions and shifts in correctly aligned regions. A region in the crystal structure of mouse cellular retinoic acid binding protein I (thin line) is compared with its model (thick line), and with the template fatty acid binding protein (broken line), (*c*) Errors in regions without a template. The C_α trace of the 112-117 loop is shown for the X-ray structure of human eosinophil neurotoxin (thin line), its model (thick line), and the template ribonuclease *A* structure (residues 111-117; broken line), (*d*) Errors due to misalignments. The N-terminal region in the crystal structure of human eosinophil neurotoxin (thin line) is compared with its model (thick line). The corresponding region of the alignment with the template ribonuclease *A* is shown. The black lines show correct equivalences, that is residues whose C_α atoms are within 5 Å of each other in the optimal least-squares superposition of the two X-ray structures. The "*a*" characters in the bottom line indicate helical residues, (*e*) Errors due to an incorrect template. The X-ray structure of α-trichosanthin (thin line) is compared with its model (thick line), which was calculated using indole-3-glycerophosphate synthase as the template.

It has been pointed out that a comparative model is frequently more distant from the actual target structure than the closest template structure used to calculate the model. However, at least for some modeling methods, this is the case only when there are errors in the template-target alignment used for modeling and when the correct structure-based template-target alignment is used for comparing the template with the actual target structure. In contrast, the model is generally closer to the target structure than any of the templates if the modeling target-template alignment is used in evaluating the similarity between the actual target structure and the template. As a result, using a model is generally better than using the template structure even when the alignment is incorrect, because the actual target structure, and therefore the correct template-target alignment, are not available in practical modeling applications.

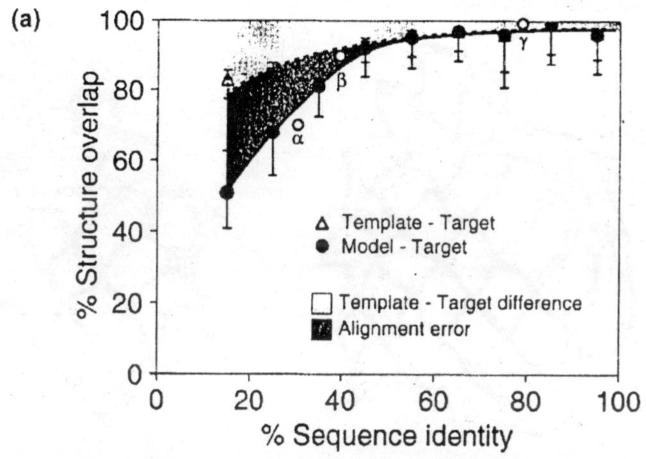

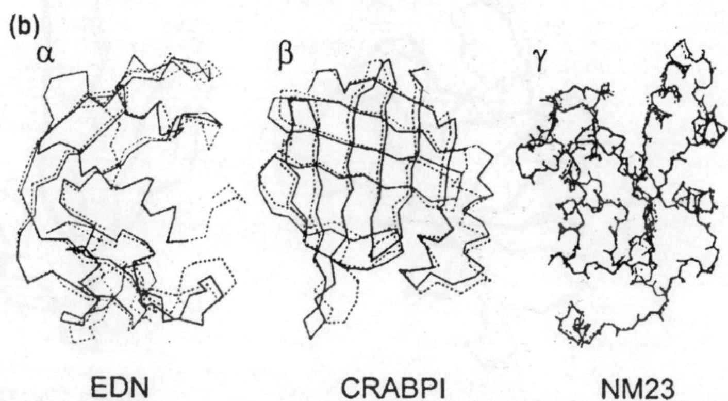

Fig. 10.8. Average model accuracy as a function of the percentage identity between the target and template sequences, (*a*) The models were calculated entirely automatically, based on single template structures. As the sequence identity between the target sequence and the template structure decreases, the average structural similarity between the template and the target also decreases (dashed line, triangles). Structure overlap is defined as the fraction of equivalent C_α atoms. For comparison of the model with the actual structure (continuous line, circles), two C_α atoms were considered equivalent if they were within 3.5 Å of each other and belonged to the same residue. For comparison of the template structure with the actual structure (dashed line, triangles), two C_α atoms were considered equivalent if they were within 3.5 Å of each other after alignment and rigid-body superposition by the ALIGN3D command in MODELLER, (*b*) Three models (solid line) compared with their corresponding experimental structures (dotted line). The models were calculated with MODELLER in a completely automated fashion before the experimental structures were available. When multiple sequence and structure information is used and the alignments are edited by hand, the models can be significantly more accurate than shown in this plot.

To put the errors in comparative models into perspective, we list the differences among structures of the same protein that have been determined experimentally. The 1 Å accuracy of main chain atom positions corresponds to X-ray structures defined at a low resolution of about 2.5 Å and with an *R*-factor of about 25%, as well as to medium resolution NMR structures determined from 10 interproton distance restraints per residue. Similarly, differences between the highly refined X-ray and NMR structures of the same protein also tend to be about 1 Å. Changes in the environment (*e.g.*, oligomeric state, crystal packing,

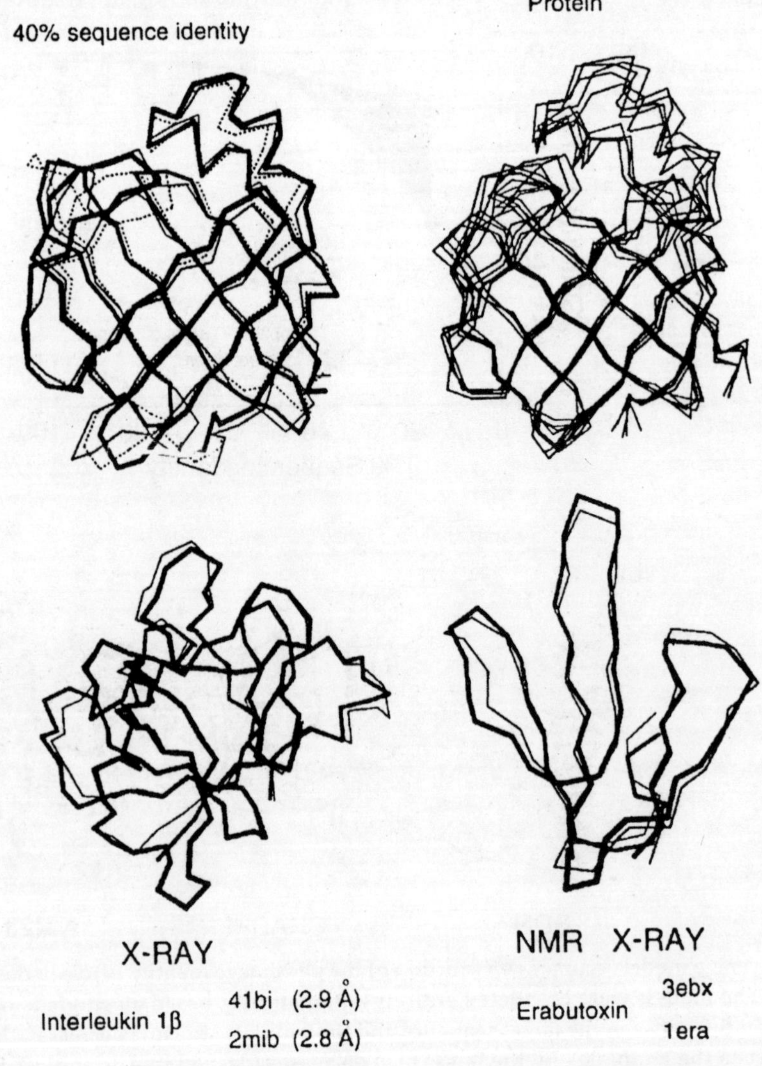

Fig. 10.9. Relative accuracy of comparative models. Upper left panel, comparison of homologous structures that share ~40% sequence identity. Upper right panel, conformations of ileal lipid-binding protein that satisfy the NMR restraints set equally well. Lower left panel, comparison of two independently determined X-ray structures of interleukin 1β. Lower right panel, comparison of the X-ray and NMR structures of erabutoxin. The figure was prepared using the program MOLSCRIPT.

solvent, ligands) can also have a significant effect on the structure. Overall, comparative modeling based on templates with more than 40% identity is almost as good as medium resolution experimental structures, simply because the proteins at this level of similarity are likely to be as similar to each other as are the structures for the same protein determined by different experimental techniques under different conditions.

However, the caveat in comparative protein modeling is that some regions, mainly loops and side chains, may have larger errors. *A* particularly informative way to test protein structure modeling methods, including comparative modeling, is provided by the biennial meetings on critical assessment of techniques for protein structure prediction (CASP). The most recent meeting was held in December 1998. Protein modelers are challenged to model sequences with unknown 3D structure and to submit their models to the organizers before the meeting. At the same time, the 3D structures of the prediction targets are being determined by X-ray crystallography or NMR methods. They become available only after the models are calculated and submitted. Thus, a bona fide evaluation of protein structure modeling methods is possible.

Model Evaluation

Essential for interpreting 3D protein models is the estimation of their accuracy, both the overall accuracy and the accuracy in the individual regions of a model. The errors in models arise from two main sources, the failure of the conformational search to find the optimal conformation and the failure of the scoring function to identify the optimal conformation. The 3D models are generally evaluated by relying on geometrical preferences of the amino acid residues or atoms that are derived from known protein structures. Empirical relationships between model errors and target-template sequence differences can also be used. It is convenient to approach an evaluation of a given model in a hierarchical manner. It first needs to be assessed if the model at least has the correct fold.

The model will have a correct fold if the correct template is picked and if that template is aligned at least approximately correctly with the target sequence. Once the fold of a model is confirmed, a more detailed evaluation of the overall model accuracy can be performed based on the overall sequence similarity on which the model is based. Finally, a variety of error profiles can be constructed to quantify the likely errors in the different regions of a model. A good strategy is to evaluate the models by using several different methods and identify the consensus between them. In addition, energy functions are in general designed to work at a certain level of detail and are not appropriate to judge the models at a finer or coarser level.

There are many model evaluation programs and servers. A basic requirement for a model is that it have good stereochemistry. The most useful programs for evaluating stereochemistry are PROCHECK, PROCHECK-NMR, AQUA, SQUID, and WHATCHECK. The features of a model that are checked by these programs include bond lengths, bond angles, peptide bond and side chain ring planarities, chirality, main chain and side chain torsion angles, and clashes between non-bonded pairs of atoms. In addition to good stereochemistry, a model also has to have low energy according to a molecular mechanics force field, such as that of CHARMM22.

However, low molecular mechanical energy does not ensure that the model is correct. Thus, distributions of many spatial features have been compiled from high resolution protein structures, and any large deviations from the most likely values have been interpreted as strong indicators of errors in the model. Such features include packing, formation of a hydrophobic core, residue and atomic solvent accessibilities, spatial distribution of charged groups, distribution of atom-atom distances, atomic volumes, and main chain hydrogen bonding.

Another group of methods for testing 3D models that implicitly take into account many of the criteria listed above involve 3D profiles and statistical potentials. These methods evaluate the environment of each residue in a model with respect to the expected environment as found in the high resolution X-ray structures. Programs implementing this approach include VERIFY3D, PROSA, HARMONY, and ANOLEA. An additional role of the model evaluation methods is to help in the actual modeling procedure. In principle, an improvement in the accuracy of a model is possible by incorporating the quality criteria into a scoring function being optimized to derive the model in the first place.

Applications of Comparative Modeling

Comparative modeling is often an efficient way to obtain useful information about the proteins of interest. For example, comparative models can be helpful in designing mutants to test hypotheses about the protein's function; identifying active and binding sites; searching for, designing, and improving ligands for a given binding site; modeling substrate specificity; predicting antigenic epitopes; simulating protein-protein docking; inferring function from calculated electrostatic potential around the protein; facilitating molecular replacement in X-ray structure determination; refining models based on NMR constraints; testing and improving a sequence-structure alignment; confirming a remote structural relationship; and rationalizing known experimental observations. Fortunately, a 3D model does not have to be absolutely perfect to be helpful in biology, as demonstrated by the applications listed above.

However, the type of question that can be addressed with a particular model does depend on the model's accuracy. At the low end of the accuracy spectrum, there are models that are based on less than 25% sequence identity and have sometimes less than 50% of their C_α atoms within 3.5 Å of their correct positions. However, such models still have the correct fold, and even knowing only the fold of a protein is frequently sufficient to predict its approximate biochemical function. More specifically, only nine out of 80 fold families known in 1994 contained proteins (domains) that were not in the same functional class, although 32% of all protein structures belonged to one of the nine superfolds]. Models in this low range of accuracy combined with model evaluation can be used for confirming or rejecting a match between remotely related proteins. In the middle of the accuracy spectrum are the models based on approximately 35% sequence identity, corresponding to 85% of the C_α atoms modeled within 3.5 Å of their correct positions.

Fortunately, the active and binding sites are frequently more conserved than the rest of the fold and are thus modeled more accurately. In general, medium resolution models frequently allow a refinement of the functional prediction based on sequence alone, because ligand binding is most directly determined by the structure of the binding site rather than its sequence. It is frequently possible to predict correctly important features of the target protein that do not occur in the template structure. For example, the location of a binding site can be predicted from clusters of charged residues, and the size of a ligand may be predicted from the volume of the binding site cleft. Medium resolution models can also be used to construct site-directed mutants with altered or destroyed binding capacity, which in turn could test hypotheses about the sequence-structure-function relationships.

Other problems that can be addressed with medium resolution comparative models include designing proteins that have compact structures without long tails, loops, and exposed hydrophobic residues for better crystallization and designing proteins with added disulfide bonds for extra stability. The high end of the accuracy spectrum corresponds to models based on 50%

sequence identity or more. The average accuracy of these models approaches that of low resolution X-ray structures (3 Å resolution) or medium resolution NMR structures (10 distance restraints per residue).

The alignments on which these models are based generally contain almost no errors. In addition to the already listed applications, high quality models can be used for docking of small ligands or whole proteins onto a given protein. We now describe two applications of comparative modeling in more detail : (1) Modeling of substrate specificity aided by a high accuracy model and (2) confirming a remote structural relationship based on a low accuracy model.

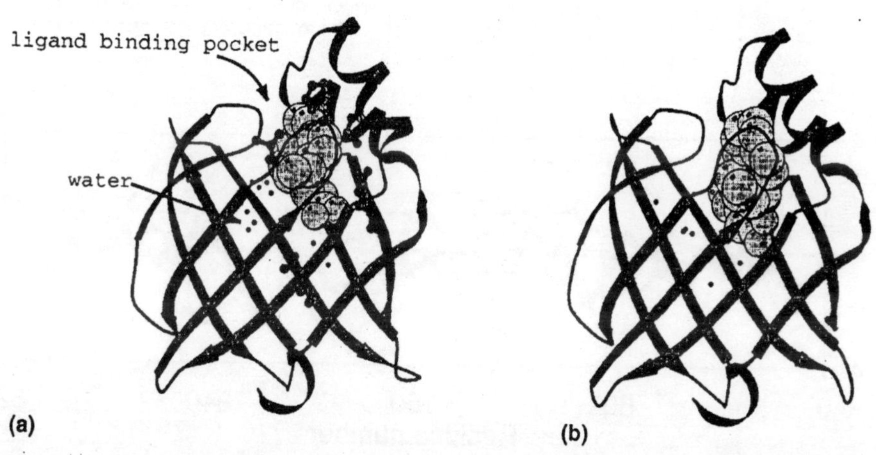

Fig. 10.6. Models of complexes between BLBP and two different fatty acids. The fatty acid ligand is shown in the CPK representation. The small spheres in the ligand-binding cavity are water molecules, (*a*) Model of the BLBP-oleic acid complex, in which the cavity is not filled, (*b*) Model of the BLBP-docosahexaenoic acid complex, in which the cavity is filled. The figure was prepared using the program MOLSCRIPT.

Ligand Specificity of Brain Lipid-Binding Protein

Brain lipid-binding protein (BLBP) is a member of the family of fatty acid binding proteins that was isolated from brain. The problem was to find out which one of the many fatty acids known to bind to fatty acid binding proteins in general is the likely physiological ligand of BLBP. To address this problem, comparative models of BLBP complexed with many fatty acids were calculated by relying on the structures of the adipocyte lipid-binding protein and muscle fatty acid binding protein, in complex with their ligands. The models were evaluated by binding and site-directed mutagenesis experiments.

The model of BLBP indicated that its binding cavity was just large enough to accommodate docosa-hexaenoic acid (DHA). Because DHA filled the BLBP binding cavity completely, it was unlikely that BLBP would bind a larger ligand. Thus, DHA was the ligand predicted to have the highest affinity for BLBP. The prediction was confirmed by the measurement of binding affinities for many fatty acids. It turned out that the BLBP-DHA interaction was the strongest fatty acid-protein interaction known to date. The binding affinities of the ligands correlated with the surface areas buried by the protein-ligand interactions, as calculated from the corresponding models, and explained why DHA had the highest affinity.

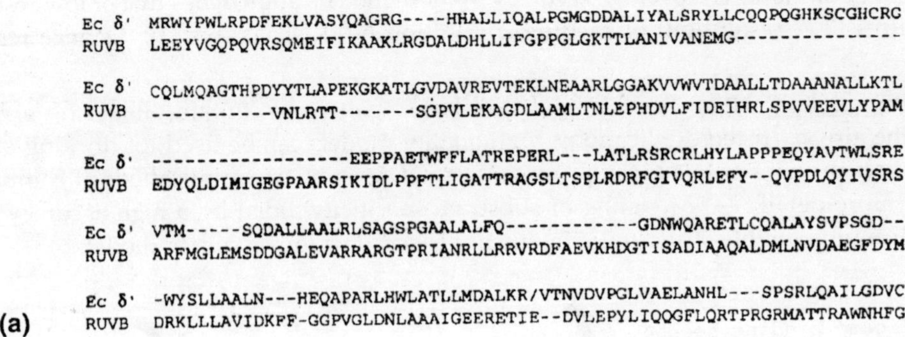

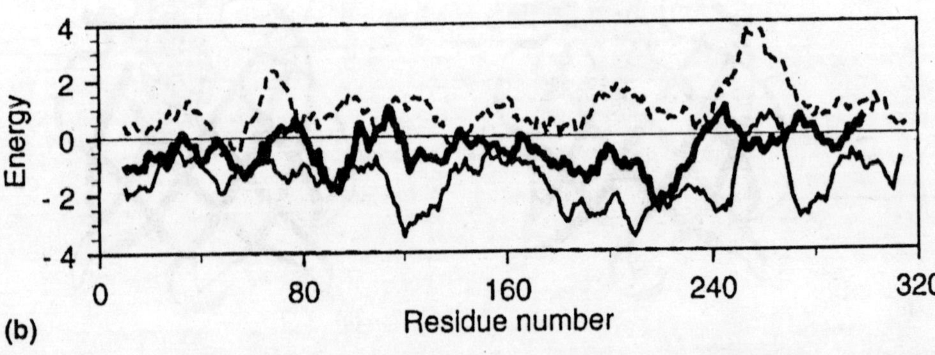

Fig. 10.11. Confirming structural similarity between the *E. coli* δ′ subunit of DNA polymerase III and RuvB. (*a*) A sequence alignment between the δ′ subunit and RuvB. (*b*) Prosall profiles for the *X*-ray structure of the δ′ subunit (thin continuous line), Z = –11.0; a model of RuvB based on its alignment to the δ′ subunit (thick line), Z = –7.3; and a test model based on an incorrect alignment (dashed line), Z = –0.9. The RuvB model based on the correct alignment has a significant Z-score and only a few positive peaks in the profile. This indicates that the model is plausible and that RuvB is indeed related structurally to the *E. coli* δ′ subunit.

This case illustrates how a comparative model provides new information that cannot be deduced directly from the template structures despite their high (60%) sequence identity to BLBP. The two templates have smaller binding sites and consequently different patterns of binding affinities for the same set of ligands. The study also illustrated how new information is obtained relative to the target-template alignment even when the similarity between the target and the template sequences is high. The volumes and contact surfaces can be calculated only from a 3D model.

Finding Proteins Remotely Related to the *E. coli* δ′ Subunit

The structure of the δ′ subunit of the clamp-loader complex of *E. coli* DNA polymerase III was determined by X-ray crystallography. Several biological considerations and extremely weak sequence patterns indicated that δ′ may be structurally related to the RuvB family of DNA helicases. However, the relationship was not possible to prove on the basis of the alignment of the corresponding sequences alone; the sequence identities ranged from only 9% to 21%. To substantiate the putative match, comparative models for several RuvB helicases were constructed using the crystal structure of the δ′ subunit as the template. The models were evaluated by calculating their PROSAII Z-scores and energy profiles. This evaluation indicated strongly that the model is plausible and that RuvB is indeed related structurally to the *E. coll* δ′ subunit.

Comparative Modeling in Structural Genomics

In a few years, the genome projects will have provided us with the amino acid sequences of more than a million proteins—the catalysts, inhibitors, messengers, receptors, transporters, and building blocks of the living organisms. The full potential of the genome projects will be realized only when we assign and understand the function of these new proteins. This will be facilitated by structural information for all or almost all proteins. This aim will be achieved by structural genomics, a focused, large-scale determination of protein structures by X-ray crystallography and nuclear magnetic resonance spectroscopy, combined efficiently with accurate, automated, and large-scale comparative protein structure modeling techniques.

Given current modeling techniques, it seems reasonable to require models based on at least 30% sequence identity, corresponding to one experimentally determined structure per *sequence family* rather than fold family. Since there are 1000-5000 fold families and perhaps about five times as many sequence families, the experimental effort in structural genomics has to deliver at least 10,000 protein domain structures. To enable the large-scale comparative modeling needed for structural genomics, the steps of comparative modeling are being assembled into a completely automated pipeline. Because many computer programs for performing each of the operations in comparative modeling already exist, it may seem trivial to construct a pipeline that completely automates the whole process. In fact, it is not easy to do so in a robust manner.

For a good reason, most of the tasks in modeling of individual proteins, including template selection, alignment, and model evaluation, are typically performed with significant human intervention. This allows the use of the best tool for a particular problem at hand and consideration of many different sources of information that are difficult to take into account entirely automatically. Because large-scale modeling can be performed only in a completely automated manner, the main challenge is to build an automated and robust pipeline that approaches the performance of a human expert as much as possible. Two applications of comparative modeling to complete genomes have been described. For the sequences encoded in the *E. coli* genome, models were built for 10–15% of the proteins using the SWISS-MODEL web server. Peitsch et al. have recently also modeled many proteins in SWISS-PROT and made the models available on their SWISS-MODEL web site.

Another large-scale modeling study was our own modeling of five prokaryotic and eukaryotic genomes. The calculation resulted in the models for substantial segments of 17.2%, 18.1.%, 19.2%, 20.4%, and 15.7% of all proteins in the genomes of *Saccharomyces cerevisiae* (6218 proteins in the genome); *Escherichia coli* (4290 proteins), *Mycoplasma genitalium* (468 proteins), *Caenorhabditis elegans* (7299 proteins, imcomplete), and *Methanococcus janaschii* (1735 proteins), respectively. An important feature of this study was an evaluation of all the models. This evaluation is important because most of the related protein pairs share less than 30% sequence identity, resulting in significant errors in the models. The models were assigned into the reliable or unreliable class by a procedure that relies on the statistical potential function from PROSAII. This allowed identification of those models that were likely to be based on correct templates and at least approximately correct alignments. As a result, 236 yeast proteins without any prior structural information were assigned to a particular fold family; 40 of these proteins did not have any prior functional annotation.

The models were also evaluated more precisely by using a calibrated relationship between the model accuracy and the percentage sequence identity on which the model is based. Almost half of the 1071 reliably modeled proteins in the yeast genome share more than approximately 35% sequence identity with their templates. All the alignments, models, and model evaluations are available in the ModBase database of comparative protein structure models.

Most recently, the combined use of PSI-BLAST with the model building and a new model evaluation allowed us to calculate reliable models for 50% of the proteins in the TrEMBL database. Large-scale comparative modeling opens new opportunities for tackling existing problems by virtue of providing many protein models from many genomes. One example is the selection of a target protein for which a drug needs to be developed. A good choice is a protein that is likely to have high ligand specificity; specificity is important because specific drugs are less likely to be toxic.

Large-scale modeling facilitates imposing the specificity filter in target selection by enabling a structural comparison of the ligand binding sites of many proteins, either human or from other organisms. Such comparisons may make it possible to select rationally a target whose binding site is structurally most different from the binding sites of all the other proteins that may potentially react with the same drug. For example, when a human pathogenic organism needs to be inhibited, a good target may be a protein whose binding site shape is different from related binding sites in all of the human proteins. Alternatively, when a human metabolic pathway needs to be regulated, the target identification could focus on that particular protein in the pathway that has the binding site most dissimilar from its human homologs.

CONCLUSION

Whereas an experimental structure or a comparative model is generally insufficient on its own to infer the biological function of a protein, it is often complementary to sequence analysis and direct experiment. Comparative modeling efficiently increases the value of sequence information from the genome projects, although it is not yet possible to model all proteins with useful accuracy. The main bottlenecks are the absence of structurally defined members in many protein families and the difficulties in detecting weak similarities, both for fold recognition and for sequence-structure alignment. The fraction of protein sequences that can be modeled with useful accuracy by comparative modeling is increasing rapidly.

The main reasons for this improvement are the increases in the numbers of known folds and the structures per fold family as well as the improvement in the fold recognition and comparative modeling techniques. It has been estimated that globular protein domains cluster in only a few thousand fold families, approximately 800 of which have already been structurally defined. Assuming the current growth rate in the number of known protein structures, the structure of at least one member of most globular folds will be determined in less than 10 years. According to this argument, comparative modeling would be applicable to most of the globular protein domains before the expected completion of the human genome project.

However, there are some classes of proteins, including membrane proteins, that will not be amenable to modeling without improvements in structure determination and modeling techniques. For example, it has been predicted that 839 (13.9%) of the yeast ORFs have at least two transmembrane helices. To maximize the number of proteins that can be modeled reliably, a concerted effort toward structural determination of the new folds by *X*-ray crystallography and nuclear magnetic resonance spectroscopy is in order. A combination of a more complete database of known protein structures with accurate modeling techniques will efficiently increase the value of sequence information from the genome projects.

STRUCTURAL BIOLOGY

Much of computational biophysics and biochemistry is aimed at making predictions of protein structure, dynamics, and function. Most prediction methods are at least in part knowledge-based rather than being derived entirely from the principles of physics. For instance, in comparative modeling of protein structure, each step in the process—from homolog identification and sequence-structure alignment to loop and side-chain modeling—is dominated by information derived from the protein sequence and structure databases. In molecular dynamics simulations, the potential energy function is based partly on conformational analysis of known peptide and protein structures and thermodynamic data. The biophysical and biochemical data we have available are complex and of variable quality and density.

We have sequences from many different kinds of organisms and sequences for proteins that are expressed in very different environments in a single organism or even a single cell. Some sequence families are very large, and some have only one known member. We have structures from many protein families, from NMR spectroscopy and from X-ray crystallography, some of high resolution and some not. These structures can be analyzed on the level of bond lengths and angles, or dihedral angles, and interatomic distances, or in terms of secondary, tertiary, and quaternary structure. Some structural features are very common, such as α-helices, and some are relatively rare, such as valine residues with backbone dihedral $\phi > 0°$. The amount of data is also increasing. The nonredundant protein sequence database available from GenBank now contains over 500,000 amino acid sequences, and there are at least 30 completed genomes from all three kingdoms of life.

The number of unique sequences in the Protein Databank of experimentally determined structures is now over 3000. The number of known protein folds is at least 400. In the next few years, the databanks will continue to grow exponentially as the *Drosophila, Arabidopsis,* corn, mouse, and human genomes are completed. Several institutions are planning projects to determine as many protein structures as possible in target genomes, such as yeast, *Mycoplasma genitalium,* and *E. coli.* To gain the most predictive utility as well as conceptual understanding from the sequence and structure data available, careful *statistical* analysis will be required.

The statistical methods needed must be robust to the variation in amounts and quality of data in different protein families and for structural features. They must be updatable as new data become available. And they should help us generate as much understanding of the determinants of protein sequence, structure, dynamics, and functional relationships as possible. In recent years, Bayesian statistics has come to the forefront of research among professional statisticians because of its analytical power for complex models and its conceptual simplicity. In the natural and social sciences, Bayesian methods have also attracted significant attention, including the fields of genetics, epidemiology, medicine, high energy physics, astrophysics, hydrology, archaeology, and economics.

Bayesian statistics have been used in molecular and structural biology in sequence alignment, remote homolog detection, threading, NMR spectroscopy, X-ray structure determination, and side-chain conformational analysis. Its counterpart, *frequentist statistics,* has in turn lost ground. To see why, we need to examine their basic conceptual frameworks. In the next section, I compare the Bayesian and frequentist viewpoints and discuss the reasons Bayesian methods are superior in both their conceptual components and their practical aspects. After that, I describe some important aspects of Bayesian statistics required for its application to protein sequence and structural data analysis. In the last section, I review several applications of Bayesian inference in molecular and structural biology to demonstrate its utility and conceptual simplicity. A useful introduction to Bayesian methods and their applications in machine learning and molecular biology can be found in the book by Baldi and Brunak.

BAYESIAN STATISTICS

Bayesian Probability Theory

The goal of any statistical analysis is inference concerning whether on the basis of available data, some hypothesis about the natural world is true. The hypothesis may consist of the value of some parameter or parameters, such as a physical constant or the exact proportion of an allelic variant in a human population, or the hypothesis may be a qualitative statement, such as "This protein adopts an α/β barrel fold" or "I am currently in Philadelphia." The parameters or hypothesis can be unobservable or as yet unobserved. How the data arise from the parameters is called the *model* for the system under study and may include estimates of experimental error as well as our best understanding of the physical process of the system. *Probability* in Bayesian inference is interpreted as the degree of belief in the truth of a statement.

The belief must be predicated on whatever knowledge of the system we possess. That is, probability is always conditional, $p(X|I)$, where X is a hypothesis, a statement, the result of an experiment, etc., and I is any information we have on the system. Bayesian probability statements are constructed to be consistent with common sense. This can often be expressed in terms of a fair bet. As an example, I might say that "the probability that it will rain tomorrow is 75%." This can be expressed as a bet: "I will bet \$3 that it will rain tomorrow, if you give me \$4 if it does and nothing if it does not." (If I bet \$3 on 4 such days, I have spent \$12; I expect to win back \$4 on 3 of those days, or \$12). At the same time, I would not bet \$3 on no rain in return for \$4 if it does not rain. This behavior would be inconsistent, since if I did both simultaneously I would bet \$6 for a certain return of only \$4. Consistent betting would lead me to bet \$1 on no rain in return for \$4. It can be shown that for consistent betting behavior, only certain rules of probability are allowed, as follows.

There are two central rules of probability theory on which Bayesian inference is based :

1. The sum rule : $p(A|I) + p(\bar{A}|I) = 1$
2. The product rule : $p(A, B|I) = p(A|B, I)\, p(B|I) = p(B|A, I)\, p(A|I)$

The first rule states that the probability of A plus the probability of not-A $(\bar{A})$ is equal to 1. The second rule states that the probability for the occurrence of two events is related to the probability of one of the events occurring multiplied by the conditional probability of the other event given the occurrence of the first event. We can drop the notation of conditioning on I as long as it is understood implicitly that all probabilities are conditional on the information we possess about the system. Dropping the I, we have the usual expression of Bayes' rule,

$$p(A, B) = p(A|B)\, p(B) = p(B|A)\, p(A) \qquad \text{...(10.1)}$$

For Bayesian inference, we are seeking the probability of a hypothesis H given the data D. This probability is denoted $p(H|D)$. It is very likely that we will want to compare different hypotheses, so we may want to compare $p(H_1|D)$ with $p(H_2|D)$. Because it is difficult to write down an expression for $p(H|D)$, we use Bayes' rule to invert the probability of $p(D|H)$ to obtain an expression for $p(H|D)$:

$$p(D) = \frac{p(D|H)\, p(H)}{p(D)} \quad ...(10.2)$$

In this expression, $p(H)$ is referred to as the prior probability of the hypothesis H. It is used to express any information we may have about the probability that the hypothesis H is true before we consider the new data D. $p(D|H)$ is the *likelihood* of the data given that the hypothesis H is true. It describes our view of how the data arise from whatever H says about the state of nature, including uncertainties in measurement and any physical theory we might have that relates the data to the hypothesis. $p(D)$ is the *marginal distribution* of the data D, and because it is a constant with respect to the parameters it is frequently considered only as a normalization factor in Eq. (2), so that $p(H|D) \propto p(D|H)\, p(H)$ up to a proportionality constant. If we have a set of hypotheses that are exclusive and exhaustive, *i.e.*, one and only one must be true, then

$$p(D) = \sum_i p(D|H_i)\, p(H_i) \quad ....(10.2\,(a))$$

$p(H|D)$ is the *posterior distribution*, which is, after all, what we are after. It gives the probability of the hypothesis after we consider the available data and our prior knowledge. With the normalization provided by the expression for $p(D)$, for an exhaustive set of hypotheses we have $\Sigma_{ip}(H_i|D) = 1$, which is what we would expect from the sum rule axiom described above.

As an example of likelihoods and prior and posterior probabilities, we give the following example borrowed from Gardner. The chairman of a statistics department has decided to grant tenure to one of three junior faculty members, Dr. A, Dr. B, or Dr. C. Assistant professor A decides to ask the department's administrative assistant, Mr. Smith, if he knows who is being given tenure. Mr. Smith decides to have fun with Dr. A and says that he won't tell her who is being given tenure. Instead, he will tell her which of Dr. B and Dr. C is going to be denied tenure. Mr. Smith does not yet know who is and who is not getting tenure and tells Dr. A to come back the next day. In the meantime, he decides that if A is getting tenure he will flip a coin and will tell A that B is not getting tenure if the coin shows heads, and that C is not getting tenure if it shows tails. If B or C is getting tenure, he will tell A that either C or B, respectively, is not getting tenure

Dr. *A* comes back the next day, and Mr. Smith tells *A* that *C* is not getting tenure. A then figures that her chances of tenure have now risen to 50%. Mr. Smith believes he has not in fact changed *A's* knowledge concerning her tenure prospects. Who is correct?

For prior probabilities, if H_A is the statement "*A* gets tenure" and likewise for H_B and H_c, we have prior probabilities $p(H_A) = p(H_B) = p(H_c\} = 1/3$. We can evaluate the likelihood of S, that Mr. Smith will say "C is not getting tenure," if H_A, H_B, or H_c is true:

$$p(S|H_A) = 0.5; \qquad p(S|H_B) = 1; \qquad p(S|H_C) = 0$$

So the posterior probability that A will get tenure based on Mr. Smith's statement is

$$p(H_A|S) = \frac{p(S|H_A)p(H_A)}{\sum_{r=A,B,C} p(S|H_r)p(H_r)} \quad ...(3)$$

$$= \frac{(1/2)\times(1/3)}{[(1/2\times 1/3)]+[1\times(1/3)]+[0\times(1/3)]} = \frac{1}{3}$$

Mr. Smith has not in fact changed *A's* knowledge, because her prior and posterior probabilities of getting tenure are both 1/3. Mr. Smith has, however, changed *A's* knowledge of *B's* prospects of tenure, which are now 2/3. Another way to think about this problem is that before Mr. Smith has told *A* anything, the probability of *B* or *C* getting tenure was 2/3. After his statement, the same 2/3 total probability applies to *B* and *C*, but now *C's* probability of tenure is 0 and *B's* has therefore risen to 2/3. *A's* posterior probability is unchanged.

Bayesian Parameter Estimation

Most often the hypothesis *H* concerns the value of a continuous parameter, which is denoted θ. The data *D* are also usually observed values of some physical quantity (temperature, mass, dihedral angle, etc.) denoted *y*, usually *a* vector. *y* may be a continuous variable, but quite often it may be a discrete integer variable representing the counts of some event occurring, such as the number of heads in a sequence of coin flips. The expression for the posterior distribution for the parameter θ given the data *y* is now given as

$$p(\theta|y) = \frac{p(y|\theta)p(\theta)}{\int_{\theta} p(y|\theta)\, p(\theta)\, d\theta} \qquad ...(10.4)$$

where the sum over hypotheses in Eq. (10.2a) has been replaced with an integral over Θ, the allowed values of the parameter θ. For example, if θ is a proportion, then the integral is from 0 to 1. The prior probability $p(\theta)$ is a *prior probability distribution*, a continuous function of θ that expresses any knowledge we have about likely values for θ before we consider the new data at hand. This probability distribution may be flat over the entire range of θ. Such a prior is referred to as uninformative (as with *A*, *B*, *C's* equal prior probability of tenure above). Or we may have some prior data so that we know that θ is more likely to be in a particular range of Θ than outside that range, so the probability should be higher in that range than outside it. This is an *informative* prior. Under ordinary circumstances, the prior should be *proper*, that is, normalized to integrate to 1 with respect to θ: $\int_{\theta} p(\theta)\, d\theta = 1$. The likelihood, by contrast, should be normalized with respect to integration over the data for given θ: $\int_{y} p(y|\theta)\, dy = 1$. If y is discrete (count data), then this is a sum over all possible *y*. The posterior distribution $p(\theta|y)$ is a continuous function of the parameter θ for known data *y*. It can be used for any inference on the value of θ desired. For instance, the probability that θ is greater than some value θ_0 is $P(\theta > \theta_0|y) = \int_{\theta_0}^{\theta_{max}} p(\theta|y)$. Ninety-five percent confidence intervals can be calculated easily by choosing θ_1 and θ_2 such that $\int_{\theta_1}^{\theta_2} p(\theta|y)\, d\theta = 0.95$. This alternative notation for parameters and data is common in the statistical literature, and we will use it throughout the rest of this chapter.

As an example of a continuous variable, we can calculate a posterior distribution for the proportion of red and green balls in a barrel sitting before us. You are asked to bet whether it is the red balls or the green balls that are more plentiful in the barrel. You have $10 in your possession, and you decide to bet in proportion to your certainty in your opinion on the red or green majority. You get a very quick look at the open barrel and estimate that the number of red and green balls look approximately equal. As a prior probability you would use a probability distribution that is a bell-shaped curve centered around θ_{red} of 0.5. (The appropriate mathematical form is a beta distribution, described below.) But your look was very brief, and you do not have a lot of confidence in this prior view. You are given a sample of balls from the barrel, which consists of 7 red balls and 3 green balls. It is clear you should bet on red, but how much? The likelihood function in this case is a binomial distribution. It gives the probability of n_{red} and n_{green} for a sample of $N = n_{red} + n_{green}$ balls, given θ_{red} and $\theta_{green} = 1 - \theta_{red}$. The results are shown in Fig. 10.13. The dashed line gives the posterior distribution of θ_{red}, the solid line gives the prior distribution, and the dotted line gives the likelihood of the data. The figure shows that the posterior is a compromise between the prior and the likelihood. We can integrate the posterior distribution to decide the amount of the bet from

$$\int_{0.5}^{1} p(\theta_{red}|n_{red} = 7; n_{green} = 3)\, d\theta_{red} = 0.72$$

So we bet $7.20 on red.

Frequentist Probability Theory

It should be noted that the Bayesian conception of probability of a hypothesis and the Bayesian procedure for assessing this probability is the original paradigm for probabilistic inference. Both Bayes and Laplace used Bayes' rule to make probability statements for hypothetical statements given observed data, $p(H|D)$, by inverting the likelihood function, $p(D|H)$ [or, more accurately, determining $p(\theta|y)$ by inverting $p(y|\theta)$]. But in the nineteenth and early

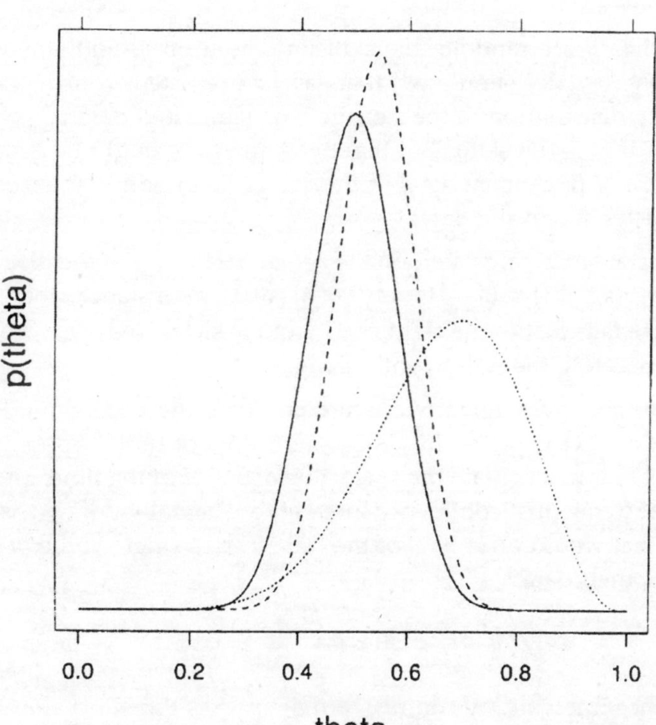

Fig. 10.12. (——) Prior, (.....) data (likelihood), and (—) posterior distributions for estimating a proportion of red and green balls in a barrel. The prior is based on a sample of 40 balls with 20 of them red, Beta. The likelihood is shown for a draw of 10 balls from the barrel, seven of them red. The posterior distribution is Beta.

twentieth centuries the idea that a hypothesis could have a probability associated with it seemed too subjective to practitioners of the new science of statistics. They were also uncomfortable with the notion of prior probabilities, especially when applied to continuous parameters, $H = \theta$. To deal with these problems, they simply removed these elements of statistical reasoning.

In frequentist statistics, probability is instead a long-run relative occurrence of some event out of an infinite number of repeated trials, where the event is a possible outcome of the trial.* A hypothesis or parameter that expresses a state of nature cannot have a probability in frequentist statistics, because after an infinite number of experiments there can be no uncertainty in the parameter left. A hypothesis or parameter value is either always true or always false. Because a hypothesis cannot have a probability, frequentist probability is restricted to inference about data given a single hypothesis or a single value for the parameter. Data can be assigned probabilities because they can have varying values due to experimental error, even under a constant hypothesis.

The usual frequentist procedure comprises a number of steps :

1. Define a procedure Π for selecting a hypothesis based on some characteristic(s) of the data, $S(y)$. S is called a *statistic*. Often this will be a *null hypothesis* that is deliberately at odds with $S(y)$. So, for instance, whereas the characteristic of the data might be the sample average $\bar{y}$ or the variance σ^2, the hypothesis H might be that the parameter θ (what y is measuring) has a value of 0.* Or if we are trying to show that two samples taken under different conditions are really different, we might define H to be the statement that the sample averages are in fact equal.
2. Because the data y are random, the statistics based on y, $S(y)$, are also random. For all possible data y (usually simulated) that can be predicted from H, calculate $p(S(y_{sim})|H)$, the probability distribution of the statistic S on simulated data y_{sim} given the truth of the hypothesis H. If H is the statement that $\theta = 0$, then y_{sim} might be generated by averaging samples of size N (a characteristic of the actual data) with variance $\sigma^2 = \sigma^2\,(y_{actual})$ (yet another characteristic of the data).
3. Compare the statistic S calculated on the actual data y_{actual} to the distribution $p(S(y_{sim})|H)$. If $\int^{\infty} S(y_{actual})\, p(S(y_{sim})|H)\, dS(y_{sim})$ is very small (<0.05, for instance), then reject the hypothesis H. If $S\,(\bar{y}_{actual})$ falls just to the right of 95% of the simulated $S(\bar{y})$, then we can conclude that $\theta > 0$ by rejecting the null hypothesis †.

If we do this over and over again, we will have done the right thing 95% of the time. Of course, we do not yet know the probability that, say, $\theta > 5$. For this purpose, confidence intervals for θ can be calculated that will contain the true value of θ 95% of the time, given many repetitions of the experiment. But frequentist confidence intervals are actually defined as the range of values for the data average that would arise 95% of the time from a single value of the parameter. That is, for normally distributed data,

$$Pr\left(\bar{y} - \frac{\sigma}{\sqrt{N}} < \theta < \bar{y} + \frac{\sigma}{\sqrt{N}}\right) = 0.95$$

Bayesian confidence intervals, by contrast, are defined as the interval in which

$$Pr\left(\theta - \frac{\sigma}{\sqrt{N}} < \bar{y} < \theta + \frac{\sigma}{\sqrt{N}}\right) = 0.95$$

The frequentist interval is often *interpreted* as if it were the Bayesian interval, but it is fundamentally defined by the probability of the data values given the parameter and not the probability of the parameter given the data.

Bayesian Methods Are Superior to Frequentist Methods

The Bayesian and frequentist theories can be considered two competing paradigms (in the sense of Kuhn of what the word "probability" means and how we should make inferential statements about hypotheses given data. It is an unusual situation in the history of science that there should be two competing views on such basic notions, both of which have sizable entrenched camps of adherents, and that the controversy has lasted so long. Bayesian views fell out of favor until the book of Jeffreys in 1939 and the work of Jaynes and Savage in the 1950s.

Since then the Bayesian camp has increased tremendously in size. Because of some of the computational difficulties in evaluating posterior distributions, the advent of Markov chain Monte Carlo methods and fast computers has vastly increased the power and flexibility of Bayesian methods. It is impossible to review the controversy in great depth here, but I will make some arguments in favor of the Bayesian view for molecular and structural biology.

The Bayesian View of Probability Corresponds to Most Scientists' Thinking

Bayesian inference is a process of taking current views of the probability that competing hypotheses about nature might be true and updating these beliefs in light of new data. It corresponds to the daily experience of scientists and nonscientists alike of judgments made on past experience and present facts. As Good has argued, all animals must be at least informal Bayesians, even non-Bayesian statisticians, because they evaluate possibilities and outcomes of their actions in light of their probabilities for success. One might argue that sane dogs are *better* Bayesians than humans, given our propensity for foolish and destructive behavior, regardless of prior experience. Bayesian methods are quite similar to most scientists' basic intuition about the nature of data and fundamental physical processes.

The use of the prior distribution formalizes what we do naturally when we restrict our view of any parameter to a certain range of values or a certain number of outcomes and when we treat outliers with suspicion. If we get a value for an experiment that yields a parameter value that seems absurd compared to previous experiments, we are likely to repeat the experiment rather than publish the absurd result. Implicitly, we set the prior distribution of the parameters that are outside our range of interest or unphysical or unlikely on some ground to 0. The remaining probability density of the parameters must lie within the range we believe to be at least remotely possible and integrates to 1 within this range. Although basic statistics textbooks spend much time discussing null hypotheses, *t*-tests, *F* tests, etc., scientists rarely use this language in assessing data and inferred values of parameters based on the data. We do not usually ask whether the value of some parameter is exactly θ_0 or simply whether it is greater than some value θ_0.

Rather we want to know what is the most likely range of values for the parameter, which is inherently a probability distribution over the parameter. By contrast, the frequentist view is often contrary to common sense and common scientific practice. The classic example of this is the *stopping* rule problem. If I am trying to decide whether a coin is fair, I can set up an experiment in two different ways. I can throw the coin some number of times *N* and then record the number of heads. Or I can keep throwing the coin until I observe some preset number of heads, n_{heads}.

Suppose one person throws the coin $N = 100$ times and observes $n_{heads} = 55$ and another person throws the coin until 55 heads are observed and in fact it takes 100 throws. To a frequentist, these are quite different experiments and result in different inferences on the fairness of the coin. This is because the experiment that would be repeated an infinite number of times to obtain p_{heads} is different in each case (in the first case, a binomial distribution is used to make the inference; in the second case, a negative binomial is used).

This seems absurd to most people, and to a Bayesian the probability of a fair coin is the same, because the data are in fact the same in each case. The previous example also highlights the controversy of the subjectivity or objectivity of the two competing views of probability. Frequentist probability theory arose because of the apparent subjectivity of Bayesian prior distributions. But it replaced this subjectivity with a set of procedures based on test statistics. Inference is based on the probability of the test statistic, calculated on *hypothetical* data consistent with the null hypothesis, being as large as or larger than the test statistic calculated on the real data. But the stopping rule example exemplifies that the nature of the repeated experiment, the *sampling space,* is in itself arbitrary, as is the functional form of the test statistic. It is the emphasis on data *not* seen that makes frequentist statistics unintuitive and gives statistics much of its reputation of being difficult for beginning students.

Bayesian Methods Perform Better

In biology we are often faced with a set of situations that demand inference based on varying amounts of data. For instance, we may try to assign a protein fold to a new protein sequence based on very remote homology to some sequence in the Protein Data Bank (PDB). In some cases, we may have many sequences related to our target sequence from the genome projects, and the multiple alignment of sequences can help us establish a homology to the PDB sequence that may in fact be quite remote. In other cases, we may have only a small number of homologs in the sequence database, and establishing a remote homology may be quite difficult.

Another example arises in side-chain conformational analysis. For some ranges of ϕ and ψ many examples exist of residues in the PDB, and reasonable estimates for the three χ_1 rotamer population can be calculated from the data. But the number of residues in some parts of the Ramachandran map is very small, but we would still like a good estimate of the three rotamer probabilities for protein folding simulations and comparative modeling. Far from being a disadvantage, the need for a prior distribution can be a distinct advantage. First, it generally forces us to consider the full range of the possibilities for the hypothesis. Second, although in the absence of any prior information we can use uninformative priors, in some cases we may choose to use informative priors.

Usually the prior corresponds in functional form to some number of data points, and we can choose to weight the prior in accordance with the strength of our belief in the prior information. Often a prior can be obtained by decoupling two or more parameters such that $p(\theta_1, \theta_2) \approx p(\theta_1)$ $p(\theta_2)$. The two factors on the right-hand side might be obtained by pooling the data of y_1 regardless of y_2 and vice versa. In any case, the posterior is always a compromise between the prior and the likelihood. If the prior represents a larger sample than the data (our data set is quite small), then the prior will dominate the data. If the data sample is large, then the prior will have little effect on the posterior, which will resemble the likelihood. In cases where there are only a few data points, frequentist methods perform badly by sacrificing good short-term behavior in favor of good long-term behavior. Another aspect in which Bayesian methods perform better than frequentist methods is in the treatment of nuisance parameters.

Quite often there will be more than one parameter in the model but only one of the parameters is of interest. The other parameter is a *nuisance* parameter. If the parameter of interest is θ and the nuisance parameter is ϕ, then Bayesian inference on θ alone can be achieved by integrating the posterior distribution over ϕ. The *marginal probability* of θ is therefore

$$p(\theta|y) = \int_{\Phi} p(\theta,\phi|y)d\phi = \frac{\int_{\phi} p(y|\theta,\phi)\,p(\theta,\phi)\,d\phi}{\int_{\theta,\phi} p(y|\theta,\phi)\,p(\theta,\phi)d\phi\,d\theta} \qquad ...(10.5)$$

In frequentist statistics, by contrast, nuisance parameters are usually treated with point estimates, and inference on the parameter of interest is based on calculations with the nuisance parameter as a constant. This can result in large errors, because there may be considerable uncertainty in the value of the nuisance parameter.

Setting Up Bayesian Models

Bayesian inference has three major components [44] :

1. Setting up the probability model for the data and parameters of the system under study. This entails defining prior distributions for all relevant parameters and a likelihood function for the data given the parameters.
2. Calculating the posterior distribution for the parameters given existing data. This calculation can sometimes be performed analytically, but in the general case it is performed by simulation.
3. Evaluating the model in terms of how well the model fits the data, including the use of posterior predictive simulations to determine whether data predicted from the posterior distribution resemble the data that generated them and look physically reasonable. Overfitting the data will produce unrealistic posterior predictive distributions.

The complexity of information that can be incorporated into the model gives Bayesian statistics much of its power.

We need a mathematical representation of our prior knowledge and a likelihood function to establish a model for any system to be analyzed. The calculation of the posterior distribution can be performed analytically in some cases or by simulation. For an analytical solution it is usually the case that we need prior distribution forms that are *conjugate* to the likelihood function. If the prior distribution and the likelihood function are conjugate, then by definition the posterior distribution will have the same mathematical form as the prior distribution. A description of the important types follows.

Binomial and Multinomial Models

Any data set that consists of discrete classification into outcomes or descriptors is treated with a binomial (two outcomes) or multinomial (three or more outcomes) likelihood function. For example, if we have y successes from n experiments, *e.g.*, y heads from n tosses of a coin or y green balls from a barrel filled with red and green balls in unknown proportions, the likelihood function is a binomial distribution :

$$p(y|\theta) = \text{Bin}\,(y|n,\theta) = \frac{n!}{y!(n-y)!}\theta^{y}(1-\theta)^{n-y} \qquad \text{...(10.6)}$$

An informative conjugate prior distribution can be formulated in terms of a beta distribution :

$$p(\theta) = \text{Beta}\,(\theta|\alpha,\beta) = \frac{\Gamma(\alpha+\beta)}{\Gamma(\alpha)\Gamma(\beta)}\theta^{\alpha-1}(1-\theta)^{\beta-1} \qquad \text{...(10.7)}$$

We can think of the beta distribution as the likelihood of α prior successes and β failures out of $\alpha + \beta$ experiments. The Γ functions in front serve as a normalization constant, so that $\int_0^1 p(\theta)\,d\theta = 1$. Note that for an integer, $\Gamma(x + 1) = x!$ The posterior distribution that results from multiplying together the right-hand sides of Eqs. (10.2) and (10.3) is also a beta distribution :

$$p(\theta|y) = p(y|\theta)p(\theta)/p(y) = \text{Beta}(\theta|\alpha+y, \beta+n-y) \qquad \text{...(10.8)}$$

$$= \frac{\Gamma(n+\alpha+\beta)}{\Gamma(\alpha+y)\Gamma(\beta+n-y)}\theta^{\alpha+y-1}\theta^{\beta+n-y1}$$

We can see that the prior and posterior distributions have the same mathematical forms, as is required of conjugate functions. Also, we have an analytical form for the posterior, which is exact under the assumptions made in the model.

It is tempting to use Eq. (10.7) to derive Eq. (10.6), because they have similar forms given the relationship of the Γ function to the factorial. But the binomial and the beta distribution are not normalized in the same way. The beta is normalized over the values of θ, whereas the binomial is normalized over the counts, y given n. That is,

$$\int_0^1 \text{Beta}\,(\theta:\alpha,\beta)\,d\theta = 1 \qquad \text{...(10.9 } a\text{)}$$

$$\sum_{y=0}^{n} \text{Bin}\,(y:n,\theta) = 1 \qquad \text{...(10.9 } b\text{)}$$

It should be noted that the expected value of θ in a beta distribution is $\alpha/(\alpha + \beta)$, and the θ, with maximum probability (the mode) is $(\alpha - 1)/(\alpha + \beta - 2)$. In terms of the expectation values, α and β behave as a total of $\alpha + \beta$ counts even though the exponents are $\alpha - 1$ and $\beta - 1$. The beta distribution is defined such that it is the expectation values and not the modes that correspond to counts of α and β.

If there are more than two outcomes, we can use the multinomial distribution for the likelihood :

$$p(y|\theta) = \left(\frac{n!}{\prod_i y_i!}\right)\prod_i \theta_i^{y_i} \qquad \text{...(10.10)}$$

where $\Sigma_{i=1}^{m} y_i = n$ for m possible outcomes of n experiments. For example, if a barrel contains red, blue, and green balls, then $m = 3$. We might make $n = 100$ draws from the barrel (returning the balls to the barrel) and get 30 green, 50 red, and 20 blue balls. The *conjugate* prior distribution is the Dirichlet distribution,

$$p(\theta) = \text{Dirichlet}(\{x_i\}) = \left(\frac{\Gamma(x_0)}{\prod_i \Gamma(x_i)}\right) \prod_{i=1}^{m} \theta_i^{x_i - 1} \quad \text{...(10.11)}$$

where $x_0 = \Sigma_{i=1} x_i$. The Dirichlet distribution can be considered a generalization of the multinomial distribution, where the counts are no longer restricted to be integers. The values of the *hyperparameters* x_i that define the prior distribution can be thought of as estimated counts for each of the m outcomes in some sample of size $x_0 = \Sigma_{i=1}^{m} x_i$. The total number of prior counts, x_0, can be scaled to any value to alter the dependence of the posterior distribution on the prior distribution. The larger x_0 is, the more precise the prior distribution is and the closer the posterior density is to values near $\theta_i = x_i/x_0$. The posterior distribution that results from Eqs. (10.10) and (10.11) is also Dirichlet with parameters $x_i + y_i$, *i.e.*,

$$p(\theta|y) = \left(\frac{\Gamma(x_0 + y_0)}{\prod_i \Gamma(x_i + y_i)}\right) \prod_i \theta_i^{x_i + y_i - 1} \quad \text{...(10.12)}$$

Again, the use of the conjugate prior distribution results in the analytical form of the posterior distribution and therefore also simple expressions for the expectation values for the θ_i, their variances, covariances, and modes :

$$E(\theta_i) = \frac{x_i + y_i}{x_0 + y_0} \quad \text{...(10.13 a)}$$

$$\text{mode}(\theta_i) = \frac{x_i + y_i - 1}{x_0 + y_0 - m} \quad \text{...(10.13 b)}$$

$$\text{var}(\theta_i) = \frac{E(\theta_i)[1 - E(\theta_i)]}{x_0 + y_0 + 1} \quad \text{...(10.13 c)}$$

$$\text{cov}(\theta_i, \theta_j) = -\frac{E(\theta_i)\, E(\theta_j)}{x_0 + y_0 + 1} \quad \text{...(10.13 d)}$$

A noninformative prior distribution could be formed by setting each x_i to 1.

Normal Models

The normal model can take a variety of forms depending on the choice of noninformative or informative prior distributions and on whether the variance is assumed to be a constant or is given its own prior distribution. And of course, the data could represent a single variable or could be multidimensional. Rather than describing each of the possible combinations, I give only the univariate normal case with informative priors on both the mean and variance. In this case, the likelihood for data y given the values of the parameters that comprise θ, μ. (the mean), and σ^2 (the variance) is given by the familiar exponential

$$p(y|\mu, \sigma) = \frac{1}{\sigma\sqrt{2\pi}} \exp\left(\frac{-(y-\mu)^2}{2\sigma^2}\right) \quad \text{...(10.14)}$$

This expression is valid for a single observation y. For multiple observations, we derive $p(y|\theta)$ from the fact that $p(y|\mu, \sigma^2) = \Pi_i\, p(y_i|\mu, \sigma^2)$. The result is that the likelihood is also normal with the average value of $y, \overline{y}$, substituted for y and σ^2/n substituted for σ^2 in Eq. (10.14). The conjugate prior distribution for Eq. (10.14) is

$$p(\mu, \sigma^2) = p(\mu|\sigma^2)\, p(\sigma^2) \qquad \text{...(10.15)}$$

$$= \frac{1}{\sigma\sqrt{2\pi}} \exp\left(\frac{-\kappa_0(\mu-\mu_0)^2}{2\sigma^2}\right) \times \left(\frac{2^{v_0/2}}{\Gamma(v_0/2)}\right) \sigma_0^{v_0/2}\, \sigma^{2-v_0} \exp\left(\frac{-v_0\, \sigma_0^2}{2\sigma^2}\right)$$

where the prior for μ is a normal distribution dependent on the value of σ^2 as well as two hyperparameters, the mean μ_0 and the scale κ_0, while the prior for σ^2 is a scaled inverse χ^2 distribution [to the right of the × sign in Eq. (10.15)] with two hyperparameters, the scale σ_0 and degrees of freedom ν_0. The posterior distribution that results from Eqs. (10.14) and (10.15) has the same form as the prior (because the prior is conjugate to the likelihood), so that [44]

$$p(\mu, \sigma^2|y) = \frac{1}{\sigma\sqrt{2\pi}} \exp\left(\frac{-\kappa_n(\mu-\mu_n)^2}{2\sigma^2}\right) \times \frac{2^{v_n/2}}{\Gamma(v_n/2)}\, \sigma_n^{v_n/2}\, \sigma^{2-v_n} \exp\left(\frac{-v_n\, \sigma_n^2}{2\sigma^2}\right) \qquad \text{...(10.16)}$$

where

$$\mu_n = \frac{1}{\kappa_n}(\kappa_0\, \mu_0 + n\overline{y}) \qquad \text{...(10.17 } a\text{)}$$

$$\kappa_n = \kappa_0 + n,\ \nu_n = \nu_0 + n \qquad \text{...(10.17 } b\text{)}$$

$$\nu_n\, \sigma_n^2 = \nu_0\, \sigma_0^2 + (n-1)\, s^2 + \frac{n\kappa_0}{\kappa_n}(\overline{y} - \mu_0)^2 \qquad \text{...(10.17 } c\text{)}$$

$$s^2 = \frac{1}{n-1}\sum_{i=1}^{n}(y_i - \overline{y})^2 \qquad \text{...(10.17 } d\text{)}$$

Although this is a complicated expression, the results can be given a simple interpretation. The data sample size is n, whereas the prior sample size is κ_0, and therefore μ_n is the weighted average of the prior "data" and the actual data, σ_n^2 is the weighted average of the prior variance (σ_0^2), the data variance (s^2), and a term from the difference in the prior and data sample means. The weight of the prior variance is represented by the degree of freedom, ν_0, while the weight of the data variance is $n - 1$.

Simulation via Markov Chain Monte Carlo Methods

In practice, it may not be possible to use conjugate prior and likelihood functions that result in analytical posterior distributions, or the distributions may be so complicated that the posterior cannot be calculated as a function of the entire parameter space. In either case, statistical inference can proceed only if random values of the parameters can be drawn from the full posterior distribution :

$$p(\theta|y) = \frac{p(y|\theta)\, p(\theta)}{\int_{\Theta} p(y|\theta)\, p(\theta)\, d\theta} \qquad \text{....(10.18)}$$

We can also calculate expected values for any function of the parameters :

$$E[f(\theta|y)] = \frac{\int_{\Theta} f(\theta)\, p(y|\theta)\, p(\theta)\, d\theta}{\int_{\Theta} p(y|\theta)\, p(\theta)\, d\theta} \qquad \text{...(10.19)}$$

If we could draw directly from the posterior distribution, then we could plot $p(\theta|y)$ from a histogram of the draws on θ. Similarly, we could calculate the expectation value of any function of the parameters by making random draws of θ from the posterior distribution and calculating

$$E[f(\theta)] \approx \frac{1}{n}\sum_{t=1}^{n} f(\theta_t) \quad \text{...(10.20)}$$

In some cases, we may not be able to draw directly from the posterior distribution. The difficulty lies in calculating the denominator of Eq. (10.18), the marginal data distribution $p(y)$. But usually we can evaluate the ratio of the probabilities of two values for the parameters, $p(\theta_t|y)/p(\theta_u|y)$, because the denominator in Eq. (10.18) cancels out in the ratio. The Markov chain Monte Carlo method proceeds by generating draws from some distribution of the parameters, referred to as the proposal distribution, such that the new draw depends only on the value of the old draw, *i.e.*, some function $q(\theta_t|\theta_{t-1})$. We accept the new draw with probability

$$\pi(\theta_t|\theta_{t-1}) = \min\left(1, \frac{p(\theta_t|y)\,q(\theta_{t-1}|\theta_t)}{p(\theta_{t-1}|y)\,q(\theta_t|\theta_{t-1})}\right) \quad \text{...(10.21)}$$

and otherwise we set $\theta_t = \theta_{t-1}$. This is the Metropolis-Hastings method, first proposed by Metropolis and Ulam [45] in the context of equation of state calculations and further developed by Hastings. This scheme can be shown to result in a stationary distribution that asymptotically approaches the posterior distribution.

Several variations of this method go under different names. The Metropolis algorithm uses only symmetrical proposal distributions such that $q(\theta_t|\theta_{t-1}) = q(\theta_{t-1}|\theta_t)$. The expression for $\pi(\theta_t|\theta_{t-1})$ reduces to

$$\pi(\theta_t|\theta_{t-1}) = \min\left(1, \frac{p(\theta_t|y)}{p(\theta_{t-1}|y)}\right) \quad \text{...(10.22)}$$

This is the form that chemists and physicists are most accustomed to. The probabilities are calculated from the Boltzmann equation and the energy difference between state t and state $t-1$. Because we are using a ratio of probabilities, the normalization factor, *i.e.*, the partition function, drops out of the equation. Another variant when θ is multidimensional (which it usually is) is to update one component at a time. We define $\theta_{t-1} = \{\theta_{t,1}, \theta_{t,2}, \ldots, \theta_{t,i-1}, \theta_{t-1,i+1}, \ldots, \theta_{t-1,m}\}$, where m is the number of components in θ. So θ_{t-i} contains all of the components except θ_i and all the components that precede the ith component have been updated in step t, while the components that follow have not yet been updated. The m components are updated one at a time with this probability :

$$\pi(\theta_{t,i}|\theta_{t-1}) = \min\left(1, \frac{p(\theta_{t,i}|y, \theta_{t-1})\,q(\theta_{t-1,i}|\theta_{t,i}, \theta_{t-1})}{p(\theta_{t-1,i}|y, \theta_{t-i})\,q(\theta_{t,i}|\theta_{t-1,i}, \theta_{t-1})}\right) \quad \text{...(10.23)}$$

If draws can be made from the posterior distribution for each component conditional on values for the others, *i.e.*, from $p(\theta_{t,i}|y, \theta_{t-i})$, then this conditional posterior distribution can be used as the proposal distribution. In this case, the probability in Eq. (10.23) is always 1, and all draws are accepted. This is referred to as Gibbs sampling and is the most common form of MCMC used in statistical analysis.

Mixture Models

Mixture models have come up frequently in Bayesian statistical analysis in molecular and structural biology as described below, so a description is useful here. Mixture models can be used when simple forms such as the exponential or Dirichlet function alone do not describe the data well. This is usually the case for a multimodal data distribution (as might be evident from a histogram of the data), when clearly a single Gaussian function will not suffice. A mixture is a sum of simple forms for the likelihood :

$$p(y|\theta) = \sum_{i=1}^{n} q_i p(y|\theta_i) \qquad ...(10.24)$$

where $\Sigma_{i=1}^{n} q_i = 1$ for the n components of the mixture. For instance, if the terms in Eq. (10.24) are normal, then each term is of the form (for a single data point y_j)

$$p(y_j|\theta_i) = \frac{1}{\sqrt{2\pi\sigma_i}} \exp\left(-\frac{(y_j - \mu_i)^2}{2\sigma_i^2}\right) \qquad (10.25)$$

so each $\theta_i = \{\mu_i, \sigma_i^2\}$.

Maximum likelihood methods used in classical statistics are not valid to estimate the θ's or the q's. Bayesian methods have only become possible with the development of Gibbs sampling methods described above, because to form the likelihood for a full data set entails the product of many sums of the form of Eq. (10.24) :

$$p(\{y_1, y_2, ..., y_N\}|\theta) = \prod_{j=1}^{N}\left(\sum_{i=1}^{n} q_i\, p(y_j|\theta_i)\right) \qquad ...(10.26)$$

Because we are dealing with count data and proportions for the values q_i, the appropriate conjugate prior distribution for the q's is the Dirichlet distribution,

$$p(q_1, q_2, ..., q_k) = \text{Dirichlet}\ (\alpha_1, \alpha_2, ..., \alpha_k)$$

where the α's are prior counts for the components of the mixture. A simplification is to associate each data point with a single component, usually the component with the nearest location (*i.e.*, μ_i). In this case, it is necessary to associate with each data point y_j a variable c_j that denotes the component that y_j belongs to. These variables c_j are unknown and are therefore called "missing data." Equation (10.26) now simplifies to

$$p(\{y_1, y_2, ..., y_N) |\theta) = \sum_{j=1}^{N} p(y_j|\theta_{c_j}) \qquad ...(10.27)$$

A straightforward Gibbs sampling strategy when the number of components is known (or fixed) is as follows.

Step 1. From a histogram of the data, partition the data into N components, each roughly corresponding to a mode of the data distribution. This defines the c_j. Set the parameters for prior distributions on the θ parameters that are conjugate to the likelihoods. For the normal distribution the priors are defined in Eq. (10.15), so the full prior for the n components is

$$p(\theta_1, \theta_2, ..., \theta_k) = \prod_{i=1}^{n} N(\mu_{0i}, \sigma_{0i}^2/\kappa_0)\ \text{Inv}\ \chi^2(V_{0i}, \sigma_{0i}^2) \qquad ....(10.28)$$

The prior hyperparameters, μ_{0i}, etc., can be estimated from the data assigned to each component. First define $N_i = \Sigma_{j=1}^{N} I(c_j = i)$, where $I(c_j - i) = 1$ if $c_j = i$ and is 0 otherwise. Then, for instance, the prior hyperparameters for the mean values are defined by

$$\mu_{0i} = \frac{1}{N_i} \sum_{j=1}^{N} I(c_j = i)\, y_j \qquad ...(10.29)$$

The parameters of the Dirichlet prior for the q's should be proportional to the counts for each component in this preliminary data analysis. So we now have a collection of prior parameters $\{\theta_{0i} = (\mu_{0i}, \kappa_{0i}, \sigma_{0i}^2, \nu_{0i})\}$ and a preliminary assignment of each data point to a component, $\{c_j\}$, and therefore the preliminary number of data points for each component, $\{N_i\}$.

Step 2. Draw a value for each $\theta_i = \{\mu_i, \sigma_i^2\}$ from the normal posterior distribution for N_i data points with average $\bar{y}_i$,

$$p(\theta_i|\{y_i\}) = N(\mu_{N_i}, \sigma_{N_i}^2)\ \text{Inv}\ \chi^2(V_{N_i}, \sigma_{N_i}^2) \qquad ...(10.30)$$

where [as in Eq. (10.17)]

$$\mu_{N_i} = \frac{1}{\kappa_{N_i}}(\kappa_{0i}\mu_{0i} + N_i\,\bar{y}_i) \qquad ...(10.31\ a)$$

$$\kappa_{N_i} = \kappa_{0i} + N_i,\ \nu_{N_i} = \nu_{0i} + N_i \qquad (10.31\ b)$$

$$\nu_{N_i}\sigma_{N_i}^2 = \nu_{0i}\sigma_{0i}^2 + (N_i - 1)s_i^2 + \frac{N_i\,\kappa_{0i}}{\kappa_{N_i}}(\bar{y}_i - \mu_{0i})^2 \qquad ...(10.31\ c)$$

$$s_i^2 = \frac{1}{N_i - 1}\sum_{k=1}^{N}(y_k - \bar{y}_i)^2 \qquad ...(10.31\ d)$$

Draw $(q_1, q_2, ..., q_n)$ from Dirichlet $(\alpha_1 + N_1, \alpha_2 + N_2, ..., \alpha_n + N_n)$, which is the posterior distribution with prior counts α_i and data counts N_i.

Step 3. Reset the c_j by drawing a random number u_j between 0 and 1 for each c_j and set c_j to i if

$$\frac{1}{Z}\sum_{i=1}^{i'-1} q_i\, p(y_j|\theta_i) < u_j \le \frac{1}{Z}\sum_{i=i'+1}^{n} q_i\, p(y_j|\theta_i) \qquad ...(10.32)$$

where $Z = \Sigma_{i=1}^{n}\ q_i\, p(y_j|\theta_i)$ is the normalization factor.

Step 4. Sum up the N_i and calculate the averages $\bar{y}_i$ from the data and the values of c_j.

Step 5. Repeat steps 2–4 until convergence.

The number of components necessary can usually be judged from the data, but the appropriateness of a particular value of n can be judged by comparing different values of n and calculating the entropy distance, or Kullback-Leibler divergence,

$$ED(g, h) = \int g(x) \ln \frac{g(x)}{h(x)} dx \qquad ...(10.33)$$

where, for instance, g might be a three-component model and h might be a two-component model. If $ED(g, h) > 0$, then the model g is better than the model h.

Explanatory Variables

There is some confusion in using Bayes' rule on what are sometimes called *explanatory variables.* As an example, we can try to use Bayesian statistics to derive the probabilities of each secondary structure type for each amino acid type, that is $p(\mu|r)$, where μ is α, β, or γ (for coil) secondary structures and r is one of the 20 amino acids. It is tempting to write $p(\mu|r) = p(r|\mu)\ p(\mu)/p(r)$ using Bayes' rule. This expression is, of course, correct and can be used on PDB data to relate these probabilities. But this is not Bayesian statistics, which relate parameters that represent underlying properties with (limited) data that are manifestations of those parameters in some way. In this case, the parameters we are after are $\theta_\mu(r) \equiv p(\mu|r)$.

The data from the PDB are in the form of counts for $y_\mu(r)$, the number of amino acids of type r in the PDB that have secondary structure μ. There are 60 such numbers (20 amino acid types × 3 secondary structure types). We then have for *each* amino acid type a Bayesian expression for the posterior distribution for the values of $\theta_\mu(r)$:

$$p(\theta(r)|y(r)) \propto p(y(r)|\theta(r))\ p(\theta(r)) \qquad ...(10.34)$$

where θ and y are vectors of three components α, β, and γ. The prior is a Dirichlet distribution with some number of counts for the three secondary structure types for amino acid type r, *i.e.*, Dirichlet $(n_\alpha(r), n_\beta(r), n_\gamma(r))$. We could choose the three $n_\mu(r)$ to be equal to some small number, say 10. Or we could set them equal to $100 \times p_\mu$, where p_μ is the proportion of each secondary structure type in the PDB. 100 would be the sample size for the prior. Because 100 is small compared to the data sample size from the PDB, the prior will have only a small influence on the posterior distribution. The likelihood is Multinom $(y_\alpha(r), y_\beta(r), y_\gamma(r) : + \theta_\alpha(r) + \theta_\beta(r) + \theta_\gamma(r))$. The posterior is Dirichlet $(\theta_\alpha(r) + \theta_\beta(r) + \theta_\gamma(r)$: $y_\alpha(r) + n_\alpha(r), y_\beta(r) + n_\beta(r), y_\gamma(r), + n_\gamma(r))$:

APPLICATIONS IN MOLECULAR BIOLOGY

Dirichlet Mixture Priors for Sequence Profiles

One of the most important goals in bipinformatics is the identification and sequence alignment of proteins that are very distantly related by descent from a common ancestor. Such remote homolog detection methods can proceed in the absence of any structural information for the proteins involved, relying instead on multiple sequence alignments of protein families. Profile methods have a long history and are based on the idea that even very distantly related proteins have particular patterns of hydrophobic and polar amino acids and some well-conserved amino acids at certain positions in the sequence. A profile consists of the probabilities of the 20 amino acids at each position in a multiple sequence alignment of a protein family.

In some cases, a "gap" amino acid is also defined to indicate the probability of a gap at each position in the alignment. Physical information derived from the protein structure, such as hydrophobicity, can also be included. One limiting factor in some cases is that there may be only a few sequences that are obviously related, and a profile built from their multiple alignment is not likely to be very accurate on statistical grounds of a small sample. For instance, for a buried residue, we might have only Val at a particular position in all of the known sequences. The question arises whether this position in an alignment of many more family members would be likely to have He and Leu as well as Val. If the conserved residue is Pro, what is the likelihood that in other related sequences this residue might be different? Bayesian statistics is ideally suited to treat the different situations we might encounter in sequence analysis within a single consistent framework.

The key is to identify prior distributions for profiles that when combined with the data (the known relatives in the family) will produce the best estimate possible of the profile that might be derived from the full sequence family from a set of known sequences of any size.

A prior distribution for sequence profiles can be derived from mixtures of Dirichlet distributions. The idea is simple : Each position in a multiple alignment represents one of a limited number of possible distributions that reflect the important physical forces that determine protein structure and function. In certain core positions, we expect to get a distribution restricted to Val, Ile, Met, and Leu. Other core positions may include these amino acids plus the large hydrophobic aromatic amino acids Phe and Trp. There will also be positions that are completely conserved, including catalytic residues (often Lys, Glu, Asp, Arg, Ser, and other polar amino acids) and Gly and Pro residues that are important in achieving certain backbone conformations in coil regions. Cys residues that form disulfide bonds or coordinate metal ions are also usually well conserved.

A prior distribution of the probabilities of the 20 amino acids at a particular position in a multiple alignment can be represented by a Dirichlet distribution. That is, it is an expression of the values of θ_r, the probabilities of each residue type r, where r ranges from 1 to 20, and $\Sigma_{r=1}^{20} \theta_r = 1$:

$$p(\theta_1, \theta_2, ..., \theta_{20} | \alpha_1, \alpha_2, ..., \alpha_{20}) = \text{Dirichlet } (\theta_1, \theta_2, ..., \theta_{20} | \alpha_1, \alpha_2, ..., \alpha_{20}) \quad ...(35)$$

$$= \left(\frac{\Gamma(\alpha_0)}{\prod_{r=1}^{20} \Gamma(\alpha_r)} \right) \prod_{r=1}^{20} \theta_r^{\alpha_r - 1}$$

α_0 $\Sigma_{r=1}^{20} \alpha_r$ represents the total number of counts that the prior distribution represents, and the α_r, the counts for each type of amino acid (not necessarily integers). Because different distributions will occur in multiple sequence alignments, the prior distribution for any position should be represented as a *mixture* of N Dirichlet distributions :

$$p(\theta_1, \theta_2, ..., \theta_{20} | \alpha_{1,1}, \alpha_{1,2}, ..., \alpha_{N,19}, \alpha_{N,20}; q_1, q_2, ..., q_N)$$

$$= \sum_{j=1}^{N} q_j \text{ Dirichlet } (\alpha_{j,1}, \alpha_{j,2}, ..., \alpha_{j,20}) \quad ...(36)$$

where the q_i represent the mixture coefficients and α_{ij} the count of the ith amino acid type in the jth component of the mixture. One can imagine components that represent predominantly a single amino acid type commonly conserved (Gly, Pro, Trp, Cys) and others that represent physical properties such as hydrophobicity and charge. We can establish such distributions from an understanding of what we expect to see in multiple sequence alignments, or the mixture can be derived from a set of sequence alignments and some optimization procedure.

Sjölander *et. al.*, describe the process assumed in their model of sequence alignments, which is how the counts for a particular position in a multiple sequence alignment would arise from the mixture Dirichlet prior :

1. A component j is chosen from among the N Dirichlet components in Eq. (10.36) according to their respective probabilities, q_j.
2. A probability distribution $(\theta_1, \theta_2, ..., \theta_{20})$ for the 20 amino acids is chosen from component j according to Dirichlet $(\alpha_{j,1}, \alpha_{j,2}, ..., \alpha_{j,20})$.

3. A count vector $(n_1, n_2, ..., n_{20})$ for the position is drawn from the multinomial distribution given $(\theta_1, \theta_2, ..., \theta_{20})$.

For a single Dirichlet component, Dirichlet $(\alpha_{j,1}, \alpha_{j,2}, ..., \alpha_{j,20})$, the expected posterior probability for amino acid type i, given the counts for the 20 amino acids in a single position of the multiple alignment is

$$\hat{p}_i = \frac{n_i + \alpha_{j,i}}{n + \alpha_{j,0}} \qquad \text{...(10.37)}$$

where n is the total of the n_i and $\alpha_{j,0}$ is the total of the $\alpha_{j,i}$.

With a mixture we have to factor in the posterior probability that component j is the correct component :

$$\hat{p}_i = \sum_{j=1}^{n} p(\vec{\alpha}_j|\vec{n}) \frac{n_i + \alpha_{j,i}}{n + \alpha_{j,0}} \qquad \text{...(10.38)}$$

where

$$p(\vec{\alpha}_j|\vec{n}) = \frac{p(\vec{n}|\vec{\alpha}_j)\, p(\vec{\alpha}_j)}{p(\vec{n})} = \frac{p(\vec{n}|\vec{\alpha}_j)\, q_j}{p(\vec{n})} \qquad \text{...(10.39)}$$

$$= q_j \frac{\Gamma(n)\,\Gamma(\alpha_{j,0})}{\Gamma(n + \alpha_{j,0})} \prod_{i=1}^{20} \frac{\Gamma(n_i + \alpha_{j,i})}{\Gamma(n_i)\,\Gamma(\alpha_{j,i})}$$

The parameters $\{\alpha_{j,i}\}$ and $\{q_j\}$ are determined from multiple sequence alignments, *e.g.*, those available from the BLOCKS [55], FSSP or HSSP databases. For each position in each alignment, there is a vector of counts, $\vec{n}_t$, and the objective is to find the set of α and q that maximizes the probability of the counts arising from the Dirichlet mixture, $\Pi_t\, p(\vec{n}_t\,|\alpha, n_t)$. As an example, we list the vectors of α for a mixture of Dirichlet distributions calculated from BLOCKS alignments by Kevin Karplus using an expectation maximization algorithm. A notation at the bottom of each column describes the type of position most likely to be represented by this mixture component (hydrophobic, charged residues, etc.). The last three components represent conserved cysteine, proline, and glycine residues. The other components represent various physical properties.

To demonstrate how the use of a Dirichlet mixture regularizes a profile derived from a small number of sequences, we used the mixture in Table 10.1 on a set of 10 nuclear hormone receptor (nhr) sequences from C. elegans and compared this with the actual data distribution from the 10 sequences and the data and regularized distributions derived from 100 nhr sequences from C. elegans. The results are shown in Table 10.2 for a short segment of nhr proteins beginning with the second conserved cysteine in the first zinc finger. Cys 21 (numbering is based on the *nhr*-25 sequence) is very conserved, which is evident in the data profiles (n_i/n) and the posterior modes of the regularized distributions ($\hat{p}_i$) derived from Eq. (10.38). For 100 sequences, the data profile and regularized profiles are quite similar, which is what we would expect from a Bayesian analysis with a large amount of data. However, when there are only 10 sequences, the regularized profiles are different from the data profiles and are in fact more similar to the profiles derived from 100 sequences. This is the intent of the Bayesian analysis—to estimate parameter values from limited data that will be as close as possible to parameter values we would calculate if we had a large amount of data.

Bayesian Sequence Alignment

The primary tool of bioinformatics is sequence alignment, and numerous algorithms have been developed for this purpose, including variants of pairwise and multiple sequence alignments, profile alignments, local and global sequence alignments, and exact and heuristic methods. Sequence alignment programs usually depend on a number of fixed parameters, including the choice of sequence similarity matrix and gap penalty parameters. The correct choice of these parameters is a much debated topic, and the choice may depend on the purpose of the alignments: to determine whether two sequences are in fact related to each other or to determine the "correct" alignment between them, whatever that may be. The Bayesian alternative to fixed parameters is to define a probability distribution for the parameters and simulate the joint posterior distribution of the sequence alignment and the parameters with a suitable prior distribution. How can varying the similarity matrix and gap penalties be justified, case by case, when these parameters in fact determine the alignment?

Similarity matrices, such as the PAM series, for instance, are parametrized as a function of evolutionary distance between the sequences to be compared. When comparing two sequences, we do not yet know the distance between them, and it is hard to say that a PAM250 or PAM 100 matrix should be used. But certainly after we look at a proposed alignment, we have a better idea of what the distance is and what the matrix should be, even if this alignment would change with a different choice of matrix. Similarly, gap penalties may be different for different sequence pairs, because some proteins may be tolerant to insertions whereas others may not, depending on the pressure of natural selection. For instance, if two proteins are orthologs, *i.e.*, they perform the same function in different species, we might expect fewer and shorter insertions and deletions. If two proteins are paralogs, *i.e.*, homologous proteins that perform different functions in one species (or potentially in different species), then there may be more insertions and deletions to account for the change in function and/or substrate. Thus the use of fixed matrix and gap penalties is not entirely justifiable, other than for convenience.

Zhu *et. al.*, and Liu and Lawrence formalized this argument with a Bayesian analysis. They are seeking a joint posterior probability for an alignment *A*, a choice of distance matrix Θ, and a vector of gap parameters, *A*, given the data, *i.e.*, the sequences to be aligned: $p(\mathbf{A}, \Theta, \Lambda|R_1, R_2)$. The Bayesian likelihood and prior for this posterior distribution is

$$p(\mathbf{A}, \Theta, \Lambda|R_1, R_2) = \frac{p(R_1, R_2|\mathbf{A}, \Theta, \Lambda)\, p(\mathbf{A}, \Theta, \Lambda)}{P(R_1, R_2)}$$

$$= \frac{p(R_1, R_2|\mathbf{A}, \Theta)\, p(A|\Lambda)\, p(\Theta)}{p(R_1, R_2)} \qquad \text{...(10.40)}$$

The second line results from the first, because Θ and Λ are independent a priori, whereas the score of the alignment **A** depends on Λ, and the likelihood of the sequences, given the alignment, depends only on the scoring matrix Θ.

The usual alignment algorithm fixes Θ and Λ as Θ_0 and Λ_0, so that the prior is 1 when $\Theta = \Theta_0$ and $\Lambda = \Lambda_0$ and is 0 otherwise. Clearly, if experience justifies this choice or some other nonuniform choice, we can choose an informative prior that biases the calculation to certain values for the parameters or limits them to some likely range. The likelihood is well defined by the alignment model defined by using a similarity matrix and affine gap penalties, so that

$$\ln p(R_1, R_2|\mathbf{A}, \Theta) = \sum_{i,j} A_{ij}\, \Psi_{R_1(i), R_2(j)} \qquad \text{...(10.41)}$$

Table 10.1. Dirichlet Mixture Components[a]

Comp, #	1	2	3	4	5	6	7	8	9	10	11	12
q	0.137	0.066	0.134	0.070	0.053	0.091	0.128	0.082	0.161	0.008	0.022	0.050
α_q	1.603	1.361	8.916	2.998	3.931	2.208	2.569	5.024	0.138	220.926	22.809	21.949
C	0.040	0.014	0.077	0.013	0.023	0.043	0.012	0.088	0.001	219.778	0.032	0.042
P	0.063	0.012	0.291	0.070	0.014	0.013	0.068	0.098	0.010	0.038	18.602	0.105
G	0.139	0.014	0.428	0.074	0.000	0.012	0.159	0.139	0.000	0.024	0.233	20.261
W	0.005	0.074	0.065	0.010	0.020	0.007	0.003	0.082	0.005	0.000	0.016	0.014
Y	0.017	0.359	0.198	0.041	0.038	0.015	0.022	0.247	0.005	0.000	0.044	0.032
F	0.014	0.406	0.144	0.029	0.203	0.057	0.012	0.293	0.005	0.000	0.071	0.010
M	0.016	0.034	0.123	0.033	0.249	0.083	0.013	0.235	0.002	0.000	0.029	0.012
I	0.035	0.053	0.245	0.051	0.378	0.637	0.018	0.599	0.003	0.045	0.188	0.009
V	0.091	0.043	0.355	0.073	0.220	0.784	0.038	0.737	0.005	0.000	0.314	0.047
L	0.044	0.116	0.348	0.092	0.530	0.280	0.034	0.644	0.008	0.000	0.301	0.059
A	0.381	0.29	0.769	0.129	0.064	0.116	0.164	0.385	0.008	0.148	0.706	0.262
T	0.211	0.018	0.730	0.132	0.051	0.091	0.136	0.343	0.010	0.145	0.285	0.049
S	0.329	0.033	0.902	0.140	0.028	0.025	0.245	0.255	0.006	0.358	0.508	0.241
H	0.012	0.059	0.253	0.091	0.016	0.006	0.057	0.100	0.008	0.000	0.095	0.046
Q	0.027	0.014	0.574	0.269	0.045	0.000	0.183	0.152	0.008	0.209	0.225	0.056
N	0.066	0.027	0.630	0.117	0.008	0.009	0.276	0.107	0.011	0.000	0.142	0.213
E	0.025	0.009	0.803	0.146	0.013	0.009	0.395	0.133	0.011	0.000	0.323	0.084
D	0.025	0.013	0.615	0.047	0.008	0.008	0.503	0.074	0.013	0.062	0.165	0.162
R	0.032	0.023	0.549	0.689	0.014	0.006	0.069	0.142	0.011	0.000	0.265	0.140
K	0.031	0.011	0.819	0.755	0.008	0.009	0.161	0.173	0.008	0.119	0.262	0.105
Description	Small, polar	Aromatic	Polar	Positive charge	Leu	Ile, Val	Negative charge	Hydro-Phobic	Polar	Cys	Pro	Gly

*This 12-component mixture was derived by Kevin Karplus (UC Santa Cruz), http://www.cse.ucsc.edu/research/compbio/dirichlets/index.html, from the BLOCKS database. The first line lists the q_j value for each component. The second line lists the total prior counts, α_0. The last line provides a rough description of each component.

Table 10.2. Raw Data and Posterior Modes from Dirichlet Mixtures for a Six Amino Acid Segment of Nuclear Hormone Receptors[a]

	Res no.	*Res. type*	*No. seq.*	*C*	*P*	*G*	*W*	*Y*	*F*	*M*	*I*	*V*	*L*	*A*	*T*	*S*	*H*	*Q*	*N*	*E*	*D*	*R*	*K*
n_i/n	21	*C*	10	100	0	0	0	0	0	0	0	0	0	0	0	0	0	0	0	0	0	0	0
$\hat{p}_i$	21	*C*	10	99	0	0	0	0	0	0	0	0	0	0	0	0	0	0	0	0	0	0	0
n_i/n	21	*C*	100	100	0	0	0	0	0	0	0	0	0	0	0	0	0	0	0	0	0	0	0
$\hat{p}_i$	21	*C*	100	100	0	0	0	0	0	0	0	0	0	0	0	0	0	0	0	0	0	0	0
n_i/n	22	*G*	10	20	0	40	0	10	10	0	0	0	0	10	0	0	0	0	0	10	0	0	0
$\hat{p}_i$	22	*G*	10	13	1	27	0	8	8	1	3	3	3	9	3	3	1	2	2	9	2	2	2
n_i/n	22	*G*	100	4	0	45	0	3	2	0	0	0	1	4	2	12	3	2	5	6	6	3	2
$\hat{p}_i$	22	*G*	100	4	0	42	0	3	2	0	0	0	1	4	3	12	3	2	5	6	6	3	3
n_i/n	23	*D*	10	0	0	20	0	0	0	0	0	0	0	0	0	0	0	10	10	30	20	0	10
$\hat{p}_i$	23	*D*	10	0	1	16	0	0	0	0	0	1	1	2	2	3	1	9	10	25	18	1	9
n_i/n	23	*D*	100	0	0	8	0	1	0	1	1	0	7	5	3	4	1	11	8	7	33	6	3
$\hat{p}_i$	23	*D*	100	0	0	8	0	1	0	1	1	0	7	5	3	5	1	11	8	7	31	6	4
n_i/n	24	*R*	10	0	30	0	0	0	0	0	0	10	0	0	10	0	0	0	0	0	0	10	40
$\hat{p}_i$	24	*R*	10	0	22	1	0	1	0	0	1	8	1	2	9	2	1	2	2	2	1	11	33
n_i/n	24	*R*	100	1	27	2	0	0	1	0	2	7	0	4	4	8	3	2	2	7	2	13	15
$\hat{p}_i$	24	*R*	100	1	25	2	0	0	1	0	2	7	0	4	4	8	3	2	2	7	2	12	15
n_i/n	25	*V*	10	0	0	10	0	0	0	0	0	10	0	60	0	20	0	0	0	0	0	0	0
$\hat{p}_i$	25	*V*	10	0	1	10	0	0	0	0	0	9	0	54	2	20	0	0	1	0	0	0	0
n_i/n	25	*V*	100	1	6	7	0	1	0	0	1	5	0	48	7	15	1	0	2	2	1	2	1
$\hat{p}_i$	25	*V*	100	1	6	7	0	1	0	0	1	5	0	45	7	15	1	1	2	3	1	2	2
n_i/n	26	*S*	10	0	0	0	0	0	10	0	0	0	0	0	10	10	40	0	20	0	10	0	0
$\hat{p}_i$	26	*S*	10	0	1	2	0	1	7	1	1	2	2	3	9	10	25	3	15	4	9	2	4
n_i/n	26	*S*	100	0	0	2	0	0	4	0	4	3	0	9	11	16	24	1	13	1	7	3	2
$\hat{p}_i$	26	*S*	100	0	0	2	0	0	4	0	4	3	0	9	11	16	22	1	13	2	7	3	3

[a]Sequence are all from the *C. elegans* genome. Sequence and sequence numbering in the second and third columns refer to the *nhr*-25 sequence.

The effect of the gap penalties occurs through the factor $p(A|\Lambda)$, which is given by

$$p(A|\Lambda) = p(A \mid \lambda_0, \lambda_e) = \frac{\lambda_0^{k_g(A)} \lambda_e^{l_g(A)-k_g(A)}}{\sum_{A'} \lambda_0^{k_g(A')} \lambda_e^{l_g(A')-k_g(A')}} \quad ...(10.42)$$

where λ_0 and λ_e are the gap opening and extension parameters, respectively, and $k_g(A)$ and $lg(A)$ are the total number and length of gaps in the alignment $\mathbf{A}$. To obtain marginal posterior distributions for the gap penalties and scoring matrix, Eq. (10.40) must be summed over all alignments. For the gap penalties alone, we need to sum over both the alignments and the scoring matrices :

$$p(\Lambda|R_1, R_2) = \sum_A \sum_\Theta \frac{p(R_1, R_2|\mathbf{A}, \Theta)\, p(A|\Lambda)\, p(\Lambda)\, p(\Theta)}{p(R_1, R_2)} \quad ...(10.43)$$

The sum in Eq. (10.43) can be obtained by a recursion algorithm used commonly in dynamic programming.

Sequence-Structure Alignment

Threading methods have been developed by many groups over a number of years and consist of three steps: (1) selecting a library of candidate folds usually derived from the Protein Data Bank (PDB) of experimentally determined protein structures; (2) selecting the most likely fold from the fold library for a protein sequence of unknown structure (the target sequence); and (3) simultaneously aligning the sequence of the target sequence to the structure (and sequence) from the fold library. Many factors can be considered in choosing the appropriate fold and aligning the sequence to the structure, including sequence similarity (with some amino acid substitution matrix such as the BLOSUM or PAM matrices); burial of hydrophobic amino acids; exposure of hydrophilic, especially charged, amino acids; *contacts* between hydrophobic amino acids (as distinct from simple burial from solvent); secondary structure predictions compared to the parent structure secondary structure; and others.

Although the methods are designed to be applicable in cases where the target sequence is not, in fact, related to the parent by descent from a common ancestor (*i.e.*, homologous), in practice most successful threadings are due to homologous relationships. Lathrop and Smith and Lathrop *et. al.*, in a series of papers, have given a Bayesian theory of sequence-structure alignment. Their goal is to derive a model for protein threading that is Bayes-optimal, that is, selecting both the fold and sequence alignment that have the highest joint probability over all possible folds and sequence alignments. The advantage of the formalism they develop is that most other threading scoring functions can be expressed within their framework. The probability they wish to calculate is $p(C, \mathbf{t}|\mathbf{a}, n)$ where C is the protein fold or "core" selected from the fold library, $\mathbf{t}$ is the alignment of the target sequence to the fold sequence and structure, $\mathbf{a}$ is the sequence of the target (assumed known), and n is its length.

Before setting up priors and likelihoods, we can factor the joint probability of the core structure choice and the alignment $\boldsymbol{t}$ by using Bayes' rule :

$$p(C, \mathbf{t}|\mathbf{a}, n) = p(\mathbf{t}|\mathbf{a}, n, C)\, p(C|\mathbf{a}, n) \quad ...(10.44)$$

[This can be thought of as follows: If $p(C, \boldsymbol{t}) = p(\boldsymbol{t}|C)\, p(C)$, then we simply add the conditionality on $\boldsymbol{a}$, n to each term to achieve Eq. (10.44).] Lathrop and coworkers provide a likelihood and prior for each term in Eq. (10.44). For the first term, considering $y = \boldsymbol{a}$ as the data and $\theta = \boldsymbol{t}$ as the parameter we are looking for, and leaving n, C where they are on the right-hand side of the vertical bar [*i.e.*, read Eq. (10.45) without the n, C, and it reads like the standard Bayesian equation $p(\theta|y) = p(y|\theta)/p(y)$], we have

$$p(\boldsymbol{t}|\boldsymbol{a}, n, C) = \frac{p(\boldsymbol{a}|\boldsymbol{t}, n, C)\, p(\boldsymbol{t}|\boldsymbol{n}, C)}{p(\boldsymbol{a}|\boldsymbol{n}, C)} \quad ...(10.45)$$

For the second term, we use the same trick: Leave n on the right-hand side of the vertical and use the Bayesian statistics equation to invert C and a :

$$p(C|\boldsymbol{a}, n) = \frac{p(\boldsymbol{a}|C, n)\, p(C|n)}{p(\boldsymbol{a}|n)} \quad ...(10.46)$$

Multiplying out Eqs. (10.45) and (10.46) and canceling the $p(\boldsymbol{a}|n, C) = p(\boldsymbol{a}|C, n)$ terms, we have

$$p(C, \boldsymbol{t}|\boldsymbol{a}, n) = p(\boldsymbol{a} \mid \boldsymbol{t}, n, C)\, \frac{p(\boldsymbol{t}|\boldsymbol{n}, C)\, p(C|n)}{p(\boldsymbol{a}|\boldsymbol{n})} \quad ...(10.47)$$

where the last term forms the prior distribution normalized by the probability of the data (*i.e.*, the sequence and its length), and the first term on the right is the likelihood function of the sequence given the alignment; its length, and the choice of core structure. What remains is to decide functional forms for each of these terms.

For $p(C|n)$, Lathrop *et. al.*, provide a noninformative prior, $p(C|n) = 1/L$, where L is the size of the library. A slightly more complicated prior could take into account that some folds are more common than others and that some folds are more common in some phyla than others. Given certain rules that constitute a proper threading, $\boldsymbol{t}$, we can establish how many threadings there can be for a given core structure C and sequence length n of the target. Their model for protein threading consists of core structures that contain only regular secondary structures (just helices and sheet strands, no loops or irregular secondary structure) and a constraint that no gap can occur within a secondary structure unit of the core structure. Under these constraints a finite number of threadings are possible, $T(n, C)$. The noninformative prior for $p(\boldsymbol{t}|n, C)$ is then just $1/T(n, C)$. A more elaborate prior for $p(\boldsymbol{t}|\boldsymbol{n}, C)$ is described in that takes into account likely loop lengths in real proteins and coverage of the core elements by residues from the sequence. Even in the absence of knowledge of the target sequence, some alignments are more likely than others. Finally, for the prior, $p(\boldsymbol{a}|\boldsymbol{n})$ is a constant and can be ignored because we are looking for relative probabilities of the choice of fold and alignment.

The likelihood function is an expression for $p(\boldsymbol{a}|\boldsymbol{t}, n, C)$, which is the probability of the sequence $\boldsymbol{a}$ (of length n) given a particular alignment $\boldsymbol{t}$ to a fold C. The expression for the likelihood is where most threading algorithms differ from one another. Since this probability can be expressed in terms of a pseudo free energy, $p(\boldsymbol{a}|\boldsymbol{t}, n, C) \propto \exp[-f(\boldsymbol{a}, \boldsymbol{t}, C)]$, any energy function that satisfies this equation can be used in the Bayesian analysis described above. The normalization constant required is akin to a partition function, such that

$$p(\boldsymbol{a}|\boldsymbol{t}, n, C) = \frac{\exp[-f(\boldsymbol{a}, \boldsymbol{t}, C)]}{\Sigma_i \exp[-f(\boldsymbol{a}, \boldsymbol{t}_i, C)]} \quad ...(10.48)$$

where the sum is over all possible alignments. The final result for the posterior distribution is

$$p(C, t|a, n) = \frac{\exp[-f(a,t,C)]}{\Sigma_i \exp[-f(a,t_i,C)]}\left(\frac{1}{LT(n,C)\,p(a|n)}\right) \quad ...(10.49)$$

The success of any method using this formalism will be entirely dependent on the function f used and the search strategy. Lathrop *et. al.*, use a branch-and-bound algorithm to search the space of legal threadings and a function f that does not depend on residue interactions. The details of these aspects are not inherently Bayesian in character and are outside the scope of this review.

APPLICATIONS IN STRUCTURAL BIOLOGY

Secondary Structure and Surface Accessibility

A common use of statistics in structural biology is as a tool for deriving predictive distributions of structural parameters based on sequence. The simplest of these are predictions of secondary structure and side-chain surface accessibility. Various algorithms that can learn from data and then make predictions have been used to predict secondary structure and surface accessibility, including ordinary statistics, information theory, neural networks and "Bayesian" methods. A disadvantage of some neural network methods is that the parameters of the network sometimes have no physical meaning and are difficult to interpret.

Unfortunately, some authors describing their work as "Bayesian inference" or "Bayesian statistics" have not, in fact, used Bayesian statistics; rather, they used Bayes' rule to calculate various probabilities of one observed variable conditional upon another. Their work turns out to comprise derivations of informative prior, distributions, usually of the form $p(\theta_1, \theta_2, ..., \theta_N) = \Pi_{i=1}^{N} p(\theta_i)$, which is interpreted as the posterior distribution for the θ_i without consideration of the joint distribution of the data variables in the likelihood.

For example, Stolorz *et. al.*, derived a Bayesian formalism for secondary structure prediction, although their method does not use Bayesian statistics. They attempt to find an expression for $(\mu|\text{seq}) = p(\text{seq}|)\, p(\mu)/p(\text{seq})$, where μ is the secondary structure at the middle position of seq, a sequence window of prescribed length. As described earlier, this is a use of Bayes' rule but is not Bayesian statistics, which depends on the equation $p(\theta|y) = p(y|\theta)\, p(\theta)/p(y)$, where y is data that connect the parameters in some way to observables. The data are not sequences alone but the combination of sequence and secondary structure that can be culled from the PDB. The parameters we are after are the probabilities of each secondary structure type as a function of the sequence in the sequence window, based on PDB data. The sequence can be thought of as an explanatory variable. That is, we are looking for

$$p(\theta_\alpha(x), \theta_\beta(x), \theta_\gamma(x)|y_\alpha(x), y_\beta(x), y_\gamma(x)) \propto p(y_\alpha(x), y_\beta(x), y_\gamma(x)|\theta_\alpha(x), \theta_\beta(x), \theta_\gamma(x)), p(\theta_\alpha(x), \theta_\beta(x), \theta_\gamma(x)) \quad ...(10.50)$$

where, for example, $\theta_\alpha(x)$ is the probability of an α-helix at the middle position of a string of amino acid positions in a protein with sequence x. Because we are dealing with count data, the appropriate functional forms for Eq. (10.50) are a multinomial likelihood and Dirichlet functions for the prior and posterior distributions. For any window of length N there are 20^N sequences possible, so there are insufficient data in the PDB for these calculations for any $N > 5$. There are three possible solutions:

1. Reduce the size of the alphabet from 20 amino acids to a smaller alphabet of, say, six (aliphatic, aromatic, charged, polar, glycine, proline).
2. Use *very* informative priors, where perhaps the prior could be based on the product of individual probabilities for each secondary structure type and amino acid type.
3. Do both of the above.

In this case, we are looking for "counts" for each secondary structure type μ for each sequence x, which might be derived from PDB data by

$$n_{\mu} \propto \prod_{i=1}^{N} p(\mu|r_i) \qquad \text{...(10.51)}$$

The constant of proportionality is based on normalizing the probability and establishing the size of the prior, that is, the number of data points that the prior represents. The advantage of the Dirichlet formalism is that it gives values for not only the modes of the probabilities but also the variances, covariances, etc.

Thompson and Goldstein improve on the calculations of Stolorz *et. al.*, by including the secondary structure of the entire window rather than just a central position and then sum over all secondary structure segment types with a particular secondary structure at the central position to achieve a prediction for this position. They also use information from multiple sequence alignments of proteins to improve secondary structure prediction. They use Bayes' rule to formulate expressions for the probability of secondary structures, given a multiple alignment. Their work describes what is essentially a sophisticated prior distribution for $\theta_{\mu}(\boldsymbol{X})$, where $\boldsymbol{X}$ is a matrix of residue counts in a multiple alignment in a window about a central position.

The PDB data are used to form this prior, which is used as the predictive distribution. No posterior is calculated with posterior = prior X likelihood. A similar formalism is used by Thompson and Goldstein [90] to predict residue accessibilities. What they derive would be a very useful prior distribution based on multiplying out independent probabilities to which data could be added to form a Bayesian posterior distribution. The work of Arnold *et al.*, is also not Bayesian statistics but rather the calculation of conditional distributions based on the simple counting argument that $p(\sigma|r) = p(\sigma, r)/p(r)$, where σ is some property, of interest (secondary structure, accessibility) and r is the amino acid type or some property of the amino acid type (hydro-phobicity) or of an amino acid segment (helical moment, etc).

Side-Chain Conformational Analysis

Analysis and prediction of side-chain conformation have long been predicated on statistical analysis of data from protein structures. Early rotamer libraries ignored backbone conformation and instead gave the proportions of side-chain rotamers for each of the 18 amino acids with side-chain dihedral degrees of freedom. In recent years, it has become possible to take account of the effect of the backbone conformation on the distribution of side-chain rotamers. McGregor *et. al.*, and Schrauber *et. al.*, produced rotamer libraries based on secondary structure. Dunbrack and Karplus instead examined the variation in χ_1 rotamer distributions as a function of the backbone dihedrals ϕ and ψ, later providing conformational analysis to justify this choice. Dunbrack and Cohen extended the analysis of protein side-chain conformation by using Bayesian statistics to derive the full backbone-dependent rotamer libraries at all values of ϕ and ψ. This library has proven useful in side-chain conformation prediction, NMR structure determination and protein design.

The PDB has grown tremendously in the last 10 years, so whereas Ponder and Richard's library was based on 19 chains in 1987, we based our most recent backbone-dependent rotamer library on 814 chains from the PDB with resolution of 2.0 Å or better, R-factor less than 20%, and mutual sequence identity less than 50%. Even though there are substantially more data for side chains than in the 1980s, the data are very uneven. Some residue types (Asp, Lys, Ser) are common, and some (Cys, Trp, Met) are not. The long side chains (Met, Glu, Gin, Lys, Arg) have either 27 or 81 possible rotamers, some of which have been seen very rarely or not at all because of the large steric energies involved. The uneven distribution of residues across the Ramachandran map is also problematic, because for prediction purposes we would like to have probabilities for the different rotamers even in unusual backbone conformations where there may be only a few cases in the PDB or again none at all.

Bayesian statistics was the most logical choice for overcoming this unevenness in data, providing a unified framework for adjusting between data-rich situations and data-poor situations where an informative prior distribution could make up for the lack of data. The parameters we are searching for are the probabilities as well as the average values of the χ angles for the χ_1, χ_2, χ_3, χ_4 rotamers given values for the explanatory variables ϕ and ψ. To distinguish between side-chain dihedrals and rotamers we denoted the χ_1 rotamer as r_1, the χ_2 rotamer as r_2, etc. For sp^3–sp^3 dihedrals, there are three well-defined rotameric positions : $\chi \sim +60°$ (g^+), $\chi \sim 180°$ (t), and $\chi \sim -60°$ (g^-). For sp^3–sp^2 dihedrals (*e.g.*, in aromatic amino acids, Asn, Asp, Glu, Gin), the situation is more complicated. This work proposed what is essentially a mixture model for side-chain conformations, although not explicitly described as such in Ref., using normal distribution functions where the μ's and σ's depend on the rotamer type and backbone conformation, as do the mixture coefficients, the amount of each component in the mixture. We denote the mixture coefficients for each region of the ϕ, ψ, map as follows :

$$\begin{aligned}\theta_{ijkl|ab} &\equiv p(r_1 = i, r_2 = j, r_3 = k, r_4 = l|\phi_a - 10° < \phi \\ &\leq \phi_a + 10°, \psi_b - 10° < \psi \leq \psi_b + 10°)\end{aligned} \quad \text{...(10.51 } a\text{)}$$

We denote the mean χ angles and their variances σ^2 similarly as $\mu^{(m)}_{ijkl|ab}$, where (m) represents the χ angle (1, 2, 3, or 4) for the $ijkl$th rotamer, and $\sigma^{2(m)}_{ijkl|ab}$ the variance. To simplify the model a bit, we set the covariances to 0. The full model is therefore

$$p(\chi_1, \chi_2, \chi_3, \chi_4|\phi_a, \psi_b) = \sum_{i=1}^{3}\sum_{j=1}^{3}\sum_{k=1}^{3}\sum_{l=1}^{3} \theta_{ijkl|ab} \prod_{m=1}^{4} N(\mu^{(m)}_{ijkl|ab}, \sigma^{2(m)}_{ijkl|ab}) \quad \text{...(10.52)}$$

where N signifies a normal distribution (Eq. 10.14). This is a likelihood function for the χ angles given their explanatory variables ϕ, ψ, *i.e.*, $p(y|\theta)$, where the y are the χ angles in each ϕ, ψ region and the θ are the mixture coefficients and the mean and variance parameters in the normal distributions. Ordinarily, which component a particular data point (a single side chain conformation and its backbone conformation) belongs to is left as missing data. However, in our case we know that our χ angles are trimodal with well-spaced modes (near +60°, 180°, and –60°), and we can assign each side chain to a particular rotamer unambiguously. We use Bayesian statistics to determine the parameters in Eq. (10.52).

Because each side chain can be identifiably assigned to a particular component, the mixture coefficients and the normal distribution parameters can be determined separately. For the $\chi_{ijkl|ab}$ we use Dirichlet priors combined with a multinominal likelihood to determine a Dirichlet posterior distribution. The data in this case are the set of counts $\{n_{ijkl|ab}\}$. We determined these counts from PDB data (lists of values for $\{\phi, \psi, \chi_1, \chi_2, \chi_3, \chi_4,\}$) by counting side chains in overlapping 20° × 20° square blocks centered on (ϕ_a, ψ_b) spaced 10° apart. The likelihood is therefore of the form

$$p(n_{ijkl|ab}\}|\theta_{ijkl|ab}) = \frac{n_{ab}!}{\prod_{i,j,k,l} n_{ijkl|ab}} \prod_{i,j,k,l} \theta_{ijkl|ab}^{n_{ijkl|ab}} \quad ...(10.53)$$

The major benefit of using Bayesian statistics for side-chain conformational data is to account for the uneven distribution of data across the Ramachandran map and the ability to determine large numbers of parameters from limited data using carefully chosen informative prior distributions. Prior distributions can come from any source, either previous data or some pooling of data that in the true posterior would be considered distinct. For instance, if we considered it likely that the distributions of two different amino acid types are likely to be very similar, we might pool their conformations to determine the prior distribution and then determine separate posterior distributions for each type by adding in the data for that type only in the likelihood.

We can use a process of building prior distributions by combining posterior distributions for subsets of the parameters. The prior distribution for the $\{\theta_{ijkl|ab}\}$ is a Dirichlet distribution where the counts, $\alpha_{ijkl|ab}$, are determined as follows :

$$\alpha_{ijkl|ab} = K\hat{\theta}_{ij|a}\,\hat{\theta}_{ij|b}\,\hat{\theta}_{kl|ij} \quad ...(10.54)$$

where K represents the total weight to the prior and the θ's come from the expectation value of the posterior distributions for these parameters. These variables have posterior distributions

$$p(\theta_{ij|a}) = \text{Dirichlet}\,(\{n_{ij|a} + \alpha_{ij|a}\}) \quad ...(10.55)$$

where the prior parameters $\{\alpha_{ij|a}\}$ are determined from the data as $\alpha_{ij|a} = K(n_{i|a}/n_a)\,(n_{j|i}/n_j)$, with K as some constant. In principle, the fractions in this expression could also be obtained from posterior distributions based on informative prior distributions, but at some point there are sufficient data to work directly from the data to determine the priors. The other factors in Eq. (10.54) are determined similarly. The χ angle distributions are also determined by deriving posterior distributions from data and prior distributions that are themselves determined from posterior distributions and their prior distributions and data.

The appropriate expressions are given in Eq. (10.17). In our previous work we did not put a prior distribution on the variances. This resulted in wildly varying values for the variances, especially for rotamers that are rare in the database. If there are only a few side chains in the count, their variance will be outweighed by the prior estimate if there is enough weight on the prior values in the Inv χ^2 distribution (the degrees of freedom, ν_0). Also, when there is only one count or no counts, a reasonable value for the variance is determined from the prior. As an example of analysis of side-chain dihedral angles, the Bayesian analysis of methionine side-chain dihedrals is given in Table 10.3 for the $r_1 = g^-$ rotamers. In cases where there are a large number of data — for example, the (3, 3, 3) rotamer — the data and posterior distributions are essentially identical. These are normal distributions with the averages and standard variations given in the table. But in cases where there are few data,

Table 10.4. Bayesian Analysis of Methionine Side Chain Dihedral Angles[a]

Dist.	r_1	r_2	r_3	N	χ_1	σ_1	χ_2	σ_2	χ_3	σ_3
data	*3*	*1*	1	24	*–80.5*	*18.7*	*77.1*	*16.7*	74.1	17.3
prior					*–78.1*	22.0	*71.9*	*11.5*	72.6	15.4
post					*–79.8*	19.5	*75.6*	15.3	73.7	16.5
data	3	*1*	1	16	*–70.4*	*27.8*	*77.1*	*29.4*	173.5	22.5
prior					*–78.1*	22.0	*73.2*	*14.0*	–175.3	28.5
post					*–73.4*	25.4	*75.6*	*24.0*	177.8	26.8
data	*3*	*1*	*3*	*7*	*–87.7*	*15.0*	*66.3*	*19.7*	*–93.5*	*29.7*
prior					*–78.1*	*22.0*	*73.8*	*15.9*	*–90.8*	*25.0*
post					*–82.1*	*19.7*	*70.7*	*17.3*	*–91.9*	*26.1*
data	3	2	1	611	–69.2	11.1	178.2	12.4	70.8	17.6
prior					–68.3	11.0	179.3	9.2	71.4	18.0
post					–69.2	11.1	178.2	12.4	70.8	17.6
data	3	2	2	332	–67.3	11.2	179.8	12.1	–175.5	23.3
prior					–68.3	11.0	179.7	9.2	–178.2	24.0
post					–67.4	11.2	179.8	12.0	–175.5	23.3
data	3	2	3	443	–67.9	10.8	–176.8	13.4	–74.5	18.7
prior					–68.3	11.0	–178.6	9.3	–73.6	19.0
post					–67.9	10.8	–176.8	13.3	–74.5	18.7
data	3	*3*	*1*	145	–65.9	10.7	*–67.7*	*14.0*	*98.2*	13.1
prior					–65.7	11.4	*–65.6*	*9.7*	*98.0*	13.2
post					–65.8	10.7	*–67.6*	*13.8*	*98.2*	13.1
data	3	3	2	157	–64.5	11.7	–66.0	13.6	168.4	25.7
prior					–65.7	11.4	–65.8	9.9	170.0	26.3
post					–64.6	11.7	–66.0	13.3	168.5	25.7
data	3	3	3	601	–66.0	11.5	–62.2	12.7	–70.0	15.9
prior					–65.7	11.4	–64.1	9.8	–70.6	16.0
post					–66.0	11.5	–62.2	12.7	–70.0	15.9

*Average data dihedrals along with the prior and posterior modes and data, prior, and posterior standard deviations are given for the *g*-rotamer. Rotamers with *syn*-pentane steric interactions are italicized.

e.g., the (3, 1, n) rotamers, the prior exhibits a strong pull on the data to yield a posterior distribution that is between their mean values. To get a feeling for the distributions involved, in Fig. 10.14 we show the data distribution along with the posterior distribution of the mean and the posterior predictive distributions for χ_3 of the (3, 1, 3), (3, 2, 3), and (3, 3, 3) rotamers. The posterior distribution for the values of $\mu_{ijk}^{(3)}$ was generated from a simulation of 1000 draws. This was achieved first by drawing values for the variance from its posterior distribution, a scaled inverse χ^2 distribution [Eqs. (10.16)] derived from the values of the prior variance and the data variance. With each value of σ^2 a draw was taken from the posterior distribution for the χ_3 angle. This distribution is normal with a variance roughly equal to σ_n^2/κ_n. For large amounts of data, for example the (3, 3, 3) rotamer, this produces a very precise estimate of the value of $\mu_{ijk}^{(3)}$. For small amounts of data, the spread in $\mu_{ijk}^{(3)}$ is larger. We can also predict future data, denoted $\bar{y}$, from the likelihood and the posterior distribution of the parameters. This is called the *posterior predictive distribution* and can be achieved by making draws from the posterior distribution and from these values, making draws from the likelihood function, *i.e.*,

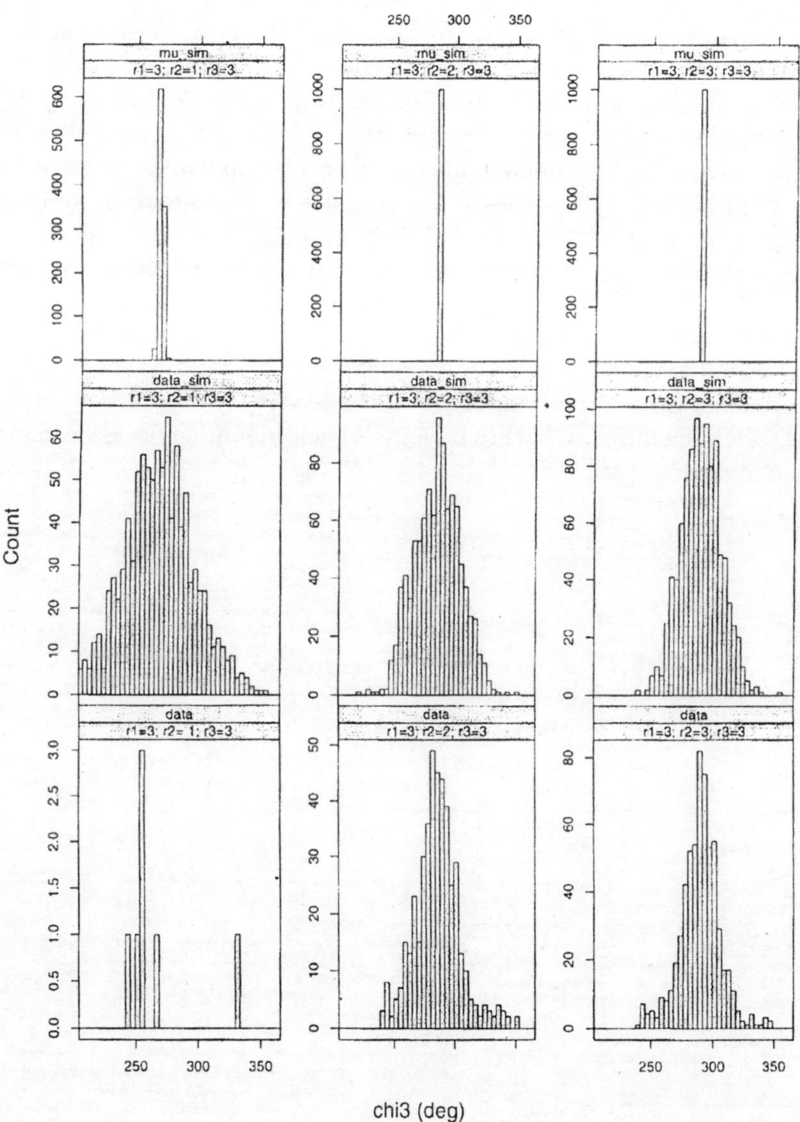

Fig. 10.14. Data distribution and draws from the posterior distribution (mu_sim) and posterior predictive distributions (data_sim) for methionine side chain dihedral angles. The results for three rotamers are shown, ($r_1 = 3$; $r_2 = 1$; $r_3 = 3$), (3, 2, 3), and (3, 3, 3). Each simulation consisted of 1000 draws, calculated and graphed with the program S-PLUS.

$$p(\bar{y}|y) = \int_{\Theta} p(\bar{y})|\theta)p(\theta|y)d\theta = \frac{\int_{\Theta} p(\bar{y}|\theta)p(y|\theta)p(\theta)\,d\theta}{\int_{\Theta} p(y|\theta)p(\theta)d\theta} \quad ...(56)$$

This distribution resembles the data closely for rotamer (3, 3, 3) but also forms a very reasonable distribution when there are only seven data points (3, 3, 1). A good posterior predictive distribution for any protein structural feature can be used in simulations of protein folding or structure prediction.

CONCLUSION

The field of statistics arose in the eighteenth and nineteenth centuries because of the need to develop good public policy based on demographic and economic data. Applications in the natural sciences were immediate, but generally natural scientists have lagged behind in their knowledge of modern statistics compared to social scientists. This is unfortunate, because many algorithms and methodologies have been developed in the last 20 years or so that make feasible sophisticated analysis of very complex data sets. Bayesian statistics has been used fruitfully in molecular and structural biology in recent years but has enjoyed more applications in genetics and clinical research and in the social sciences. Bayesian methods are particularly useful in modeling complex data, where the distribution of information may be uneven or hierarchical. This is true not only of the sequence and structure databases described in this chapter but also of more recently developed experimental methods such as DNA microarrays for analyzing mRNA expression levels over many thousands of genes. The computational challenges for this kind of data are immense. Particularly now, when the influx of data in biology is overwhelming, Bayesian statistical analysis promises to be an important tool.

MOLECULAR BIOMECHANICS 11

Computer-aided drug design (CADD) has enjoyed a long and successful history in its development and applications. Even before the computer age, chemists and medicinal chemists used their intuition and experience to design molecules with desired biological profiles. However, as in practically all areas of chemistry, rapid accumulation of vast amounts of experimental information of biologically active molecules naturally led to the development of theoretical and computational tools for the rational analysis and design of bioactive molecules. The power of these tools rests solidly on a variety of well-established scientific disciplines including theoretical and experimental chemistry, biochemistry, biophysics, pharmacology, and computer science.

This unique combination of underlying scientific disciplines makes CADD an indispensable complementary tool for most types of experimental research in drug design and development. The advancement of drug design as a quantitative, analytical methodology should be, perhaps, attributed to Professor Corwin Hanch, who in the early 1960s initiated the development of the quantitative structure-activity relationships (QSAR) method (reviewed in Ref. 1). Early QSAR studies made extensive use of various physicochemical properties (such as the octanol-water distribution coefficient, which serves as a measure of hydrophobicity) and chemical topology based molecular descriptors to arrive at reliable models correlating structure and activity. Beginning in the early 1970s, CADD started to employ computer graphics or, more specifically, molecular graphics. This led to the development of three-dimensional molecular modeling, which can be defined as the generation, manipulation, calculation, and prediction of realistic chemical structures and associated physicochemical and biological properties.

The application of molecular modeling approaches in the drug design area culminated in the development of the first three-dimensional (3D) pharmacophore assignment method, the active analog approach (AAA). This was quickly followed by the development of pharmacophore-based database searching methodologies and the establishment of 3D QSAR approaches such as comparative molecular field analysis (CoMFA), which remains one of the most popular methods of CADD today. Finally, beginning in the early 1980s, a growing number of drug design applications started to use experimental structural information about macromolecular drug targets, i.e., proteins and nucleic acids.

Modern methods of computer-aided drug design fall into two major categories: ligand-based and receptor-based methods. The former methods, which include QSAR, various pharmacophore assignment methods, and database searching or mining, are based entirely on experimental structure-activity relationships for receptor ligands or enzyme inhibitors. Their application in the last three or four decades led to several drugs currently on the market. The structure-based design methods, which include docking and advanced molecular simulations, require structural information about the receptor or enzyme that is available from X-ray crystallography, nuclear magnetic resonance (NMR) techniques, or protein homology model building. This strategy became popular only recently owing to rapid advances in structure elucidation methods and has already led to several promising drug candidates and even marketed drugs such as several approved inhibitors of the HIV protease for the treatment of AIDS [8].

It is practically impossible to review all major developments, applications, approaches, and caveats of the various drug design methods in one chapter. Several monographs and numerous publications have been devoted to CADD, and many specialized reviews addressing various aspects of molecular modeling as applied to drug design have been published in recent years in the ongoing series *Reviews in Computational Chemistry*. We decided to concentrate in this chapter on one outstanding problem of modern chemistry and theoretical biology, that of ligand-receptor recognition, which is the underlying mechanism of action of most drugs. Using this problem as a molecular modeling paradigm, we introduce major molecular modeling concepts and tools that one can apply to drug design. To provide a practical example of a drug design project and make the discussion somewhat didactic, we consider the dopaminergic ligand-receptor system, which has been the focus of our research in recent years. We believe that by considering one particular drug design problem we can better illustrate the power and limitations of underlying techniques and at the same time provide a practical introduction to the major concepts of molecular modeling as applied to drug design.

RELATIONSHIPS BETWEEN EXPERIMENTAL AND THEORETICAL APPROACHES TO STUDYING DRUG-RECEPTOR INTERACTIONS

The ultimate goal of molecular modeling as a pharmacological and medicinal chemistry tool is to predict, in advance of any laboratory testing, novel biologically active compounds. Molecular modeling research starts from the analysis of experimental observables of drug-receptor interaction. This interaction leads to the formation of the ligand-receptor complex followed by the conformational change of the receptor, which constitutes the putative mechanism of signal transduction. One can obtain either ligand-receptor binding constants from direct measurements in vitro or the biological activity data measured on isolated tissues or on the whole organism. Thus, experimental observables of ligand-receptor systems include the chemical structure of the ligands, biological activity or binding constants, receptor primary sequence, and, in some cases, receptor 3D structure.

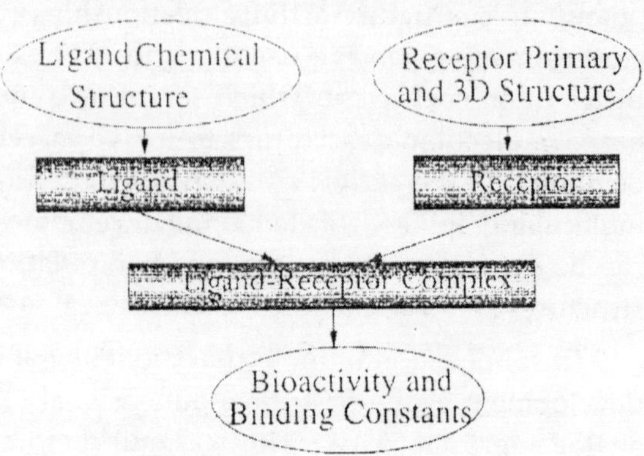

Fig. 11.1. A flow chart of experimental information about drug-receptor interaction. The objects of experimental and theoretical investigation are boxed, and the experimental information is circled.

Molecular modeling uses the experimental observations as input data to programs that afford rational analysis of structure-activity relationships. As mentioned above, there are two main approaches to drug design: ligand-based and structure-based methods. First, if the receptor structure has been characterized by either X-ray crystallography or NMR, computer graphics methods can be used to design novel compounds based on complementarity between the surfaces of the ligand and the receptor binding site. Energy calculations may also be applied to quantitate and refine the graphics design. Second, if the receptor structure is unknown, one is left to infer the shape of the receptor binding site. Several receptor mapping techniques have been developed, such as AAA and CoMFA. The idea of these approaches is to construct a pharmacophore that integrates all the important structural and physicochemical features of the active compounds required for their successful binding to a receptor. Although the pharmacophore itself has significant predictive power as far as rational drug design is concerned, the design can be greatly facilitated if the receptor model is available. In many cases structural information about the receptor can be obtained experimentally by the means of X-rays or NMR. However, in many instances, especially for transmembrane receptors such as G-protein coupled receptors (GPCR), experimental structural determination is difficult if not impossible. Thus, theoretical structure modeling remains the only possible source of structural information about many receptors.

Recent successes in deciphering primary sequences of many receptors and developing several computerized protein-modeling tools enabled researchers to devise putative three-dimensional models of these receptors from their primary sequences. The development of molecular docking methodologies (which allow one to bring together the ligand and the receptor and "dock" the former into the latter) complete the round of computer representations of natural processes of ligand-receptor interaction (*i.e.*, a phar-macophore-based conformationally constrained ligand is brought into the vicinity of the modeled receptor binding site via docking).

The interaction between ligands and their receptors is clearly a dynamic process. Once the static model of ligand-receptor interaction has been obtained, the stability of ligand-receptor complexes should be evaluated by means of molecular dynamics simulations.

One can see that at this point only two types of experimental data are used for the analysis of a ligand-receptor system: ligand chemical structure (from which one generates the 3D pharmacophore conformation) and the primary receptor sequence (which is used to generate the 3D receptor model). The most important property of the receptors, however, is their ability to discriminate ligands on the basis of their chemical structure; we quantify this as ligand binding constants. The ability to reproduce binding constants, or at least their relative order, is the most sensitive

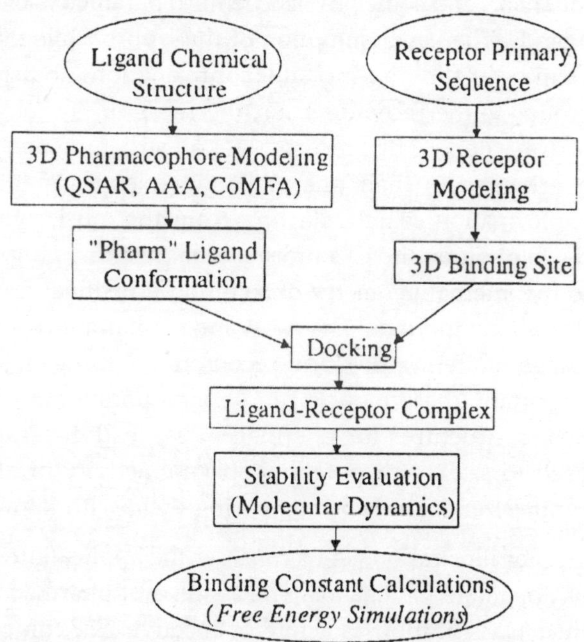

Fig. 11.2. A flow chart of theoretical modeling drug-receptor interaction and relation to experiment. The objects of theoretical investigation are in rectangles, and the experimental information is in the ovals.

test of any putative receptor model. The technique that, in principle, can provide such data is the free energy simulations/thermodynamic cycle approach, a method that relies on full atomistic representations of both ligand and receptor in the molecular model. This approach is very robust but requires substantial computational resources, which limits its practical applicability for drug design. Nevertheless, in several reported cases, it was able to reproduce accurately the experimental binding constants.

The successful realistic modeling of ligand-receptor interactions can only be based on a combined approach incorporating several computer-assisted modeling tools. As outlined above, these tools may include ligand-based "negative image" receptor modeling, QSAR, primary structure-based "real" receptor modeling, docking, structure refinement (molecular dynamics), and relative binding constant calculations. We now discuss these techniques in the order in which one can proceed while working on a receptor model.

COMPUTATIONAL APPROACHES TO MODELING LIGAND-RECEPTOR INTERACTIONS

Ligand-Based Approaches

The advent of faster computers and the development of new algorithms led in the late 1970s and early 1980s to the development and active use of 3D negative receptor image modeling tools such as AAA. In this approach, first proposed by Marshall *et. al.*, the researcher infers the size, the shape, and some physicochemical parameters of the receptor binding site by modeling receptor ligands. The key assumption of this approach is that all the active receptor ligands adopt, in the vicinity of the binding site, conformations that present common pharmacophoric functional groups to the receptor. Thus, the receptor is thought of as a negative image of the generalized pharmacophore that incorporates all structural and physicochemical features of all active analogs overlapped in their pharmacophoric conformations. The latter are found in the course.of a conformation search, starting from the most rigid active analogs and proceeding with more flexible compounds. Conformational restraints imposed by more rigid compound(s) with respect to the internal geometry of common functional groups are used to facilitate the search for more flexible compounds. The AAA and similar approaches have been successfully applied to negative image modeling of many receptors, yielding important leads for rational drug design. Most important, these models can be incorporated into a common database providing the source of ligand structures for pharmaceutical lead generation. Furthermore, the same database can be used to search for potential specific activity or cross-reactivity of independently designed or synthesized ligands by comparing them with known 3D pharmacophores.

For illustration, let us consider the application of AAA to an important pharmacological class of dopaminergic ligands. The analysis of pharmacological activity of D, receptor ligands leads to their classification as active or inactive. Six compounds were chosen as active on the basis of three criteria: They had affinity for the D_1 receptor ($K_{0.5}$ < 300 nM); they could increase cAMP synthesis in rat striatal membranes to the same degree as dopamine; and this increase could be blocked completely by the D_1 antagonist SCH23390. The compounds that met these criteria were dopamine, DHX, SKF89626, SKF82958, A70108, and A77636. As shown in Table 11.1, all of these compounds caused complete activation of dopamine-sensitive adenylate cyclase in this preparation and had $K_{0 \cdot 5}$ values ranging from 267 nM (dopamine) to 0.9 nM (A70108).

The molecular modeling studies with this set of dopaminergic ligands involved the following analytical steps :

1. The tentative pharmacophoric elements of the D_1 receptor were determined on the basis of known structure-activity relationships.
2. A rigorous conformational search on the active compounds was performed to determine their lowest energy conformation(s).

Table 11.1. Pharmacological Analysis of DHX and Related Compounds.

Drug	*D_1 affinity $K_{0.5}$ (nM)*	*D_2 affinity $K_{0.5}$ (nM)*	*Adenylate cyclase EC50 (nM)*	*Max. stimulation of adenylate cyclase (% vs. DA)*
Dopamine	267	36	5000	100
(+) DHX	2.3	43.8	30	120
SKF89626	61	142	700	120
SKF82958	4	73	491	94
A70108	0.9	41	1.95	96
A77636	31.7	1290	5.1	92
cis-DHX	$>10^3$	$>10^3$	$>10^4$	17
N-Propyl-DHX	326	27	$>10^4$	36
N-Allyl-DHX	328	182	$>10^4$	32
Ro 21-7767	477	61	$>10^4$	22
N-Benz-5, 6-DTN	$>10^3$	$>10^3$	$>10^3$	38
N-Benz-6, 7-ADTN	$>10^3$	335	$>10^4$	25

3. Conformationally flexible superimposition of these compounds was done to determine their common (pharmacophoric) conformation.
4. Similar conformational analyses were performed for inactive compounds, and inactive compounds in pharmacophoric conformations were superimposed with the active compounds to determine steric limitations in the active site. Where appropriate, the geometry of each inactive molecule was obtained by modifying the chemical structure of the relevant active analogs followed by the energy minimization of the resulting structure.
5. Finally, an evaluation was made of excluded receptor volume and shape as spatial equivalents of the volume and shape of the pharmacophore.

All molecular modeling studies were performed with the multifaceted molecular modeling software package SYBYL (Tripos Associates Inc., St. Louis, MO).

Based on the experimental SAR data, the following functional groups of agonists were defined as key elements of the D_1 agonist pharmacophore : the two hydroxyl groups of the catechol ring, the nitrogen, and (except for dopamine) an accessory hydrophobic group (*e.g.*, the aromatic ring in dihydrexidine or SKF82958). Thus, the task of molecular modeling analysis was to identify a pharmacophoric conformation for each compound in which these key pharmacophoric elements were spatially arranged in the same way for all active compounds.

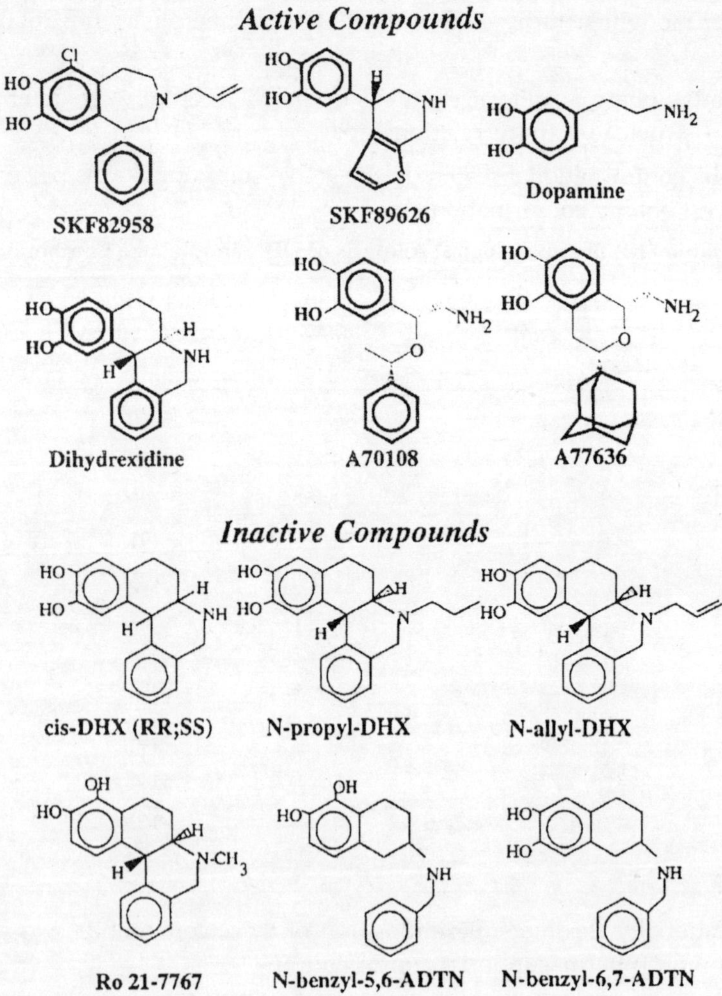

Fig. 11.3. The chemical structures of the ligands used in the molecular modeling study of the D_1 dopamine receptor. The ligands were divided into two groups (active and inactive) based on their pharmacological properties. The hypothesized pharmacophoric elements are shown in bold.

(a) Lowest Energy Conformations of Active Compounds. Construction of the Pharmacophore. The evaluation of the D_1 agonist phars*(b) D_1 Receptor Mapping and Agonist Pharmacophore Development*. The "pharm" configurations of the active molecules also were used to map the volume of the receptor site available for ligand binding. The steric mapping of the D_1 receptor site, using the MVolume routine in SYBYL, involved the construction of a pseudoelectron density map for each of the active analogs superimposed in their pharmacophore conformations. A union of the van der Waals density maps of the active compounds defines the *receptor excluded volume*.

The essential feature of the AAA is a comparison of active and inactive molecules. A commonly accepted hypothesis to explain the lack of activity of inactive molecules that possess the pharmacophoric conformation is that their molecular volume, when presenting the pharmacophore, exceeds the receptor excluded volume. This additional volume apparently is filled by the receptor and is unavailable for ligand binding; this volume is termed the *receptor essential volume*. Following this approach, the density maps for each of the inactive compounds (in their "pharm" conformations superimposed with that of active compounds) were constructed; the difference between the combined inactive compound density maps and the receptor excluded volume represents the receptor essential volume. These receptor-mapping techniques supplied detailed topographical data that allowed a steric model of the D_1 receptor site to be proposed.

These modeling efforts relied upon dihydrexidine as a structural template for determining molecular geometry because it was not only a high affinity full agonist, but it also had limited conformational flexibility relative to other more flexible, biologically active agonists. For all full agonists studied (dihydrexidine, SKF89626, SKF82958, A70108, A77636, and dopamine), the energy difference between the lowest energy con-former and those that displayed a common pharmacophore geometry was relatively small (<5 kcal/mol). The pharmacophoric conformations of the full agonists were also used to infer the shape of the receptor binding site. Based on the union of the van der Waals density maps of the active analogs, the excluded receptor volume was calculated. Various inactive analogs (partial agonists with D_1 $K_{0.5}$ > 300 nM) subsequently were used to define the receptor essential volume (*i.e.*, sterically intolerable receptor regions). These volumes, together with the pharmacophore results, were integrated into a three-dimensional model estimating the D_1 receptor active site topography.

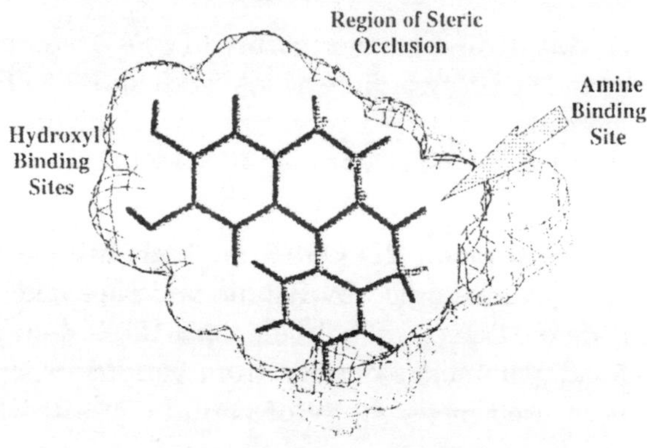

Fig. 11.4. Excluded volume for the D, agonist pharmacophore. The mesh volume shown by the black lines is a cross section of the excluded volume representing the receptor binding pocket. Dihydrexidine (see text) is shown in the receptor pocket. The gray mesh represents the receptor essential volume of inactive analogs. The hydroxyl binding, amine binding, and accessory regions are labeled, as is the steric occlusion region.

Fig. 11.4 represents the typical result of the application of the active analog approach. Based on the steric description of the essential receptor volume, new ligands can be designed that fit geometrical descriptions of the pharmacophore. Indeed, recent research based on the active analog approach led to the design of a novel highly selective dopamine agonist dihydrexidine.

Quantitative Structure-Activity Relationship Method

The quantitative structure-activity relationship (QSAR) approach was first introduced by Hansch and coworkers on the basis of implications from linear free energy relationships (LFERs) in general and the Hammett equation in particular. It is based upon the assumption that the difference in physicochemical properties accounts for the differences in biological activities of compounds. According to this approach, the changes in physicochemical properties that affect the biological activities of a set of congeners are of three major types : electronic, steric, and hydrophobic. These structural properties are often described by Hammett electronic constants,

Verloop STERIMOL parameters, hydrophobic constants, etc. The quantitative relationships between biological activity (or chemical property) and the structural parameters could be conventionally obtained by using multiple linear regression (MLR) analysis.

The typical QSAR equation is

$$\log\left(\frac{1}{C}\right) = b_0 + \sum_i b_i D_i$$

where C is drug concentration, b_0 is a constant, b_i is a regression coefficient, and D_i is a molecular descriptor such as the Hansch π parameter (hydrophobicity) or the Hammett a parameter (electron-donating or accepting properties), or molar refractivity (MR), which describes the volume and electronic polarizability, etc. The fundamentals and applications of this method in chemistry and biology have been summarized by Hansch and Leo. This traditional QSAR approach has generated many useful and, in some cases, predictive QSAR equations and has led to several documented drug discoveries.

Many different approaches to QSAR have been developed since Hansch's seminal work. These include both two-dimensional (2D) and 3D QSAR methods. The major differences among these methods can be analyzed from two viewpoints : (1) the structural parameters that are used to characterize molecular identities and (2) the mathematical procedure that is employed to obtain the quantitative relationship between a biological activity and the structural parameters.

Most of the 2D QSAR methods are based on graph theoretic indices, which have been extensively studied by Randic and Kier and Hall. Although these structural indices represent different aspects of molecular structures, their physicochemical meaning is unclear. . On the other hand, parameters derived from various experiments through, chemometric methods have also been used in the study of peptide QSAR, where partial least square (PLS) analysis has been employed.

With the development of accurate computational methods for generating 3D conformations of chemical structures, QSAR approaches that employ 3D descriptors have been developed to address the problems of 2D QSAR techniques, *e.g.*, their inability to distinguish stereoisomers. The examples of 3D QSAR include molecular shape analysis (MSA), distance geometry and Voronoi techniques.

Perhaps the most popular example of 3D QSAR is CoMFA. Developed by Cramer *et. al.*, CoMFA elegantly combines the power of molecular graphics and partial least squares (PLS) technique and has found wide applications in medicinal chemistry and toxicity analysis (see Ref. 38 for several excellent reviews). This method is one of the most recent developments in the area of ligand-based receptor modeling. This approach combines traditional QSAR analysis and three-dimensional ligand alignment central to AAA into a powerful 3D QSAR tool. CoMFA correlates 3D electrostatic and van der Waals fields around sample ligands typically overlapped in their pharmacophoric conformations with their biological activity. This approach has been successfully applied to many classes of ligands.

CoMFA methodology is based on the assumption that since, in most cases, the drug-receptor interactions are noncovalent, the changes in biological activity or binding constants of sample compounds correlate with changes in electrostatic and van der Waals fields of these molecules. To initiate the CoMFA process, the test molecules should be structurally aligned in their pharmacophoric conformations; the latter are obtained by using, for instance, the AAA described above. After the alignment, steric and electrostatic fields of all molecules are sampled with a probe atom, usually an sp^3 carbon bearing a +1 charge, on a rectangular grid that encompasses structurally aligned molecules. The values of both van der Waals and electrostatic interaction between the probe atom and all atoms of each molecule are calculated in every lattice point on the grid using the force field equation described above and entered into the CoMFA QSAR table. This table thus contains thousands of columns, which makes it difficult to analyze; however, application of special multivariate statistical analysis routines such as partial least squares analysis, cross-validation, and bootstrapping ensures the statistical significance of the final CoMFA equation. A cross-validated R^2 (q^2) that is obtained as a result of this analysis serves as a quantitative measure of the quality and predictive ability of the final CoMFA model.

The statistical meaning of the q^2 is different from that of the conventional r^2; a q^2 value greater than 0.3 is considered significant. Recent trends in both 2D and 3D QSAR studies have focused on the development of optimal QSAR models through variable selection. This implies that only a subset of available descriptors of chemical structures that are the most meaningful and statistically significant in terms of correlation with biological activity is selected. The optimum selection of variables is achieved by combining stochastic search methods with correlation methods such as MLR, PLS analysis, or artificial neural networks (ANN). More specifically, these methods employ either generalized simulated annealing, genetic algorithms, or evolutionary algorithms as the stochastic optimization tool. Since the effectiveness and convergence of these algorithms are strongly affected by the choice of a fitting function, several such functions have been applied to improve the performance of the algorithms.

It has been demonstrated that combinations of these algorithms with various chemometric tools have effectively improved the QSAR models compared to those without variable selection.

The variable selection approaches have also been adopted for region selection in the area of 3D QSAR. For example, GOLPE was developed using chemometric principles. q^2-GRS.0000000000000000000000000000000000000000 was developed based on independent CoMFA analyses of small areas (or regions) of near-molecular space to address the issue of optimal region selection in CoMFA analysis. More recently, a genetic algorithm based sampling of 3D regions of CoMFA fields was implemented. These methods have been shown to improve the QSAR models compared to the original CoMFA technique.

Most of the QSAR techniques (both 2D and 3D) assume the existence of a linear relationship between a biological activity and molecular descriptors, which may be an adequate assumption for relatively small datasets (dozens of compounds). However, the fast collection of structural and biological data, owing to recent development of combinatorial chemistry and high throughput screening technologies, has challenged traditional QSAR techniques. First, 3D methods may be computationally too expensive for the analysis of a large volume of data, and in some cases, an automated and unambiguous alignment of molecular structures is not achievable, Second, although existing 2D techniques are computationally efficient, the assumption of linearity in the SAR may not hold true, especially when a large number of structurally diverse molecules are included in the analysis. Thus, several nonlinear QSAR methods have been proposed in recent years. Most of these methods are based on either artificial neural network (ANN) or machine learning techniques. Such applications, especially when combined with variable selection, represent fast-growing trends in modern QSAR research.

It is important not to overinterpret the QSAR models in terms of the underlying mechanism of the ligand-receptor interaction or their value for the design of novel compounds with high biological activity. Successful QSAR models present a statistically significant correlation between chemical features of compounds (descriptors) and biological activity, which does not imply any particular mechanism of drug action. In some instances, such as when CoMFA or simple linear regression QSAR models are used with clear-cut physicochemical descriptors (*e.g.*, van der Waals and electrostatic molecular fields or hydrophobicity), the models could be formally interpreted in terms of the structural modifications required to increase the biological activity. However, it is essential in all cases to avoid forecasting biological activity for compounds too structurally different from those in the training set.

Structure-Based Drug Design

Ligand-based methods of drug design provide only limited information about the stereochemical organization of the receptor binding site. Detailed knowledge of the active site geometry can be obtained experimentally by means of X-ray crystallography or NMR; unfortunately, despite rapid progress in practical applications of these techniques, the structures of very few receptors and ligand-receptor complexes are available experimentally. In the absence of any experimental information on the 3D structure of many receptors, predictions from primary sequence remain the only means of generating the receptor structure. Obviously, the knowledge of the active site geometry significantly facilitates the design of new drugs by providing real spatial and chemical limitations on the structures of newly designed molecules. We briefly review in the following subsections first the methods of molecular simulations applicable to the analysis of known ligand-receptor complexes and then the current state of the art in the area of modeling GPCRs, which include dopamine receptors. Excellent reviews of methods and successes in the area of structure-based drug design were published in the early 1990s.

Ligand-Receptor Docking

In ligand-receptor docking, the ligand is brought (manually or in some automated way) into the vicinity of the binding site and oriented so that electrostatic and van der Walls interactions (which correspond to Coulomb and dispersion terms, respectively, in molecular mechanics energy expressions) between ligand and receptor are optimized. Pharmaco-phoric conformation of ligands is frequently used in these studies. Earlier docking programs allowed the docking of only a rigid ligand into a rigid receptor, *i.e.*, no conformational change of either receptor or ligand was permitted as the latter approached the former. Clearly, this is not the way the interaction occurs in nature, because both ligand and receptor are relatively flexible molecules that adjust their conformation to each other in the process of binding to maximize steric and chemical complementarity. Flexible docking calculations are very difficult and require an extremely fast computer. Nevertheless, the consideration of both ligand and receptor active site flexibility is becoming part of modern docking approaches. Further descriptions of docking algorithms can be found in several publications.

Molecular Dynamics

Once the model of a ligand-receptor complex is built, its stability should be evaluated. Simple molecular mechanics optimization of the putative ligand-receptor complex leads only to the identification of the closest local minimum. However, molecular mechanics optimization of molecules lacks two crucial properties of real molecular systems: temperature and, consequently, motion. Molecular dynamics studies the time-dependent evolution of coordinates of complex multimolecular systems as a function of inter- and intramolecular interactions. Because simulations are usually performed at normal temperature (~300 K), relatively low energy barriers, on the order of kT (0.6 kcal), can be easily overcome. Thus if the starting configuration of the whole system (*i.e.*, drug-receptor complex) resulting from docking is separated from the more stable configuration by such a low barrier, molecular dynamics will take the system over the barrier. Molecular simulations may identify more stable, therefore more realistic, conformational states of ligand-receptor complexes. Furthermore, they may provide unique information about conformational changes of the receptor due to ligand binding. They may shed light on the intimate mechanisms of receptor activation that currently cannot be studied by any other technique. Finally, molecular simulations frequently incorporate solvent and thus allow the inclusion of solvent effects in the consideration. Unfortunately, due to the inherently very short elementary simulation step size, ~2 fs, this technique is presently limited to relatively short total simulation times, on the order of hundreds of picoseconds to nanoseconds. These limitations are mainly due to the fact that available computer power is still inadequate for significantly longer simulation times.

Binding Constant Calculation

The combination of free energy simulation (FES) and the thermodynamic cycle (TC) approach is the most promising modern technique for calculating relative ligand-receptor binding constants by simulating ligand-receptor interaction. Experimentally, equilibrium free energies of binding of two ligands to the same receptor are evaluated in two independent experiments, according to the following scheme.

$$\text{Ligand 1} + \text{Receptor} \Leftarrow \Delta G_1^o \Rightarrow \text{Ligand1/Receptor}$$

$$\text{Ligand2} + \text{Receptor} \Leftarrow \Delta G_2^o \Rightarrow \text{Ligand2/Receptor}$$

To calculate the relative binding constants of the two ligands, Ligand1 and Ligand2, we construct the following cyclic scheme,

$$\begin{array}{ccc} \text{Ligand1} + \text{Receptor} \Rightarrow & \Delta G_1^o & \Rightarrow \text{Ligand1/Receptor} \\ \Downarrow & & \Downarrow \\ \Delta G_3^o & & \Delta G_4^o \\ \Downarrow & & \Downarrow \\ \text{Ligand2} + \text{Receptor} \Rightarrow & \Delta G_2^o & \Rightarrow \text{Ligand2/Receptor} \end{array}$$

where ΔG_1^o and ΔG_2^o correspond to the experimental binding free energies of Ligand1 and Ligand2, respectively, and ΔG_3^o and ΔG_4^o correspond to the free energy of the formal transformation of the chemical structure of Ligand1 into that of Ligand2 in solution and in the binding site, respectively. From a thermodynamic viewpoint, the foregoing scheme represents a closed thermodynamic cycle that consists of four transformations: the binding of Ligandl in solution to the receptor (ΔG_1^0); the chemical transformation of bound Ligandl to bound Ligand2 (ΔG_4^0); the chemical transformation of Ligand1 in solution to Ligand2 in solution (ΔG_3^o); and the binding of Ligand2 in solution to the receptor (ΔG_2^o). Using the thermodynamic cycle (TC) relationship, $\Delta\Delta G_{\text{cycle}}^o = 0$, one obtains the *difference* in the free energy of binding of Ligand1 and Ligand2, *i.e.*, $(\Delta G_1^o - \Delta G_2^o)$. Thus,

$$\Delta\Delta G^o_{\text{cycle}} = \Delta G^o_1 + \Delta G^o_4 - \Delta G^o_3 - \Delta G^o_2 = 0$$

so

$$\Delta\Delta G^o_{\text{binding}} = \Delta G^o_1 - \Delta G^o_2 = \Delta G^o_3 - \Delta G^o_4$$

The latter two free energies of chemical transformation are computed in the course of FES. The advantage of this approach is that it avoids calculations of ligand-receptor binding free energy per se, *i.e.*, ΔG^o_1 and ΔG^o_2, which would be extremely computationally intensive, without sacrificing the theoretical rigor of calculation of binding constants from molecular simulations of ligand-receptor interaction. Faster free energy simulation approaches based on the linear response theory have been introduced to improve the computational efficiency of free energy simulations. Despite its computational intensity, the FES-TC approach finds important and growing applications in the analysis of ligand-receptor interactions and drug design.

Chemical Informatics and Drug Design

Combinatorial chemical synthesis and high throughput screening have significantly increased the speed of the drug discovery process. However, it is still impossible to synthesize all of the library compounds in a reasonably short period of time. As many as 3000^3 (2.7×10^{10}) compounds can be synthesized from a molecular scaffold with three different substitution positions when each of the positions has 3000 different substituents. If a chemist could synthesize 1000 compounds per week, 27 million weeks (~0.5 million years) would be required to synthesize all of these compounds.

Furthermore, many of these compounds can be structurally similar to each other, and chemical information contained in the library can be redundant. Thus, there is a need for rational library design (*i.e.*, rational selection of a subset of building blocks for combinatorial chemical synthesis) so that a maximum amount of information can be obtained while a minimum number of compounds are synthesized and tested. Similarly, there is a closely related task in computational database mining, *i.e.*, rational sampling of a subset of compounds from commercially available or proprietary databases for biological testing. Thus, new challenges to computational drug design in the context of combinatorial chemistry include information management, rational library design, and database analysis. These are the research topics of chemoinformatics—a new area of computational chemistry.

Chemoinformatics (or cheminformatics) deals with the storage, retrieval, and analysis of chemical and biological data. Specifically, it involves the development and application of software systems for the management of combinatorial chemical projects, rational design of chemical libraries, and analysis of the obtained chemical and biological data. The major research topics of chemoinformatics involve QSAR and diversity analysis. The researchers should address several important issues. First, chemical structures should be characterized by calculable molecular descriptors that provide quantitative representation of chemical structures. Second, special measures should be developed on the basis of these descriptors in order to quantify structural similarities between pairs of molecules. Finally, adequate computational methods should be established, for the efficient sampling of the huge combinatorial structural space of chemical libraries.

There are two types of experimental combinatorial chemistry and high throughput screening research directions : targeted screening and broad screening. The former approach involves the design and synthesis of chemical libraries with compounds that are either analogs of some active compounds or can specifically interact with the biological target under study. This is desired when a lead optimization (or evolution) program is pursued. On the other hand, a broad screening project involves the design and synthesis of a large array of maximally diverse chemical compounds, leading to diverse (or universal) libraries that are then tested against a variety of biological targets. This design strategy is suited for lead identification programs. Thus, two categories of computational tools should be developed and validated to meet the needs of the two different types of projects.

In a targeted screening project, computational library design involves the selection of a subset of chemical building blocks from an available pool of chemical structures. This subset of selected building blocks affords a limited virtual library with a high content of compounds similar to a lead molecule. Molecular similarity is quantified by using a chosen set of molecular descriptors and similarity metrics. Building blocks can also be chosen such that the resulting virtual library could have a high percentage of compounds that are predicted to be active from a preconstructed QSAR model. When the structure of the biological target is known, one can select building blocks such that the resulting library compounds are stereochemically complementary to the binding site structure of the underlying target. Other approaches to targeted library design using different criteria have also been considered. Similar approaches have long been used in targeted database mining that were based on the principle of either molecular similarity or structure-based drug design. In abroad screening project, computational library design or database mining involves the selection of a subset of compounds that are optimally diverse and representative of available classes of compounds, leading to a nonredundant chemical library or a set of nonredundant compounds for biological testing. Reported methods can be generally classified into several categories.

1. Cluster sampling methods, which first identify a set of compound clusters, followed by the selection of several compounds from each cluster.
2. Grid-based sampling, which places all the compounds into a low-dimensional descriptor space divided into many cells and then chooses a few compounds from each cell.
3. Direct sampling methods, which try to obtain a subset of optimally diverse compounds from an available pool by directly analyzing the diversity of the selected molecules.

Many reports have been published that address various aspects of diversity analysis in the context of chemical library design and database mining.

SUMMARY AND CONCLUSIONS

We have reviewed several methods of computer-aided drug design that are formally divided into ligand-based and receptor-based methods. Ligand-based methods of analysis are used widely because they are not very computationally intensive and afford rapid generation of QSAR models from which the biological activity of newly designed compounds can be predicted. Most of the existing methods require the generation of a 3D pharmacophore hypothesis (i.e., unique 3D arrangements of important functional groups common to all or the majority of the receptor ligands). In many cases, when the receptor ligands are not very structurally diverse and include conformationally rigid compounds, a pharmacophore can be generated in a reasonably unbiased and unique way, using either automated (*e.g.*, DISCO) or semiautomated pharmacophore prediction methods. The pharmacophore hypothesis can help medicinal chemists gain an insight into the key interactions between ligand and receptor when a 3D ligand-receptor complex structure has not been determined. It can be used as the search query for 3D database mining, and this approach has been demonstrated as very productive for lead compound discovery. It can also be used for 3D QSAR analysis, grouping together the compounds that follow the same binding mode and indicating the possible 3D alignment rules. However, this task of unique pharmacophore generation becomes less feasible for more structurally diverse and/or conformationally flexible compounds. In general, in the absence of detailed structural information about the receptor binding site, any pharmacophore inferred from only the ligand structure remains hypothetical.

Structure-based design methods present an appealing alternative to more traditional approaches to CADD. Molecular docking algorithms can generate fairly accurate orientations of known or designed receptor ligands in the active site (*e.g.*, GRID, DOCK). However, their application in practice is limited by the availability of macromolecu-lar target structure. The protein and nucleic acid structural databases have been growing exponentially over the past decade, with the size of the Protein Data Bank (PDB) currently exceeding 10,000 entries. Nevertheless, many of these entries are highly homologous proteins, and the structure of many pharmacologically important molecules such as transmembrane receptor proteins cannot be resolved experimentally at the present time.

Furthermore, predictions of the binding affinity of receptor ligands are fast but rather inaccurate because the scoring functions used in docking calculations are not robust enough. Alternatively, these predictions can be fairly accurate but very computationally intensive when free energy simulation methods are used. Therefore, because of these practical limitations, ligand-based design methods will probably remain the main arena of CADD efforts, supplemented, when possible, by structure-based approaches.

We have shown that successful molecular modeling and the design of new drugs can be achieved by integration of different computational chemistry and molecular modeling techniques. Owing to the rapid increase in computational power due to advances in both hardware and software, molecular modeling has become an important integral part of multidisciplinary efforts in the design and synthesis of new potent pharmaceuticals. Practical knowledge of these techniques and their limitations is a necessary component of modern research in drug discovery.

CONFORMATIONAL SAMPLING 12

The goal of conformational analysis is to shed light on conformational characteristics of flexible biomolecules and to gain insight into the relationship between their flexibility and their function. Because of the importance of this approach, conformational analysis plays a role in many computational projects ranging from computer-aided drug design to the analysis of molecular dynamics simulations and protein folding. In fact, most structure-based drug design projects today use conformational analysis techniques as part of their toolchest. As will be discussed earlier, in structure-based drug design a rational effort is applied to identifying potential drug molecules that bind favorably into a known three-dimensional (3D) binding site, the structure of which was determined through X-ray crystallography, NMR spectroscopy, or computer modeling. Because such an effort requires, among other things, structural compatibility between the drug candidate and the binding site, computational methods were developed to "dock" ligands into binding sites. These docking calculations are used for screening large virtual molecular libraries, saving both time and money.

However, although docking is fairly straightforward with rigid molecules, it becomes significantly more complicated when flexible molecules are considered. This is because flexible molecules can adopt many different conformations, each of which may, in principle, lead to successful docking. Although there are a few "flexible docking" approaches that account for flexibility during the docking process itself, most docking applications rely on conformational analysis to deal with this problem (*e.g.*, by generating a multitude of molecular conformations that are docked separately into the binding site).

The importance of conformational analysis in the context of drug design extends beyond computational docking and screening. Conformational analysis is a major tool used to gain insight for future lead optimization. Furthermore, even when the 3D structure of the binding site is unknown, conformational analysis can yield insights into the structural characteristics of various drug candidates. In a different context, conformational analysis is essential for the analysis of molecular dynamics simulations. As discussed earlier the direct output of a molecular dynamics simulation is a set of conformations ("snapshots") that were saved along the trajectory. These conformations are subsequently analyzed in order to extract information about the system. However, if, during a long simulation, the molecule moves from one conformation class to another, averaging over the whole simulation is likely to be misleading. Conformational analysis allows one to first identify whether such drastic conformational transitions have occurred and then to focus the analysis on one group of conformations at a time.

In view of their importance it is not surprising that conformation sampling and analysis constitute a very active and innovative field of research that is relevant to biomolecules and inorganic molecular clusters alike. The following sections offer an introduction to the main methodologies that are used as part of a conformational analysis study. These are arranged according to the three main steps applied in such studies : (1) conformation sampling, (2) conformation optimization, and (3) conformational analysis.

CONFORMATION SAMPLING

Conformation sampling is a process used to generate the collection of molecular conformations that will later be analyzed. Ideally, all locally stable conformations of the molecule should be accounted for in order for the conformational analysis to be complete. However, owing to the complexity of proteins and even fairly small peptides it is impractical to perform such an enumeration. The number of locally stable conformations increases so fast with the molecular size that the task of full enumeration becomes formidable. Even enumerating all possible (ϕ, ψ), conformations of a protein backbone rapidly becomes intractable. As a result, most conformational studies must rely on sampling techniques. The basic requirement from such sampling procedures is that the resulting conformational sample ("ensemble") will be representative of the system as a whole. This means that in most biomolecular studies a "canonical" ensemble, characterized by a constant temperature, is sought. Therefore sampling methods that were designed for canonical ensembles and that guarantee "detailed balance" are especially suitable for this task. Two such methods are high temperature molecular dynamics and Monte Carlo simulations. However, because of the complexity and volume of biomolecular conformational space, other sampling techniques, which do not adhere to the canonical ensemble constraint, are also often employed.

High Temperature Molecular Dynamics

Molecular dynamics simulations, which were discussed earlier, are among the most useful methods for sampling molecular conformational space. As the simulation proceeds, the classical trajectory that is traced .is in fact a subset of the molecular conformations available to the molecule at that energy (for microcanonical simulations) or temperature (for canonical simulations). Assuming that the ergodic hypothesis holds, an infinitely long MD trajectory will cover all of conformational space. The problem with room temperature MD simulations is that a shorter finite-time trajectory is not likely to sample all of conformational space. Even a nanosecond MD trajectory will most likely be confined to limited regions of conformational space (*a*). The room temperature probability of crossing high energy barriers is often too small to be observed during a finite MD simulation.

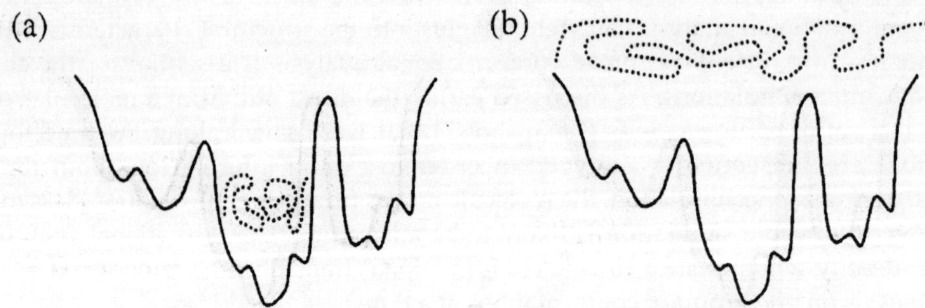

Fig. 12.1. A schematic view of (*a*) a low temperature simulation that is confined by high energy barriers to a small region of the energy landscape and (*b*) a high temperature simulation that can overcome those barriers and sample a larger portion of conformational space.

A common solution that allows one to overcome the limited sampling by MD simulations at room temperature is simply to raise the temperature of the simulation. The additional kinetic energy available in a higher temperature simulation makes crossing high energy barriers more likely and ensures a broad sampling of conformational space. In raising the simulation temperature to 1000 K or more, one takes advantage of the fact that chemical bonds cannot break in most biomolecular force fields.

Namely, the fact that bonds are modeled by a harmonic potential means that regardless of the simulation temperature these bonds can never spontaneously break, and the chemical integrity of the molecule remains intact. The effect of the unrealistically high temperatures employed is primarily to "shake" the system and allow the molecule to cross high energy barriers.

There is no definite rule regarding what temperature is "high temperature" in this context, as this depends on the character of the underlying energy landscape. Temperatures on the order of 1000 K are often used for sampling the conformations of peptides and proteins, because this temperature is below the temperature at which unwanted cis-trans transitions of the peptide bond frequently occur. In other cases, such as for sampling the conformations of a ligand bound in a protein's active site, much lower temperatures must be used. Otherwise the ligand will dissociate and the simulation will sample the conformations of an unbound ligand rather than those of the bound ligand.

The main advantage of using MD for conformation sampling is that information of molecular forces is used to guide the search process into meaningful regions of the potential. A disadvantage associated with this sampling technique is the fact that high temperature simulations sample not only the regions of interest at room temperature but also regions that are inaccessible to the molecule at room temperature. To overcome this problem the sampled conformations have to be energy-minimized or preferably annealed before being considered as sampled conformations. These methods will be discussed later.

Monte Carlo Simulations

Monte Carlo search methods are stochastic techniques based on the use of random numbers and probability statistics to sample conformational space. The name "Monte Carlo" was originally coined by Metropolis and Ulam during the Manhattan Project of World War II because of the similarity of this simulation technique to games of chance. Today a variety of Monte Carlo (MC) simulation methods are routinely used in diverse fields such as atmospheric studies, nuclear physics, traffic flow, and, of course, biochemistry and biophysics. In this section we focus on the application of the Monte Carlo method for conformational searching.

In performing a Monte Carlo sampling procedure we let the dice decide, again and again, how to proceed with the search process. In general, a Monte Carlo search consists of two steps (1) generating a new "trial conformation" and (2) deciding whether the new conformation be accepted or rejected.

Starting from any given conformation we "roll the dice," *i.e.*, we let the computer choose random numbers, to decide what will be the next trial conformation. The precise details of how these moves are constructed vary from one study to another, but most share similar traits. For example, assuming that the search proceeds via polypeptide torsion moves, choosing a new trial conformation could include the following steps. First, roll the dice to randomly select an amino acid position along the polypeptide backbone. Second, randomly select which of the several rotatable bonds in that amino acid will be modified (*e.g.*, the ϕ, ψ or χ torsion angles). Finally, randomly select a new value for this torsion angle from a predefined set of values. In this example it took three separate random selections to generate a new trial conformation. Multiple torsion moves as well as Cartesian coordinate moves are among the many possible variations on this procedure.

Once a new "trial conformation" is created, it is necessary to determine whether this conformation will be accepted or rejected. If rejected, the above procedure will be repeated, randomly creating new trial conformations until one of them is accepted. If accepted, the new conformation becomes the "current" conformation, and the search process continues from it. The trial conformation is usually accepted or rejected according to a temperature-dependent probability of the Metropolis type,

$$p = \begin{cases} e^{\beta \Delta U}, & e^{-\beta \Delta U} < 1 \\ 1, & e^{-\beta \Delta U} \geq 1 \end{cases} \quad \text{or } p = \min[1, e^{-\beta \Delta U}] \qquad \text{...(12.1)}$$

where $\beta = 1/kT$ and ΔU is the change in the potential energy. This means that if the energy of the new trial conformation is lower than that of the current conformation, $\Delta U < 0$, it is always accepted. But even if the energy of the trial conformation is higher than the current energy, $\Delta U > 0$, there is a certain probability, proportional to the Boltzmann factor, that it will be accepted. To find out whether a higher energy trial conformation is accepted, a random number r in the range is selected and compared to the Metropolis probability defined in Eq. (1). If $r < p$, the conformation is accepted; otherwise it is rejected. This acceptance probability satisfies the principle of detailed balance, ensuring that if the process continues for a long enough time then a stationary solution will be achieved.

In Monte Carlo simulations, just as in MD simulations, temperature plays an important role. In general, MC simulations tend to move toward low energy states. However, at high temperatures (small β values) there is a significant probability of climbing up energy slopes, allowing the search process to cross high energy barriers. This probability becomes significantly smaller at low temperatures, and it vanishes altogether in the limit of $T \rightarrow 0$, where the method becomes equivalent to a minimization process. Thus, high temperature MC is often used to sample broad regions of conformational space.

As stated above, MC simulations are popular in many diverse fields. Their popularity is due mainly to their ease of use and their good convergence properties. Nonetheless, straightforward and application of MC methods to biomolecules is often problematic due to very low acceptance ratios, which significantly reduce the efficiency of the method. The reason for the low acceptance ratio, *i.e.*, the ratio between accepted MC moves and total MC trial moves, is the compact character of most biomolecules. This means that many move attempts end up "bumping" into other parts of the molecule and are rejected because of the clash. This is true in particular of moves defined in Cartesian coordinates. To partially overcome this problem it is recommended that torsion move-sets be used. Advanced applications of MC methods for conformation sampling often involve various techniques for enhanced sampling.

A special comment must be added regarding the random number generator. Because the MC process is driven by random numbers, it is sensitive to the quality of the random number generator that is being used. The random number generator is supposed to generate uniformly distributed random numbers in the range. But in fact the computer does not generate random numbers at all. It uses a deterministic algorithm to generate a pseudorandom series of numbers with a finite periodicity. A high quality algorithm will have a long enough periodicity that the observed distribution does indeed appear random. However, there are many so-called random number generators on the market that are anything but random. It is good practice to check the random number generator prior to using it in an actual MC simulation. A simple histogram of 10,000 numbers or more will easily show whether or not the resulting distribution is uniform.

Genetic Algorithms

A genetic algorithm (GA) evolves a population of possible solutions through genetic operations, such as mutations and crossovers, to a final population of low energy conformations that minimize the energy function (the fitness function). For the purpose of confor-mational sampling, the translational, rotational, and internal degrees of freedom are encoded into "genes," which are represented by the real number values of those degrees of freedom. Each individual conformation, named a "chromosome," consists of a collection of genes and is represented by the appropriate string of real numbers. A fitness value (energy) is assigned to each chromosome.

The GA evolutionary process iterates through the following two steps: (1) Generation of a children population from a parent population by means of genetic operators, and (2) a generation update step. The process starts with a random population of initial chromosomes. A new population is generated from the old one by the use of genetic operators. The two most fundamental operators are schematically depicted in Fig. 12.2. The *mutation* operator (Fig. 12.2(*a*)) changes the value of a randomly selected gene by a random value, depending on the type of gene. The crossover operator (Fig. 12.2(*b*)) exchanges a set of genes between two parent chromosomes, creating two children with genes from both parents. Additional genetic operators, such as the *multiple simultaneous mutation* operator or a *migration* operator, which moves individual chromosomes from one subpopulation to another, may also be used.

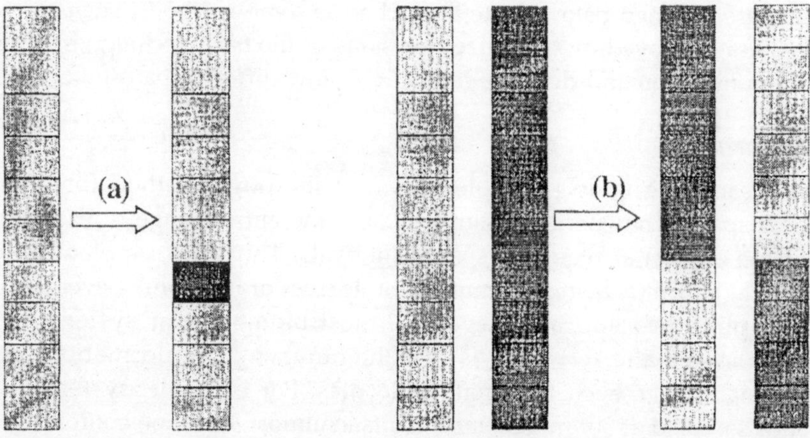

Fig. 12.2. Genetic operators used to create a population of children chromosomes from a population of parent chromosomes, (*a*) Single-point mutation. A gene to be mutated is selected at random, and its value is modified, (*b*) One-point crossover. The crossover point is selected randomly, and the genes are exchanged between the two parents. Two children are created, each having genes from both parents.

Parents are selected for breeding based on their relative fitness and an external evolutionary "pressure," which directs the process to favor breeding by parents with higher fitness. Following each breeding cycle the chromosome population is updated. A common update scheme is to replace all but the most fit "elite" chromosomes of the original set. For example, a "generation update with an elitism of 2" indicates that all but the two most fit parent chromosomes are replaced. The iteration through the two steps of the evolutionary process continues until convergence or until the maximum number of steps is reached.

Application of GA to conformational searching typically uses the potential energy as the fitness function. The degrees of freedom encoded in the "genes" are usually only a subset of coordinates that are most significant for the conformational search. These may include backbone dihedral angles, side chain dihedral angles, or any other coordinate that may be considered useful. Application of GA to docking problems requires that the three center-of-mass coordinates and the three rotational Euler angles also be encoded in genes.

An advantage of GA for conformational analysis, besides its elegance, is that it is very easy to code and run. In addition, GA usually requires fewer iterations than either MD or MC to generate a large population of low energy conformations. On the other hand, because in every iteration whole populations are propagated, each iteration takes significantly more CPU time than in either of the other two methods. Furthermore, in many complex cases GA is known for its slow convergence.

A successful application of GA to conformation sampling is, for example, as a part of flexible docking. It should be noted, however, that none of the three sampling methods discussed above, MD, MC, and GA, was shown to outperform the other two in any general way. In fact, a comparison of the three methods in the context of flexible docking showed similar efficiency for all three [12], although specific advantages are likely to exist for particular applications.

Other Search Methods

Many additional search methods have been devised in addition to the basic three just discussed. A few of them are outlined below. Note that whereas some methods, such as parallel tempering and J-walking, are improved or specialized versions of the basic techniques, other methods, such as systematic enumeration and distance geometry, follow different paths altogether.

Systematic Enumeration

Systematic sampling is, in principle, the most thorough method for searching molecular conformational space. The energy is sampled over the entire range of each degree of freedom (typically torsion angles) at regularly spaced intervals. Thus, the sampled conformations lie on an N-dimensional lattice (N being the number of degrees of freedom). Several problems make the systematic sampling procedure irrelevant to most biomolecular systems. A major problem associated with systematic searching is that the number of conformations generated rapidly becomes extremely large even for small molecules. For example, systematic search with six rotatable bonds sampled at 30° increments results in almost 3 million conformations. The number of samples can be reduced by limiting the range of rotation for a symmetrical substituent (*e.g.*, 0-180° for a phenyl group) and/or increasing the rotation step size (all staggered conformations of a saturated $C - C$ bond can be sampled at 120° increments). Another problem arises from the fact that a systematic search is likely to generate a large number of unphysical conformations, in which one part of the molecule crosses over another part (*e.g.*, a protein backbone that crosses over itself). As a result of the these limitations, systematic search is applied only to small molecules such as small peptides or ligands in the context of flexible docking.

"Pruning" is a sophisticated way to reduce the computational requirements associated with systematic search. Pruning takes advantage of the fact that all conformations along a given branch of the search tree that are downstream from a high energy conformation will also have high energies and need not be calculated. For example, if an irrevocable clash such as the backbone folding over itself occurs after the fifth angle in an eight-angle systematic search, there is no need to continue sampling the three remaining angles. Pruning can be performed using geometrical constraints that must be satisfied for all conformations. An energy cutoff is a simplistic filter that can be applied at the end of the search to reduce the number of conformations stored. With this filter, conformations with energies above the cutoff are discarded. This filter, of course, does not reduce the number of calculations but does help in managing and analyzing the data.

Distance Geometry

Distance geometry is a general method for building conformational models of complex molecular systems based on a set of distance constraints. It is a purely geometric technique that generates structures that satisfy the given set of constraints without requiring a starting conformation or an energy function. The distance constraints, which are the input to the method, can be qualitative or approximate. As such they are defined by upper and lower bounds and often into a matrix form. The distance geometry approach converts, or "embeds," the uncertain distance constraints into 3D Cartesian coordinates.

In the basic metric matrix implementation of the distance constraint technique one starts by generating a distance bounds matrix. This is an $N \times N$ square matrix (N the number of atoms) in which the upper bounds occupy the upper diagonal and the lower bounds are placed in the lower diagonal. The matrix is filled by information based on the bond structure, experimental data, or a hypothesis. After smoothing the distance bounds matrix, a new distance matrix is generated by random selection of distances between the bounds. The distance matrix is converted back into a 3D conformation after the distance matrix has been converted into a metric matrix and diagonalized. A new distance matrix randomly generated within the same bounds will result in another possible conformation that satisfies the model, and so forth. The tighter the bounds, the more restricted the search. Naturally, distance geometry is very useful in situations in which many distance constraints are known, in particular for suggesting models that agree with NMR data. This method is also useful for pharmacophore modeling based on known bioactive molecules. However, for general-purpose conformational searching this method is less appropriate unless the search space is limited by known constraints (such as active site data in a docking context). Computationally, distance geometry is limited to moderately sized systems (up to several thousand atoms) because it requires computationally expensive matrix manipulation.

Parallel Tempering and J-Walking

Energy barriers that confine the search to limited regions of conformational space restrict both molecular dynamics and Monte Carlo simulations, preventing them from reaching ergodicity. As discussed earlier, this limitation is often overcome by raising the simulation temperature. The higher temperature allows the simulation to overcome high energy barriers and extend the search space. However, it also causes the simulation to sample regions of conformational space that are irrelevant and inaccessible to room temperature molecules. The methods of parallel tempering and J-walking address the problem of how to raise the simulation temperature without wasting computational effort on inaccessible conformations.

The idea behind both parallel tempering and J-walking is to incorporate information obtained by ergodic high temperature simulations into low temperature simulations. By periodically passing information from high temperature simulations to low temperature simulations, these methods allow the low temperature system to overcome the barriers between separate regions. In practice, both methods, which can be applied in both MC and MD simulations, require propagation of at least two (and often more) parallel simulations—one at the desired low temperature and another at.a high temperature. For simplicity we shall refer to the Monte Carlo implementation of these techniques; the molecular dynamics implementation is similar in nature.

In J-walking the periodic MC trial probability for a simulation at temperature T is taken to be a Boltzmann distribution at a high temperature, T_J ($\beta_J = 1/kT_J$), The jumping temperature, T_J, is sufficiently high that the Metropolois walk can be assumed to be ergodic. This results in the acceptance probability,

$$p = \min[1, e^{-(\beta-\beta_J)\Delta U}] \quad \text{...(12.2)}$$

where $\beta = 1/kT$, $\beta_J = 1/kT_J$, and ΔU is the change in the potential energy. In practice, at temperature T the trial moves are taken from the Metropolis distribution of Eq. (12.1) about 90% of the time, with jumps using Eq. (12.2) attempted about 10% of the time. The jumping conformations are generated with a Metropolis walk at temperature T_J. Several methods have been devised to overcome correlations between the low and high temperature walks.

In parallel tempering, the high and low temperature walkers exchange configurations, unlike in J-walking, in which only the high temperature walker feeds conformations to the low temperature walker. By exchanging conformations, parallel tempering satisfies detailed balance, assuming the two walks are long enough. Parallel tempering simulations can be performed for more than two temperatures. It should be noted that the gaps between adjacent temperatures must be chosen so that the exchanges are accepted with sufficient frequency. This means that the distribution characteristics of the two temperatures must overlap so that a conformation generated at one temperature will have a significant probability of being accepted at the other temperature. If the temperature gap is too large, the likelihood of any conformation being accepted by the other simulation is very small.

CONFORMATION OPTIMIZATION

The structures generated by the various sampling methods often correspond to transient conformations. It is a desirable practice to bring these conformations to nearby local minima, which represent locally stable conformations, before further analysis is performed.. The most widely used methods for this purpose are the various minimization techniques. These vary in accuracy and computational efficiency and are useful in many applications as well as conformational analysis. Minimization methods are used as a tool (often alongside other computational tools) in model building, preparation of the initial structure for molecular dynamics, preparation of structures for normal mode analysis, and more. Because of their importance we go into some detail in discussing the different minimization approaches. Another conformation optimization approach is simulated annealing. In the context of conformational analysis, simulated annealing is often used to optimize conformations generated by high temperature MD or MC simulations. It is also often used as an independent method for minimization and identification of the global minimum in a complex energy landscape.

Minimization

A drop of water that is placed on a hillside will roll down the slope, following the surface curvature, until it ends up in the valley at the bottom of the hill. This is a natural minimization process by which the drop minimizes its potential energy until it reaches a local minimum. Minimization algorithms are the analogous computational procedures that find minima for a given function. Because these procedures are "downhill" methods that are unable to cross energy barriers, they end up in local minima close to the point from which the minimization process started (Fig. 12.3 *a*). It is very rare that a direct minimization method will find the global minimum

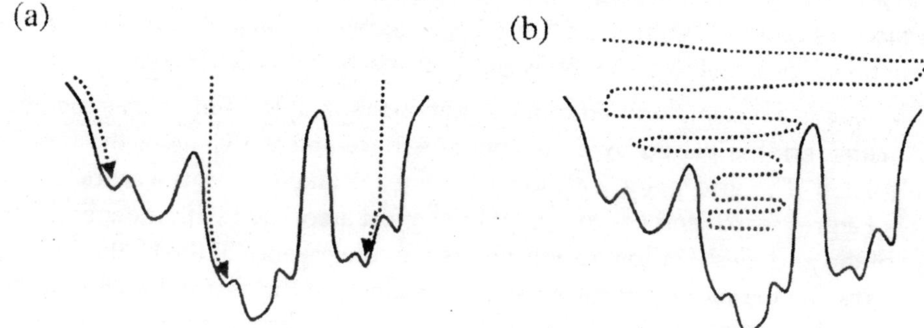

Fig. 12.3. A schematic representation of two optimization schemes, (*a*) Minimization, which leads to the closest local minimum; (*b*) simulated annealing, which can overcome intermediate energy barriers along its path.

of the function. This can be easily understood from our water drop analogy. The drop rolls down the slope until it reaches the smallest of hollows and stops there. It will not cross the nearest rise even if the valley on the other side of that rise is much deeper than the hollow it is in now.

In conformational analysis, minimization reduces the potential energy of a given conformation, typically by relieving local strains in the structure. However, the minimized structure will not stray far from the initial conformation. This is why the minimization process is considered an optimization technique and not a search method. Global optimization—namely, reaching the lowest point on the energy surface—cannot be addressed by straightforward minimization.

The goal of all minimization algorithms is to find a local minimum of a given function. They differ in how closely they try to mimic the way a drop of water or a small ball would roll down the slope, following the surface curvature, until it ends up at the "bottom." Consider a Taylor expansion around a minimum point X_0 of the general one-dimensional function $F(X)$, which can be written as

$$F(X) = F(X_0) + (X - X_0)F'(X_0) + \frac{1}{2}(X - X_0)^2 F''(X_0) + \ldots \qquad \ldots(12.3)$$

where F' and F'' are the first and second derivatives of the function. The extension to a multidimensional function requires replacement of the variable X by the vector X and replacement of the derivatives by the appropriate gradients. Minimization algorithms are classified according to the amount of information about the function that is being used, represented by the highest derivative used by the algorithm. Algorithms that use only the value of the function and no derivative are called "order 0" algorithms. Algorithms that use the first derivative—the slope—of the function are denoted as "order 1" algorithms. Finally, "order 2" algorithms are algorithms that take advantage of the second derivative—the curvature—of the function. As in most computational techniques, higher order methods that use more information about the function are generally more accurate at the price of being computationally more expensive, taking more time and more resources than the lower order methods. This section offers a brief discussion of the most common minimization algorithms.

Order 0 Algorithms

Order 0 minimization methods do not take the slope or the curvature properties of the energy surface into account. As a result, such methods are crude and can be used only with very simple energy surfaces, *i.e.*, surfaces with a small number of local minima and monotonic behavior away from the minima. These methods are rarely used for macro-molecular systems.

Grid searching is an example of an order 0 minimization algorithm. In this method a regular grid (*e.g.*, a cubic grid) is placed over the energy surface and the value of the function at each node is calculated. The grid point with lowest energy is taken to represent the real minimum. The quality of a grid search and the computational effort associated with it depend, of course, on the density of the grid mesh. In some applications the convergence of the method is improved by gradually increasing the density of the grid in the vicinity of the best point of an earlier, coarser grid. This method is inefficient for large molecules and can easily converge to a false minimum. Grid searching as a minimization technique is rarely used in biomolecular studies.

Another order 0 method is the *downhill simplex method*. A *simplex* is a geometrical element consisting, in N dimensions, of $N + 1$ points (vertices) and all their interconnecting line segments, polygonal faces, etc. In two dimensions, a simplex is a triangle; in three dimensions it is a tetrahedron. The downhill simplex method starts with $N + 1$ points, defining an initial simplex. The energy of each point is evaluated, and the method moves the point of the simplex from where the function is largest through the opposite face of the simplex to a lower point. Usually such "reflections" conserve the volume of the simplex, but when it can do so it will expand or contract the simplex to minimize the function. The minimization proceeds by taking the highest point in the new simplex and reflecting it through the opposite face and so forth. This process is schematically described in Fig. 12.4. The simplex method is sometimes used for crude placement of starting conformations, for example in the context of docking.

Order 1 Algorithms

Order 1 algorithms, which represent a fair balance between accuracy and efficiency, are the most commonly used minimization methods in macromolecular simulations. As indicated by their name, these algorithms use the gradient of the function to direct the minimization process toward the nearest local minimum. They use information about the *slope* of the function but do not include information about its *curvature*. Thus, to compensate for the deficiency of curvature data, all order 1 minimization methods employ a stepwise iterative scheme. The iterations are used for recalculating the gradient and correcting the minimization approach pattern following changes in the direction of the slope.

In general, order 1 methods iterate over the following equation in order to perform the minimization until it converges or until it reaches a preset maximum number of steps :

$$\hat{r}_k = \hat{r}_{k-1} + \lambda_k \hat{S}_k \qquad \text{...(12.4)}$$

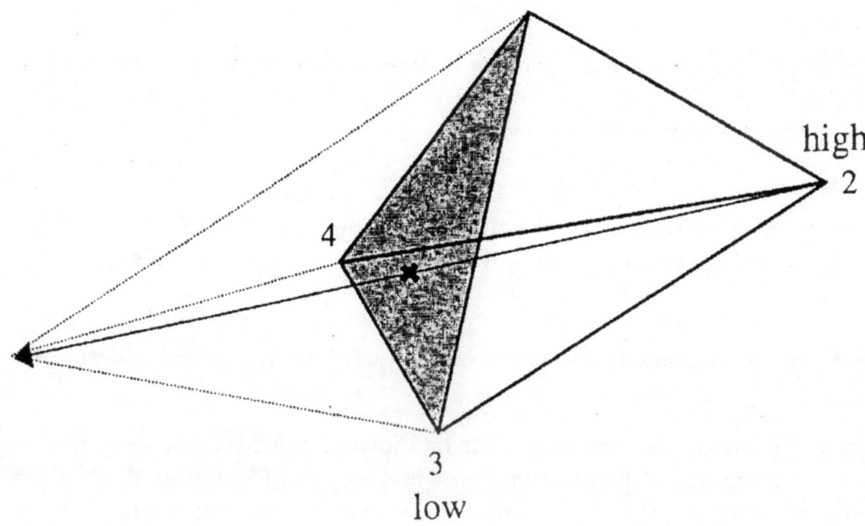

Fig. 12.4. A representative step in the downhill simplex method. The original simplex, a tetrahedron in this case, is drawn with solid lines. The point with highest energy is reflected through the opposite triangular plane (shaded) to form a new simplex. The new vertex may represent symmetrical reflection, expansion, or contractions along the same direction.

where r_k is the new position at step k, r_{k-1} is the position at the previous step $k - 1$, λ_k is the size of the step to be taken at step k, and S_k is the direction of that step. Various methods differ in the way they choose the step size and the step direction, *i.e.*, in the way they try to compensate for the lack of curvature information. For example, the *steepest descent* method uses a variable step size for this purpose, whereas the *conjugated gradients* method takes advantage of its memory of previous steps. These two popular algorithms are described next.

(*a*) *Steepest Descent*. Steepest descent (SD), which is the simplest minimization algorithm, follows the direction of the instantaneous gradients calculated at each iteration. This means that once the gradient of the energy function, g_k, at the current position is calculated, the minimization step is taken in the direction opposite to it (*i.e.*, in the direction of the force),

$$\hat{S}_k = -\hat{g}_k = -\nabla U(r) \quad ...(12.5)$$

The step size, λ_k, is adjusted at each iteration to compensate for the lack of curvature information. If the energy of the new conformation is lower than that of the previous one, the algorithm assumes that the approach direction is correct and increases the step size λ_K. by a small factor (often by the factor 1.2) to improve efficiency. However, if the energy of the new conformation turns out to higher, indicating that the real path has curved away from the current minimization direction, the step size λ_k is decreased (often by the factor 0.5) to allow the algorithm to correct the direction more efficiently. Because finite step sizes are used, the method does not flow smoothly down to the minimum but rather jitters around it. Furthermore, SD's imprecise approach to the minimum usually means that the method does not converge to one point and gets stuck in a limit cycle around the minimum. This means that SD often gets close to the minimum but rarely reaches it. Despite the relatively poor convergence of SD it is an efficient minimization procedure that is very useful for crude local optimization, such as relieving bad contacts in an initial structure before a dynamics simulation or as the first phase of a more complex minimization scheme.

(*b*) *Conjugated Gradients.* To compensate for the missing curvature information, the conjugated gradients (CG) method makes use of its "memory" of gradients calculated in previous steps. The first step is taken in the direction of the force,

$$\hat{S}_1 = -\hat{g}_1 \qquad ...(12.6)$$

but for all subsequent iterations, $k > 1$, the direction of the minimization step is determined as a weighted average of the current gradient and the direction taken in the previous iteration, *i.e.*,

$$\vec{S}_k = -\vec{g}_k + b_k \vec{S}_{k-1} \qquad ...(12.7)$$

The weight factor b_k is calculated as a ratio of the squares of the current and previous gradients.

$$b_k = |g_k|^2/|g_{k-1}|^2 \qquad ...(12.8)$$

For quadratic surfaces of dimension n, it can be shown that the method of conjugated gradients is very efficient, converging on the nth step. Nonetheless, even for nonquadratic surfaces such as molecular energy surfaces, the CG method converges much better than SD. A numerical disadvantage of the gradient memory employed in CG is that it accumulates numerical errors along the way. This problem is usually overcome by restarting the minimization process every so often; that is, at given intervals the gradient memory is wiped out by setting b_k equal to zero.

Order 2 Algorithms

Order 2 minimization algorithms, which use the second derivative (curvature) as well as the first derivative (slope) of the potential function, exhibit in many cases improved rate of convergence. For a molecule of N atoms these methods require calculating the $3N \times 3N$ Hessian matrix of second derivatives (for the coordinate set at step k).

$$[\boldsymbol{H}_k]_{ij} = \frac{\partial^2 U(\boldsymbol{r})}{\partial r_i\, \partial r_j} \qquad ...(12.9)$$

in addition to the $3N$ vector of first derivatives $\boldsymbol{g}_k$ discussed above.

(*a*) *Newton-Raphson.* The Newton-Raphson minimization method is an order 2 method based on the assumption that, near the minimum, the energy can be approximated by a quadratic function. For a one-dimensional case, assuming that $F(X) = a + bX + cX^2$, the Newton-Raphson minimization can be formulated as

$$X^* = X - F'(X)/F''(X) \qquad ...(12.10)$$

where X^* is the minimum, X is the current position, and F' and F'' are the first and second derivatives at the current position. Namely, for a quadratic function, this algorithm finds the minimum in a single step. This conclusion can be generalized to the multidimensional case :

$$\vec{S}_k = -\boldsymbol{H}_k^{-1}\, \vec{g}_k \qquad ...(12.11)$$

For nonquadratic but monotonic surfaces, the Newton-Raphson minimization method can be applied near a minimum in an iterative way.

There are several reasons that Newton-Raphson minimization is rarely used in macromolecular studies. First, the highly nonquadratic macromolecular energy surface, which is characterized by a multitude of local minima, is unsuitable for the Newton-Raphson method. In such cases it is inefficient, at times even pathological, in behavior. It is, however, sometimes used to complete the minimization of a structure that was already minimized by another method. In such cases it is assumed that the starting point is close enough to the real minimum to justify the quadratic approximation. Second, the need to recalculate the Hessian matrix at every iteration makes this algorithm computationally expensive. Third, it is necessary to invert the second derivative matrix at every step, a difficult task for large systems.

(*b*) *Adopted Basis Newton-Raphson.* A derivative method that is suited for large systems such as proteins is the *adopted basis Newton-Raphson* (ABNR) algorithm. Instead of calculating the full multidimensional curvature (*i.e.*, the full Hessian matrix) at each minimization step, the ABNR method limits its calculation to a small subspace of dimension *s* in which the system has made the most progress in past moves. The idea is to add curvature information only in those directions where it is likely to contribute most. This way the system moves in the best direction in the restricted subspace. Because the dimensionality *s* of the subspace is taken to be between 4 and 10, ABNR is an efficient minimization method comparable in CPU time to the order 1 methods such as the conjugated gradient. ABNR is also comparable to CG in terms of convergence. Since the method is not self-starting, the first steps are taken with an order 1 method, usually the steepest descent method.

Minimization Protocol

The foregoing discussion highlights the fact that the different minimization algorithms have relative strengths and weaknesses. As an example, Fig. 12.5 compares the results of minimizing the same protein with the steepest descent (SD) and conjugated gradients (CG) methods. It is evident that although initially SD reduces the energy faster than CG, in the long run the latter outperforms the former. A similar result is obtained when CG is replaced by ABNR. A detailed comparison between the various minimization algorithms applied to a peptide and a protein is given in Ref. 25.

To optimize the minimization procedure it is usually best to combine several algorithms into a single minimization protocol, taking advantage of their relative strengths. A good minimization scheme will usually start with SD and then use CG or ANBR to finish the job. The number of steps to be used in each phase depends on the goal of the minimization and on the character of the system. When high quality minimization is required, the minimization can be completed with Newton-Raphson. The termination criterion is usually defined in terms of the gradient RMS (GRMS), which is defined as the root-mean-square of all 3*N* gradients (or forces).

Simulated Annealing

Simulated annealing is a popular method that is often used for global optimization, i.e., finding the global minimum of a potential energy surface. The method takes its name from the natural annealing process in which a glass or a metal is first heated and then slowly cooled into a stable low energy state. The key factor in this process is slow cooling. If the cooling is done too fast, the materials will end up in unstable brittle states. Alternalively phrased, heating up the system shakes and rattles the molecule around the energy landscape, infusing it with thermal energy, analogous to kT, enabling it to jump out of its initial local minimum. The gradual cooling that is subsequently applied decreases the amplitude of these shakes, bringing the details of the energy surface back into focus and causing the system to slowly settle down to a lower energy minimum. Fig. 12.3*b* is a schematic representation of the simulated annealing process. In fact, we see that simulated annealing bridges between the high temperature conformational sampling simulations that are insensitive to the energy barriers and the low temperature situations, which are sensitive to the details of the energy landscape.

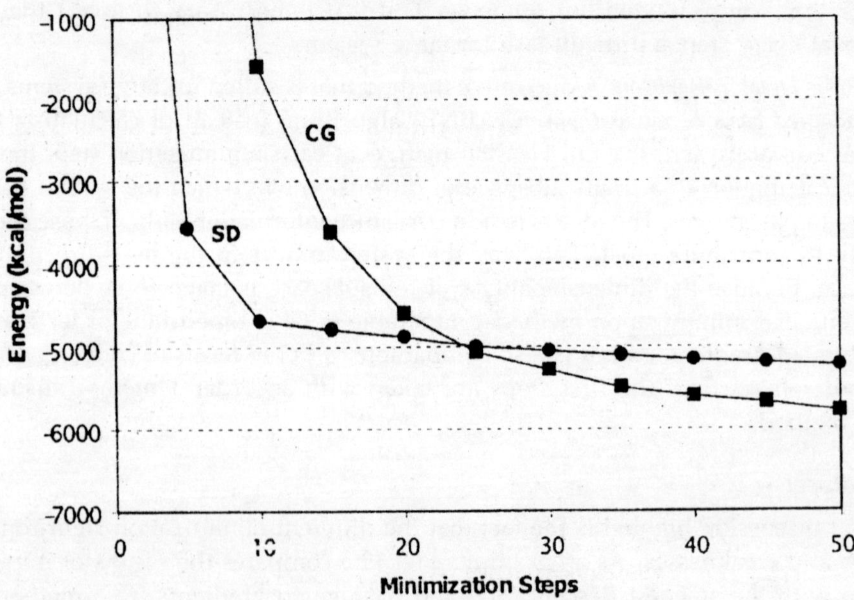

Fig. 12.5. A comparison of steepest descent (SD) minimization and conjugated gradients (CG) minimization of the same protein.

Simulated annealing can be easily implemented in both molecular dynamics and Monte Carlo simulations. In molecular dynamics, the temperature is controlled through coupling to a heat bath; with simulated annealing, the temperature of the bath is decreasing gradually. In Monte Carlo the trial move is accepted or rejected according to a temperature-dependent probability of the Metropolis type. In simulated annealing MC, the temperature used in the acceptance probability is gradually decreased. It should be noted that it is not necessary to anneal all the way to 0 *K*, because once the kinetic energy *kT* gets below the characteristic barrier height, a significant change cannot occur. Thus, many simulated annealing protocols cool to room temperature (or somewhat below) and are followed by a local minimization algorithm to remove the excess energy.

Specific implementations vary in cooling schedules, initial temperatures, the possibility of repeated heating "spikes," etc. Although simulated annealing is often considered a global optimization method, this is not the case when biomolecules are concerned. It can be shown that in systems characterized by a broad distribution of energy scales, a simulated annealing trajectory (either MD or MC) will have to be extremely long before it is able to find the global minimum. Since the energy landscape of proteins is broadly distributed and rough, this means that simulated annealing is an inefficient and infeasible strategy for protein folding. Nonetheless, even with biomolecules, simulated annealing remains a very useful method for local optimization. Its advantage is that, unlike direct minimization, which takes the molecule only as far as the nearest local minimum, simulated annealing is able to locate lower local minima further away from the initial conformation. An example of the application of this method can be found in Ref. 27. In the context of conformational analysis, simulated annealing is often used in conjunction with a high temperature sampling simulation. Each of the molecular structures generated by the high temperature simulation is first annealed back to room temperature before it is included in the conformational sample and subjected to further analysis.

CONFORMATIONAL ANALYSIS

To extract the conformational properties of the molecule that is being studied, the conformational ensemble that was sampled and optimized must be analyzed. The analysis may focus on global properties, attempting to characterize features such as overall flexibility or to identify common trends in the conformation set. Alternatively, it may be used to identify a smaller subset of characteristic low energy conformations, which may be used to direct future drug development efforts. It should be stressed that the different conformational analysis tools can be applied to any collection of molecular conformations. These may be generated by the above sampling techniques but can also have an experimental origin, such as NMR models or different X-ray structures of the same molecule (or analogous molecules).

Similarity Measures

A similarity measure is required for quantitative comparison of one structure with another, and as such it must be defined before the analysis can commence. Structural similarity is often measured by a root-mean-square distance (RMSD) between two conformations. In Cartesian coordinates the RMS distance d_{ij} between conformation i and conformation j of a given molecule is defined as the minimum of the functional

$$d_{ij} = \left[\frac{1}{N}\sum_{k=1}^{N} |r_k^{(i)} - r_k^{(j)}|^2\right]^{1/2} \qquad ...(12.12)$$

where N is the number of atoms in the summation, k is an index over these atoms, and $r_k^{(i)}, r_k^{(j)}$ are the Cartesian coordinates of atom k in conformations i and j. The minimum value of Eq. (12.12) is obtained by an optimal superposition of the two structures. The resulting RMS distances are usually compiled into a distance matrix Δ, where the elements Δ_{ij} are the RMS distances between conformations i and j.

Since the summation in Eq. (12.12) may be on any subset of atoms, it can be fine-tuned to best suit the problem at hand. The summation may be over the whole molecule, but it is very common to calculate conformational distances based only on non-hydrogen "heavy" atoms or, in the case of proteins, even based on only the backbone C_α atoms. Alternatively, in a study related to drug design one may consider, for example, focusing only on atoms that make up the pharmacophore region or that are otherwise known to be functionally important.

The conformational distance does not have to be defined in Cartesian coordinates. For comparing polypeptide chains it is likely that similarity in dihedral angle space is more important than similarity in Cartesian space. Two conformations of a linear molecule separated by a single low barrier dihedral torsion in the middle of the molecule would still be considered similar on the basis of dihedral space distance but will probably be considered very different on the basis of their distance in Cartesian space. The RMS distance is dihedral angle space differs from Eq. (12.12) because it has to take into account the 2π periodicity of the torsion angle,

$$d_{ij} = \left\{ \frac{1}{N} \sum_{k=1}^{N} \min[(\theta_k^{(i)} - \theta_k^{(j)})^2, (2\pi - \theta_k^{(i)} + \theta_k^{(j)})^2 \right\}^{1/2} \quad ...(12.13)$$

where N is the number of dihedral angles in the summation and $\theta_k^{(j)}, \theta_k^{(ij)}$ are the values of the dihedral angle θ_k in the two structures. As with the Cartesian distance, any appropriate subset of dihedral angles may be used, ranging from only the backbone ϕ, ψ, angles to a full set that includes all the side chain χ angles.

It is up to the researcher to decide whether to use a Cartesian similarity measure or a dihedral measure and what elements to include in the summation. It should be stressed that while the RMS distances perform well and are often used, there are no restrictions against other similarity measures. For example, similarity measures that emphasize chemical interactions, hydrophobicity, or the relative orientation of large molecular domains rather than local geometry may serve well if appropriately used.

Cluster Analysis

The distance matrix Δ, which holds the relative distances (by whatever similarity measure) between the individual conformations, is rarely informative by itself. For example, when sampling along a molecular dynamics trajectory, the Δ matrix can have a block diagonal form, indicating that the trajectory has moved from one conformational basin to another. Nonetheless, even in this case, the matrix in itself does not give reliable information about the size and shape of the respective basins. In general, the distance matrix requires further processing.

Cluster analysis is a common analytical technique used to group conformations. This approach highlights structural similarity, as defined by the distance measure being used, within a conformational sample. Starting from one selected conformation (often that of the lowest energy), all conformations that are within a given cutoff distance from this structure are grouped together into the first cluster C_1. Next, one of the conformations that were not grouped into the first cluster is selected, and a new cluster is formed around it. The process continues until all the conformations in the sample are assigned to a cluster C_i. This process often generates overlapping clusters, that is, clusters with nonzero intersection $C_i \cap C_j \neq 0$. The overlapping clusters are typically treated in one of two ways: (1) Group together the overlapping clusters C_i and C_j to form a single large cluster that is their union $C_i \cup C_j$ (Fig. 12.6*a*) or (2) make the overlapping clusters disjoint by removing their intersection $C_i \cap C_j$ from one of the clusters, typically the one that started with a higher energy conformation. Since the optimal cutoff distance by which to cluster the conformations is not a priori known, cluster analysis is usually performed hierarchically. Starting with a short cutoff the analysis is repeated again and again, each time with a larger cutoff distance. The results are often represented as a dendogram.

In many conformational studies, cluster analysis is used as a way to focus future effort on a small set of characteristic conformations. One conformation, typically the lowest energy one, is picked from each of the highly populated conformational clusters. The resulting small number of distinctly different conformations are then used as starting points for further computational analysis (such as free energy simulations) or as a basis for generating a pharmacological hypothesis used for directing future drug development. It should be noted, however, that conformational clusters generated by the above procedure do not necessarily represent the correct basin structure of the underlying energy landscape.

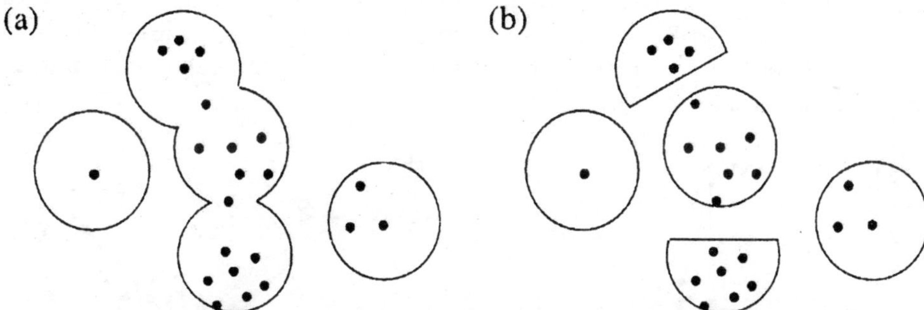

Fig. 12.6. A schematic representation of two clustering methods, in which each point represents a single molecular conformation and the circles are the similarity cutoff distances used to define the clusters, (*a*) Three clusters are defined when overlapping clusters are grouped together, (*b*) Five clusters are defined when the overlaps are removed from one of the overlapping clusters.

It has been shown that similar conformations that belong to adjacent energy basins separated by high energy barriers are incorrectly grouped together by the straightforward cluster analysis.

Principal Component Analysis

An inherent problem associated with conformational analysis is the high dimensionality of the molecular conformational spaces. An N-atom molecule has $3N$ degrees of freedom, and its corresponding conformational space is $(3N - 6)$-dimensional. As a result, even relatively small molecules have very large conformational spaces, making them difficult to analyze. For example, a small heptapeptide may have a 100-dimensional or even 150-dimensional conformational space, depending on its precise amino acid composition. Principal component analysis (PCA) is a computational tool that reduces the effective dimensionality of molecular conformational spaces while retaining an accurate representation of the interconformational distances. This task is accomplished by projecting the original multidimensional data onto an optimal low-dimensional subspace, allowing visual inspection of conformational spaces and of dynamic trajectories that traverse these spaces. Principal component analysis (PCA) was introduced to protein simulations under the name quasi-harmonic analysis by Ichiye and Karplus and is becoming widely used for a variety of applications involving sampling and visualization of conformational spaces.

How does principal component analysis work? Consider, for example, the two-dimensional distribution of points shown in Fig. 12.7*a*. This distribution clearly has a strong linear component and is closer to a one-dimensional distribution than to a full two-dimensional distribution. However, from the one-dimensional projections of this distribution on the two orthogonal axes X and Y you would not know that. In fact, you would probably conclude, based only on these projections, that the data points are homogeneously distributed in two dimensions. A simple axes rotation is all it takes to reveal that the data points are preferentially spread along one dimension, as reflected by the broad distribution along one of the new axes and a narrow distribution along the other in Fig. 12.7*b*. The above procedure is what is done by PCA. Starting from a multidimensional distribution of data points (in our case, of molecular conformations), PCA performs a similarity transformation on the original axes to find a new set of axes that best fits the data. The first new axis is selected such that the variance of the distribution along it is the largest possible. The second axis is placed orthogonal to the first in the direction of the second largest variance of the distribution, and so forth. In this new axes set it is usually possible to identify a low-dimensional subspace that captures most of the relative distances between individual conformations.

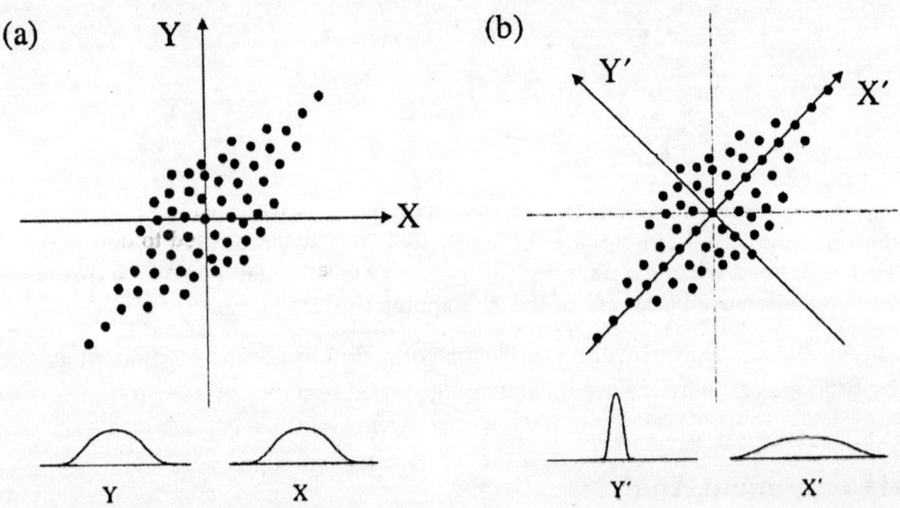

Fig. 12. 7 (*a*). A two-dimensional distribution of points and their one-dimensional projections on the original axes. Judging just from the 1D projections one would probably conclude that the original distribution is homogeneously distributed in two dimensions, (*b*) The same distribution of points and their 1D projections on the new axes set obtained by PCA by using a similarity transformation. The new 1D projections highlight the strong 1D character of the distribution.

In general, two related techniques may be used: principal component analysis (PCA) and principal coordinate analysis (PCoorA). Both methods start from the $n \times m$ data matrix **M**, which holds the m coordinates defining n conformations in an m-dimensional space. That is, each matrix element M_{ij} is equal to q_{ij}, the jth coordinate of the ith conformation. From this starting point PCA and PCoorA follow different routes.

Principal component analysis (PCA) takes the m-coordinate vectors q associated with the conformation sample and calculates the square $m \times m$ M^TM matrix, reflecting the relationships between the *coordinates*. This matrix, also known as the covariance matrix C, is defined as

$$C = \langle (q - \langle q \rangle)(q - \langle q \rangle)^T \rangle \qquad ...(12.14)$$

where the averaging is over the conformation sample (in Cartesian space $m = 3N$ for an N-atom molecule). The covariance matrix C is diagonalized to obtain the eigenvectors that capture most of the variation in atomic position fluctuations.

Principal coordinate analysis (PCoorA), on the other hand, operates on the square $n \times n$ MM^T matrix, reflecting the relationships between the *conformations*. The elements of this matrix, also known as the distance matrix Δ, are distances d_{ij} between two conformations i andy j [such as those defined in Eqs. (12.12) and (12.13)]. Since the distances d_{ij} can also be obtained from the $n \times n$ matrix A of latent roots (eigenvectors), one can use this matrix for the projection, defining $A_{ij} = -1/2\, d_{ij}^2$ and $A_{ij} = 0$ (for $i, j = 1, 2, ..., n$). To guarantee that the matrix A has a zero root (and thus guarantee that it corresponds to a real configuration) it is "centered," so that the sum of every row and the sum of every column of A is zero. This centering, which does not alter the distances d_{ij}, is defined as

$$A^*_{ij} = A_{ij} - \langle\langle A_{ij}\rangle_i - \langle A_{ij}\rangle_j + 2\langle A_{ij}\rangle_{ij} \quad ...(12.15)$$

where $\langle\cdot\rangle_k$ is the mean over all specific indices $k = i, j, ij$. The centered matrix $A*$ is diagonalized using standard matrix algebra to obtain the latent eigenvectors and the diagonal matrix of eigenvalues. The resulting eigenvalues (normalized) give the percentage of the projection of the original distribution on the new coordinate set, and the eigenvectors (scaled by their corresponding eigenvalues) give the new coordinates of the original points in the new axes set.

It should be stressed that PCA and PCoorA are dual methods that give the same analytical results. Using one or the other is simply a matter of convenience, whether one prefers to work with the covariance matrix C or with the distance matrix Δ.

As stated earlier, the main motivation for using either PCA or PCA is to construct a low-dimensional representation of the original high-dimensional data. The notion behind this approach is that the effective (or essential, as some call it) dimensionality of a molecular conformational space is significantly smaller than its full dimensionality ($3N$-6 degrees of freedom for an N-atom molecule). Following the PCA procedure, each new axis k is associated with a normalized eigenvalue λ_k that reflects the relative weight of that axis in reproducing the original data. An axis with a high λ_k value is significant for the projection, whereas axes with small λ_k values are insignificant. By sorting the new axes according to their λ_k weight it is possible to select a small subset of effective coordinates that capture most of the conformational relationships of the original high-dimensional space. The quality of such a projection can be estimated by the average difference between conformation distances reconstructed in the low s-dimensional subspace $d_{ij}^{(s)}$ and the original distance d_{ij}. The reconstructed distances are defined as

$$d_{ij}^{(s)2} = \sum_{k=1}^{s} (Q_{ik} - Q_{jk})^2 \quad ...(12.16)$$

where Q_{ik} is the coordinate of the ith conformation along the kth new (principal) axis. It can be shown that the average deviation of the distances in s dimensions from the exact distances is given by the sum of the first s eigenvalues,

$$\langle d_{ij}^2 - d_{ij}^{(s)2}\rangle_{ij} = -\sum_{k=1}^{s} \lambda_k \quad ...(12.17)$$

Thus by summing the normalized eigenvalues of the first s dimensions one can judge the quality of a projection onto that subspace.

Fortunately, it was found that in polypeptide systems the effective dimensionality of conformational spaces is significantly smaller than the dimensionality of the full space, with only a few principal axes contributing to the projection. In fact, in many cases a projection quality of 70-90% can be achieved in as few as three dimensions, opening the way for real 3D visualization of molecular conformational space. Fig. 12.8 shows a 3D visualization of the conformational spaces of two hexapeptides, $(Ala)_6$ and cyclic-$(Ala)_6$, jointly projected on the same principal coordinate set. The comparison shows that the conformation volume occupied by the linear peptide is about 10 times larger than the conformation volume occupied by its conformationally constrained counterpart. Quantifying relative flexibility of analogous peptides through joint principal projections was shown to be useful, for example, in predicting their relative bioactivity.

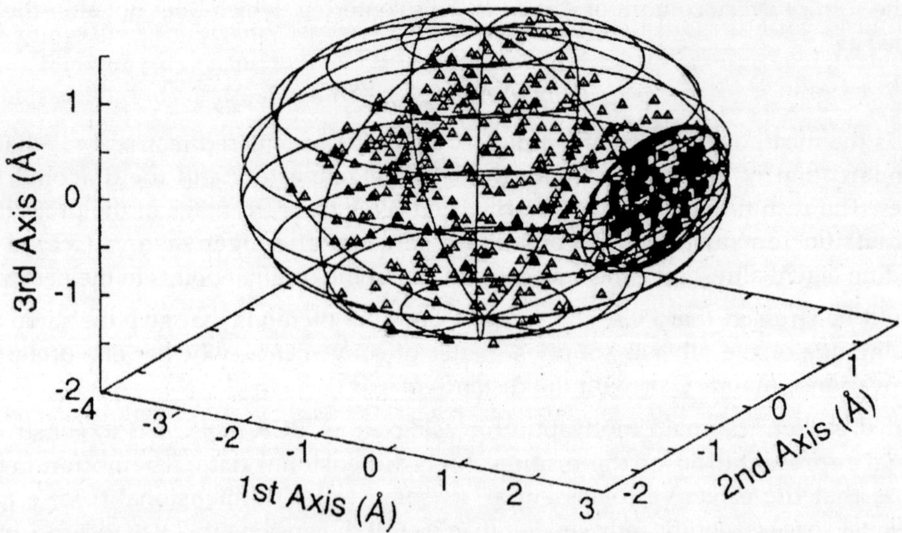

Fig. 12.8. A joint principal coordinate projection of the occupied regions in the conformational spaces of linear $(Ala)_6$ (triangles) and its conformational constraint counterpart, cyclic-$(Ala)_6$ (squares), onto the optimal 3D principal axes. The symbols indicate the projected conformations, and the ellipsoids engulf the volume occupied by the projected points. This projection shows that the conformational volume accessible to the cyclic analog is only a small subset of the conformational volume accessible to the linear peptide.

CONCLUSION

In this chapter we surveyed a variety of computational methods that contribute to the "conformational analysis" of complex molecules. These include methods for searching and sampling the molecular conformation space, methods for local optimization of the sampled conformations, and basic analytical techniques. In practice, many variations of the basic methodologies are reported as researchers continuously try to improve and enhance these procedures. The different methods are often used in conjunction to form a complete conformational analysis study of a bimolecular system of interest. However, each of the different procedures can also be used separately as part of computational studies with other goals. The need for these analytical techniques is to a large extent brought about by the continuous increase is simulation times, which generates more data than ever before, requiring systematic ways to interpret and represent them.

INDEX

D

E

F

P

S

V

W

X

❑❑❑